ENCYCLOPAEDIA OF
NANO PHYSICS

ENCYCLOPAEDIA OF NANO PHYSICS

Hugh Watson

ANMOL PUBLICATIONS PVT. LTD.
NEW DELHI - 110 002 (INDIA)

ANMOL PUBLICATIONS PVT. LTD.
H.O.: 4374/4B, Ansari Road, Darya Ganj,
New Delhi-110 002 (India)
Ph.: 23278000, 23261597
B.O.: No. 1015, Ist Main Road, BSK IIIrd Stage
IIIrd Phase, IIIrd Block,
Bangalore - 560 085 (India)
Visit us at: www.anmolpublications.com

Encyclopaedia of Nano Physics

First Published, 2006

ISBN 81-261-2983-2

PRINTED IN INDIA

Printed at Mehra Offset Press, Delhi.

Contents

Preface

This book titled "Encyclopaedia of Nano Physics" deals briefly with the subject areas of nano-optics, nanotube, nanofibre, nano-sensor, nanowire and nanoribbon. Keeping in mind the applications and utilities of nano physics for meeting the contemporary global energy requirements, it seems that this subject will become the core area of future nanoscience and nanotechnology research. Nanostructured materials do not represent a new phenomenon. However, the ability to probe, manipulate, understand and engineer matter at atomic scales has only recently come within our grasp. Developments such as the invention of the Scanning Tunneling Microscope in 1981 have since made nanoscale science a reality. Today, the term nanotechnology is used to describe specific construction and manipulation performed on an atomic and molecular scale. Generally, systems with a maximum size of less than a hundred nanometers are classified as nanotechnology.

In physics, the prefix "nano" is used to describe a billionth of a unit, so a nanometer is a billionth of a meter. This is the equivalent of approximately five to ten atoms laid end-to-end. These dimensions are also reflected in the English translation of the Greek word "nanos"—"dwarf". A single human hair is around ten thousand times thicker than a nanometer, and another impressive comparison – a nanoparticle is to a soccer ball as a soccer ball is to the Earth. But nanotechnology is not just microtechnology on a smaller scale; it is much more than a mere improvement on previous forms of technology. Nanotechnology is a completely new production process in a molecular world, opening up possibilities far beyond what we have known before. Nature provides the inspiration for this new technology – from time immemorial, nanomachines have been at work in all organic cells, building our world of plants, animals and people from a store of chemical elements. The discovery in recent years of the mechanisms for accessing individual building blocks of matter, and with it the increasing understanding of how these building blocks organize themselves, has sparked off the industrial conquest of the nanosphere throughout the world. For instance, Nanotube is real. Strictly speaking, any tube with nanoscale dimensions, but generally used to refer to carbon nanotubes (a commonly mentioned non-carbon variety is made of boron nitride), which are sheets of graphite rolled up to make a tube. The dimensions are variable (down to 0.4 nm in diameter) and you can also get nanotubes within nanotubes, leading to a distinction between multi-walled and single-walled nanotubes. Apart from remarkable tensile strength, nanotubes exhibit varying electrical properties (depending on the way the graphite structure spirals around the tube, and other factors), and can be insulating, semiconducting or conducting (metallic). Carbon nanotubes are considered by some to be fullerenes (strictly speaking, this is probably correct, but they are more commonly treated as distinct) and are sometimes referred to as buckytubes. Microelectromechanical systems (MEMS) are generally used to refer to systems that can respond to a stimulus or create physical forces (sensors and actuators) and that have dimensions on the micrometer scale. They are almost exclusively made using the same lithographic techniques that are used to make silicon chips for computing. Miniature accelerometers are the most successful product in this field and are used to trigger air bags in cars. When such systems can be made with nanoscale dimensions they can be classified as NEMS. The success story of the giant magnetoresistance effect (GMR) shows in what a short space of time nanotechnology can make it from research laboratory to complete market penetration.

This book tends to cover the aforementioned subject areas in their broadest possible terms. All useful supplementary research and reference tools have been included for readers' further investigations in the subject area of Nano Physics.

—*Editor*

1
Introduction to Nano Physics

Molecular Nanotechnology & Quantum Dot Lasers

Thorough, inexpensive control of the structure of matter based on molecule-by-molecule control of products and byproducts; the products and processes of molecular manufacturing. Technology as we know it is a product of industry, of manufacturing and chemical engineering. Industry as we know it takes things from nature—ore from mountains, trees from forests—and coerces them into forms that someone considers useful. Trees become lumber, then houses. Mountains become rubble, then molten iron, then steel, then cars. Sand becomes a purified gas, then silicon, then chips. And so it goes. Each process is crude, based on cutting, stirring, baking, spraying, etching, grinding, and the like. Trees, though, are not crude: to make wood and leaves, they neither cut, grind, stir, bake, spray, etch, nor grind. Instead, they gather solar energy using molecular electronic devices, the photosynthetic reaction centers of chloroplasts. They use that energy to drive molecular machines—active devices with moving parts of precise, molecular structure—which process carbon dioxide and water into oxygen and molecular building blocks. They use other molecular machines to join these molecular building blocks to form roots, trunks, branches, twigs, solar collectors, and more molecular machinery. Every tree makes leaves, and each leaf is more sophisticated than a spacecraft, more finely patterned than the latest chip from Silicon Valley. They do all this without noise, heat, toxic fumes, or human labor, and they consume pollutants as they go. Viewed this way, trees are high technology. Chips and rockets aren't. Trees give a hint of what molecular nanotechnology will be like, but nanotechnology won't be biotechnology because it won't rely on altering life. Biotechnology is a further stage in the domestication of living things. Like selective breeding, it reshapes the genetic heritage of a species to produce varieties more useful to people. Unlike selective breeding, it inserts new genes. Like biotechnology—or ordinary trees—molecular nanotechnology will use molecular machinery, but unlike biotechnology, it will not rely on genetic meddling. It will be not an extension of biotechnology, but an alternative or a replacement. Molecular nanotechnology could have been conceived and analyzed—though not built—based on scientific knowledge available forty years ago. Even today, as development accelerates, understanding grows slowly because molecular nanotechnology merges fields that have been strangers: the molecular sciences, working at the threshold of the quantum realm, and mechanical engineering, still mired in the grease and crudity of conventional technology. Nanotechnology will be a technology of new molecular machines, of gears and shafts and bearings that move and work with parts shaped in accord with the wave equations at the foundations of natural law. Mechanical engineers don't design molecules. Molecular scientists seldom design machines. Yet a new field will grow—is growing today—in the gap between. That field will replace both chemistry as we know it and mechanical engineering as we know it. And what is manufacturing today, or modern technology itself, but a patchwork of crude chemistry and crude machines?

Next chapters will paint a concrete picture of molecular machines and molecular manufacturing, but for now analogy will serve. Picture an automated factory, full of conveyor belts, computers, rollers, stampers, and swinging robot arms. Now imagine something like that factory, but a million times smaller and working a million times faster, with parts and workpieces of molecular size. In this factory, a "pollutant" would be a loose molecule, like a ricocheting bolt or washer, and loose molecules aren't tolerated. In many ways, the factory is utterly unlike a living cell: not fluid, flexible, adaptable, and fertile, but rigid, preprogrammed and specialized. And yet for all of that, this microscopic molecular factory emulates life in its clean, precise molecular construction.

Advanced molecular manufacturing will be able to make almost anything. Unlike crude mechanical and chemical technologies, molecular manufacturing will work from the bottom up, assembling intricate products from the molecular building blocks that underlie everything in the physical world. Nanotechnology will bring new capabilities, giving us new ways to make things, heal our bodies, and care for the environment. It will also bring unwelcome advances in weaponry and give us yet more ways to foul up the world on an enormous scale. It won't automatically solve our problems: even powerful technologies merely give us more power. As usual, we have a lot of work ahead of us and a lot of hard decisions to make if we hope to harness new developments to good ends. The main reason to pay attention to nanotechnology now, before it exists, is to get a head start on understanding it and what to do about it.

Semiconductor lasers are key components in a host of widely used technological products, including compact disk players and laser printers, and they will play critical roles in optical communication schemes. The basis of laser operation depends on the creation of nonequilibrium populations of electrons and holes, and coupling of electrons and holes to an optical field, which will stimulate radiative emission. Calculations carried out in the early 1970s by C. Henry (Dingle and Henry 1976) predicted the advantages of using quantum wells as the active layer in such lasers: the carrier confinement and nature of the electronic density of states should result in more efficient devices operating at lower threshold currents than lasers with "bulk" active layers. In addition, the use of a quantum well, with discrete transition energy levels dependent on the quantum well dimensions (thickness), provides a means of "tuning" the resulting wavelength of the material. The critical feature size—in this case, the thickness of the quantum well—depends on the desired spacing between energy levels. For energy levels of greater than a few tens of millielectron volts (meV, to be compared with room temperature thermal energy of 25 meV), the critical dimension is approximately a few hundred angstroms. Although the first quantum well laser, demonstrated in 1975, was many times less efficient than a conventional laser (van der Ziel et al. 1975), the situation was reversed by 1981 through the use of new materials growth capabilities (molecular beam epitaxy), and optimization of the heterostructure laser design (Tsang 1982).

Even greater benefits have been predicted for lasers with quantum dot active layers. Arakawa and Sakaki (1982) predicted in the early 1980s that quantum dot lasers should exhibit performance that is less temperaturedependent than existing semiconductor lasers, and that will in particular not degrade at elevated temperatures. Other benefits of quantum dot active layers include further reduction in threshold currents and an increase in differential gain—that is, more efficient laser operation (Asada et al. 1986). The key concepts in the laser operation. Stimulated recombination of electron-hole pairs takes place in the GaAs quantum well region, where the confinement of carriers and of the optical mode enhance the interaction between carriers and radiation. In particular, note the change in the electronic density of states, as a function of the "dimensionality" of the active layer.

The population inversion (creation of electrons and holes) necessary for lasing occurs more efficiently as the active layer material is scaled down from bulk (3-dimensional) to quantum dots (0-dimensional). However, the advantages in operation depend not only on the absolute size of the nanostructures in the active region, but also on the uniformity of size. A broad distribution of sizes "smears" the density of states, producing behavior similar to that of bulk material. Thus, the challenge in realizing quantum dot lasers with operation superior to that shown by quantum well lasers is that of forming high quality, uniform quantum dots in the active layer. Initially, the most widely followed approach to forming quantum dots was through electron beam lithography of suitably small featured patterns (~300 Å) and subsequent dryetch transfer of dots into the substrate material. The problem that plagued these quantum dot arrays was their exceedingly low optical efficiency: high surface-to-volume ratios of these nanostructures and associated high surface recombination rates, together with damage introduced during the fabrication itself, precluded the successful formation of a quantum dot laser.

Density of electronic states as a function of structure size. With the demonstration of the high optical efficiency self-assembled formation of quantum dots, formed without need of external processing and having the natural overgrowth of cladding material (which addressed issues of surface recombination), there ensued a marked increase in quantum dot laser research. The first demonstration of a quantum dot laser with high threshold density was reported by Ledentsov and colleagues in 1994. Bimberg et al. (1996) achieved improved operation by increasing the density of the quantum dot structures, stacking successive, strain-aligned rows of quantum dots and therefore achieving vertical as well as lateral coupling of the quantum dots. In addition to utilizing their quantum size effects in edge-emitting lasers, self-assembled quantum dots have also been incorporated within vertical cavity surface-emitting lasers. A partial summary of the work and achievements in quantum dot lasers. As with the demonstration of the advantages of the quantum well laser that preceded it, the full promise of the quantum dot laser must await advances in the understanding of the materials growth and optimization of the laser structure. Although the self-assembled dots have provided an enormous stimulus to work in this field, there remain a number of critical issues involving their growth and formation: greater uniformity of size, controllable achievement of higher quantum dot density, and closer dot-todot interaction range will further improve laser performance. Better understanding of carrier confinement dynamics and capture times, and better evaluation of loss mechanisms, will further improve device characteristics. It should be noted that the spatial localization of carriers brought about by the quantum dot confinement may play a role in the "anomalous" optical efficiency of the GaN-based materials, which is exceptional in light of the high concentration of threading dislocations (~ 108 - 1010 cm-2) that currently plague this material system. The localization imposed by the perhaps natural nanostructure of the GaN materials may make the dislocation largely irrelevant to the purely optical (but not to the electrical) behavior of the material.

The first synthesis and characterization of carbon nanotubes were reported by Iijima from NEC in late 1991. The initial theoretical study of their electronic structure was soon followed with the work by Dresselhaus and coworkers at MIT (Dresselhaus et al. 1992; Saito et al. 1992a; Saito et al. 1992b). Since then, the fabrication of nanotubes has been improved by several groups, and methods other than arc discharge have been explored. The main issues are to separate the nanotubes from other forms of carbon also produced in the fabrication process and to increase the yield of singlewalled nanotubes (SWNT) for potential applications. Following Iijima's work, macroscopic quantities of MWNT were produced with an improved arc discharge method by Ebbesen and coworkers at NEC (Tsukuba) (Ebbesen and Ajayan 1992).

In the following year, that same group produced what is considered to be among the best SWNT material generated so far; over 70% of the volume of material was nanotubes bundled together into crystalline ropes of metallic character (Thess et al. 1996). Also in 1996, a group from the Chinese Academy of Science used chemical vapor deposition to produce a 50 mm thick film of nanotubes that were highly aligned perpendicular to the surface (Li et al. 1996). Progress in recent years leads one to predict that it will indeed be possible to produce high quality carbon nanotubes in macroscopic quantities needed for many of the applications outlined below. Nanotube bundles form a low density material and are expected to have high stiffness and axial strength as a result of their seamless cylindrical graphitic structure.

It is therefore predicted that they can be used to fabricate a material with better mechanical properties than the present carbon fiber materials. Information about the mechanical properties of nanotubes has been gathered recently by a study of the thermal vibrations of a single attached to a substrate (Treacy et al. 1996). Ebbesen's group at the Princeton NEC Research Institute found that nanotubes have an exceptionally high Young's modulus (~ 2 x 109 Pa) (Treacy et al. 1996). In order to reach a better understanding of the mechanical properties and intrinsic limitations of nanotubes, Bernholc's group from North Carolina State University theoretically studied the behavior of nanotubes beyond the linear Hooke's law and the nature of the defects leading to dislocations and fractures (Yakobson et al. 1996; Nardelli et al. 1998). Nanotubes are highly polarizable nanoscale straws, a property that confers on them the capacity to ingest inorganic elements by nanocapillarity (Pederson and Broughton 1992). As a result, it has been conjectured that they could be used as minute molds to shape nanometer-sized quantum wires and as miniature test tubes. Ajayan and coworkers at NEC (Tsukuba) have first shown that lead can be introduced into carbon nanotubes (Ajayan and Iijima 1993).

The efficiency of their process is low, and prior removal of the caps from the ends of the nanotubes is expected to improve the situation (Tsang et al. 1993). More information about the mechanism of NT filling was obtained by Pascard and coworkers from the École Polytechnique in France by studying the propensity to form nanowires for 15 encapsulated metal elements (Guerret-Plécourt et al. 1994). Finally, external decoration of nanotubes with metal atoms has been demonstrated and is predicted to have applications in catalysis (Satishkumar et al. 1996). The primary methods of nanotube fabrication and the institutions engaged in specific methods of nanotube fabrication. Early theoretical studies already showed that the electronic properties of nanotubes strongly depend on their diameter and their chirality leading to metallic or semiconducting structures (Saito et al. 1992c). It was conjectured that these properties can be used to construct nanoscale electronic devices. While theoretical studies were promptly published, it was only in 1996 that Ebbesen and coworkers (1996) at the Princeton NEC Research Institute presented reliable four-point probe conductivity measurements on MWNT, confirming the theoretical predictions. In 1997, two groups, one at Lawrence Berkeley National Laboratory (Bockrath et al. 1997) and the second at Delft University (Tans et al. 1997) in the Netherlands showed that conductivity through nanotubes is controlled by low dimensional effects such as resonant tunneling and single-electron charging effects. Hall effect measurements at the École Polytechnique Fédérale de Lausanne (EPFL) in Switzerland have shown that hole transport is predominant in electronic conductance (Baumgartner et al. 1997). Despite these and other very recent and encouraging efforts such as those studying the mean free path of carriers in nanotubes, the conduction mechanism is still only partially understood (Petit et al. 1997).

For this reason and possibly others related to quantum confinement effects, nanotubes are expected to form outstanding field-emitting materials. In 1995, the Rice University group showed that nanotubes emit electrons very efficiently when immersed in an electric field and irradiated by a laser to remove their cap (Rinzler et al. 1995). They attribute their observation to the unraveling of an atomic wire of carbon atoms. Efficient field emission was also obtained from carefully aligned nanotubes by de Heer and coworkers (1995) at EPFL, whereas Collins and coworkers from the University of California at Berkeley (Collins and Zettl 1996, 1997), have used randomly oriented nanotubes with similar results. Efficient field-emitting material is highly desirable for the production of field-emission displays and microwave tubes. Recently, a group from Mie University in Japan has built a cathode ray tube (CRT) using nanotube field emitters (Saito et al. n.d.). In this work, the layers of nanotubes were cut out from the soot produced in an arc discharge chamber. This fabrication method is presently not compatible with industrial production requirements, and more progress must be made before this effort can be translated into an industrial product. The electrical and field emission properties of nanotubes, with the representative institutions pursuing these studies. In summary, macroscopic amounts of good quality nanotubes can presently be fabricated by several groups around the world, and the theoretical understanding of the electronic structure and related properties of nanotubes has reached a very good level. However, despite the fact that many potential applications are mentioned in the literature over and over again, only the outstanding field emission properties of nanotubes have achieved realization in practical devices. One of the main obstacles simply remains the controlled manipulation of nanoscale objects. In this respect it seems that the generation of self-aligned structures is a path to explore further, especially after the encouraging successes reported in the literature in 1997-98.

Nanostructured Materials Behavior

Bulk nanostructured materials are defined as bulk solids with nanoscale or partly nanoscale microstructures. This category of nanostructured materials has historical roots going back many decades but has a relatively recent focus due to new discoveries of unique properties of some nanoscale materials. Early in the century, when "microstructures" were revealed primarily with the optical microscope, it was recognized that refined microstructures, for example, small grain sizes, often provided attractive properties such as increased strength and toughness in structural materials. A classic example of property enhancement due to a refined microstructure—with features too small to resolve with the optical microscope—was age hardening of aluminum alloys. The phenomenon, discovered by Alfred Wilm in 1906, was essentially explained by Merica, Waltenberg, and Scott in 1919 (Mehl and Cahn 1983, 18), and the microstructural features responsible were first inferred by the X-ray studies of Guinier and Preston in 1938. With the advent of transmission electron microscopy (TEM) and sophisticated X-ray diffraction methods it is now known that the fine precipitates responsible for age hardening, in Al-4% Cu alloys, for example, are clusters of Cu atoms— Guinier-

Preston (GP) zones—and the metastable partially coherent q' precipitate (Silcock et al. 1953-54; Cohen 1992). Maximum hardness is observed with a mixture of GPII (or q") (coarsened GP zones) and q' with the dimensions of the q' plates, typically about 10 nm in thickness by 100 nm in diameter.

Therefore, the important microstructural feature of age-hardened aluminum alloys is nanoscale. There are a number of other examples of nanoscale microstructures providing optimized properties. The critical current density JC of commercial superconducting Nb3Sn is controlled by grain size and is inversely proportional to grain size, with grain sizes of 50-80 nm providing high values of JC (Scanlan et al. 1975). The field of nanocrystalline (or nanostructured, or nanophase) materials as a major identifiable activity in modern materials science results to a large degree from the work in the 1980s of Gleiter and coworkers (Gleiter 1990), who synthesized ultrafine-grained materials by the in situ consolidation of nanoscale atomic clusters. The ultrasmall size (< 100 nm) of the grains in these nanocrystalline materials can result in dramatically improved—or different—properties from conventional grain-size (> 1 μm) polycrystalline or single crystal materials of the same chemical composition. This is the stimulus for the tremendous appeal of these materials. While there are a number of bulk properties that may be dramatically changed when the microstructure is nanoscale, this chapter focuses on those for which the recent work with nanostructured materials has been most extensive. These are (1) the mechanical properties of nanostructured materials for a variety of potential structural applications, and (2) ferromagnetic materials with nanoscale microstructures for potential applications as soft magnetic materials and permanent magnet materials, and for other special applications such as information storage, magnetoresistance spin valves, and magnetic nanocomposite refrigerants. Other bulk applications such as hydrogen storage are discussed briefly.

The great interest in the mechanical behavior of nanostructured materials originates from the unique mechanical properties first observed and/or predicted for the materials prepared by the gas condensation method. Among these early observations/predictions were the following:

- lower elastic moduli than for conventional grain size materials—by as much as 30 - 50%
- very high hardness and strength—hardness values for nanocrystalline pure metals (~ 10 nm grain size) are 2 to 7 times higher than those of larger grained (>1 mm) metals
- a negative Hall-Petch slope, i.e., decreasing hardness with decreasing grain size in the nanoscale grain size regime
- ductility—perhaps superplastic behavior—at low homologous temperatures in brittle ceramics or intermetallics with nanoscale grain sizes, believed due to diffusional deformation mechanisms

While some of these early observations have been verified by subsequent studies, some have been found to be due to high porosity in the early bulk samples or to otherartifacts introduced by the processing procedures. The following summarizes the author's understanding of the state of the art of the mechanical behavior of nanostructured materials, as determined from the literature, presentations at the U.S. workshop (Siegel et al. 1998), and the WTEC panel's site visits in Japan and Europe.

Elasticity

Early measurements of the elastic constants on nanocrystalline (nc) materials prepared by the inert gas condensation method gave values, for example for Young's Modulus, *E*, that were significantly lower than values for conventional grain size materials. While various reasons were given for the lower values of *E*, it was suggested by Krstic and coworkers (1993) that the presence of extrinsic defects—pores and cracks, for example—was responsible for the low values of *E* in nc materials compacted from powders. This conclusion was based on the observation that nc NiP produced by electroplating with negligible porosity levels had an *E* value comparable to fully dense conventional grain size Ni (Wong et al. 1994, 85). Subsequent work on porosity-free materials has supported these conclusions, and it is now believed that the intrinsic elastic moduli of nanostructured materials are essentially the same as those for conventional grain size materials until the grain size becomes very small, e.g., < 5 nm, such that the number of atoms associated with the grain boundaries and triple junctions becomes very large. For nanocrystalline Fe prepared by mechanical attrition and measured by a nano-indentation technique. Thus, for most nanostructured materials (grain size > 10 nm), the elastic moduli are not unique properties and not a "negative."

Hardness and Strength

Hardness and strength of conventional grain size materials (grain diameter, d > 1 mm) is a function of grain size. For ductile polycrystalline materials the empirical Hall-Petch equation has been found to express the grain-size dependence of flow stress at any plastic strain out to ductile fracture. In terms of yield stress, this expression is so = si + kd-1/2, where so = yield stress, si = friction stress opposing dislocation motion, k = constant, and d = grain diameter. Similar results are obtained for hardness, with Ho = Hi + kd-1/2. To explain these empirical observations, several models have been proposed, which involve either dislocation pileups at grain boundaries or grain boundary dislocation networks as dislocation sources. In all cases the Hall-Petch effect is due to dislocation motion/generation in materials that exhibit plastic deformation.

Ratio of the Young's (*E*) and shear (G) moduli of nanocrystalline materials to those of conventional grain size materials as a function of grain size. The dashed and solid curves correspond to a grain boundary thickness of 0.5 and 1 nm, respectively (Shen et al. 1995). Most of the mechanical property data on nc materials have pertained to hardness, although some tensile test data are becoming available. Several recent reviews have summarized the mechanical behavior of these materials (Siegel and

Fougere 1994, 233–261; Siegel 1997; Morris and Morris 1997; Weertman and Averback 1996, 323–345). It is clear that as grain size is reduced through the nanoscale regime (< 100 nm), hardness typically increases with decreasing grain size and can be factors of 2 to 7 times harder for pure nc metals (10 nm grain size) than for large-grained (> 1 μm) metals.

The experimental results of hardness measurements, summarized previously, show different behavior for dependence on grain size at the smallest nc grains (< 20 nm), including (a) a positive slope ("normal" Hall-Petch behavior), (b) essentially no dependence (~ zero slope), and (c) in some cases, a negative slope (Siegel and Fougere 1994, 233–261; Siegel 1997; Morris and Morris 1997; Weertman and Averback 1996, 323–345). Most data that exhibit the negative Hall-Petch effect at the smallest grain sizes have resulted from nc samples that have been annealed to increase their grain size. It is suggested that thermally treating nanophase samples in the as-produced condition may result in such changes in structure as densification, stress relief, phase transformations, or grain boundary structure, all of which may lead to the observed negative Hall-Petch behavior (Siegel and Fougere 1994, 233–261). Only a few cases of negative Hall-Petch behavior have been reported for as-produced nanocrystalline samples with a range of grain sizes. These include electrodeposited nc alloys and devitrified nc alloys (Erb et al 1996, 93-110; Alves et al. 1996). Nanocrystalline thin films with grain sizes £ 6 nm are also observed to exhibit a negative Hall-Petch effect (Veprek 1998). While it seems likely that in many cases the observed negative Hall-Petch slopes are due to artifacts of the specimen preparation methods, it is also likely that conventional dislocation-based deformation is not operable in nanocrystalline materials at the smallest grain sizes (< ~30 nm). At these grain sizes, theoretically, mobile dislocations are unlikely to occur; nor have they been observed in *in situ* TEM deformation experiments (Siegel and Fougere 1994, 233–261; Milligan et al 1993; Ke et al. 1995). Thus, the hardness, strength, and deformation behavior of nanocrystalline materials is unique and not yet well understood.

Ductility and Toughness

It is well known that grain size has a strong effect on the ductility and toughness of conventional grain size (> 1 mm) materials. For example, the ductile/brittle transition temperature in mild steel can be lowered about 40°C by reducing the grain size by a factor of 5. On a very basic level, mechanical failure, which limits ductility, is an interplay or competition between dislocations and cracks (Thomson 1996, 2208–2291). Nucleation and propagation of cracks can be used as the explanation for the fracture stress dependence on grain size (Nagpal and Baker 1990). Grain size refinement can make crack propagation more difficult and therefore, in conventional grain size materials, increase the apparent fracture toughness. However, the large increases in yield stress (hardness) observed in nc materials suggest that fracture stress can be lower than yield stress and therefore result in reduced ductility. The results of ductility measurements on nc metals are mixed and are sensitive to flaws and porosity, surface finish, and method of testing (e.g., tension or compression testing). In tension, for grain sizes < 30 nm, essentially brittle behavior has been observed for pure nanocrystalline metals that exhibit significant ductility when the grain size is conventional. Elongation to failure in tension vs. grain size for some nanocrystalline metals and alloys. In some metals, Cu for example, ductile behavior is observed in compression, along with yield strengths about twice those observed in tension. While it is likely that the flaws and porosity present in many nc samples seriously affect the results of mechanical tests and may be partly responsible for the asymmetry of results in compression compared to tension tests, the nature of the deformation process in terms of shear banding may also be important.

The above behavior is presumably due to the inability of usual dislocation generation and motion to occur at these smallest nc grain sizes. An intriguing suggestion based on early observations of ductile behavior of brittle nc ceramics at low temperatures is that brittle ceramics or intermetallics might exhibit ductility with nc grain structures (Karch et al. 1987; Bohn et al. 1991). Karch and colleagues (1987) observed apparent plastic behavior in compression in nc CaF2 at 80°C and nc TiO2 at 180°C. These observations were attributed to enhanced diffusional creep providing the plasticity at these temperatures, where conventional grain-size materials would fail in the elastic regime. It was assumed that diffusional creep was responsible for the plasticity; observations were rationalized, with boundary diffusion dominating the behavior such that the strain (creep) rate is defined as de dt = BsWDDb d3kT where s is the applied stress, W the atomic volume, d the grain size, k the Boltzmann constant, T the temperature, B a constant, and Db the grain boundary diffusion coefficients. Going from a grain size of 1 μm to 10 nm should increase de/dt by 106 or more if Db is significantly larger for nc materials. However, these results on nc CaF2 and nc TiO2 have not been reproduced, and it is believed that the porous nature of these samples was responsible for the apparent ductile behavior. In addition, the idea of unusually high creep rates at low temperatures has been refuted. Recent creep measurements of nc Cu, Pd, and Al-Zr at moderate temperatures by Sanders et al. (1997) find creep rates comparable to or lower than corresponding coarse-grain rates. The creep curves at low and moderate homologous temperatures (.24 – .48 TM) could be fit by the equation for exhaustion (logarithmic) creep. One explanation is that the observed low creep rates are caused by the high fraction of low energy grain boundaries in conjunction with the limitation on dislocation activity by the small grain sizes. In sum, the predicted ductility due to diffusional creep in nc brittle ceramics or intermetallics at temperatures significantly less than 0.5 TM has not been realized.

Superplasticity

Superplasticity is the capability of some polycrystalline materials to exhibit very large tensile deformations without necking or fracture. Typically, elongations of 100% to > 1000% are considered the defining features of this phenomenon. As grain size is decreased it is found that the temperature is lowered at which superplasticity occurs, and the strain rate for its occurrence is increased. As discussed previously, Equation 1 suggests that creep rates might be enhanced by many orders of magnitude and superplastic behavior might be observed in nc materials at temperatures much lower than 0.5 TM. As mentioned above, actual creep experiments have not borne out this prediction, but instead have shown creep rates (Equation 1) comparable to or lower than those in coarse-grained samples of the same material. This is presumably why little enhancement in ductility or superplastic behavior has been observed for nc materials at temperatures < 0.5 TM. However, there is evidence of enhancement of superplastic behavior in nc materials at temperatures > 0.5 TM. Superplasticity has been observed at somewhat lower temperatures and at higher strain rates in nc materials. The evidence for tensile superplasticity is limited and observed typically at temperatures greater than 0.5 TM and in materials that exhibit superplasticity in coarser grain sizes (1–10 μm). For example, Mishra et al. (1997) observed superplastic behavior in nc Pb-62%Sn at 0.64 TM and nc Zn–22%Al at 0.52 to 0.60 TM. However, Salishekev et al (1994) observed superplastic behavior in submicron—200 nm—Ti and several Ti and Ni base alloys. Here, superplasticity (190% elongation, m = 0.32) was observed in Ti at 0.42 TM. This was at a temperature 50°C lower than for 10 μm grain size Ti. The flow stress for the 200 nm Ti at 550°C was 90 MPa, compared to 120 MPa for 10 μm Ti at 600°C. Very recently, Mishra and Mukherjee (1997) have observed superplastic behavior in Ni3Al with a 50 nm grain size at temperatures of 0.56 to 0.60 TM to strains of 300 - 600%, but with unusual stress-strain behavior and significant apparent strain-hardening. These new results suggest very different mechanisms may be causing superplastic behavior in these nc materials.

Mechanical Properties

While there are still only limited data on the mechanical behavior— especially tensile properties—of nc materials, some generalizations may be made regarding the deformation mechanisms. It is likely that for the larger end of the nanoscale grain sizes, about 50 - 100 nm, dislocation activity dominates for test temperatures < 0.5 TM. As grain size decreases, dislocation activity apparently decreases. The essential lack of dislocations at grain sizes below 50 nm is presumably the result of the image forces that act on dislocations near surfaces or interfaces. The lack of dislocations in small, confined spaces such as single-crystal whiskers has been known for many years (Darken 1961). Creation of new dislocations is also made difficult as the grain size reaches the lower end of the nanoscale (< 10 nm). Stresses needed to activate dislocation sources, such as the Frank-Read source, are inversely proportional to the distance between dislocation pinning points. Since nanoscale grains will limit the distance between such pinning points, the stresses to activate dislocation sources can reach the theoretical shear stress of a dislocation-free crystal at the smallest grain sizes (~ 2 nm). Thus, at the smallest grain sizes we may have new phenomena controlling deformation behavior. It has been suggested that such phenomena may involve grain boundary sliding and/or grain rotation accompanied by short-range diffusion-assisted healing events (Siegel 1997). Several examples of deformation by shear banding have been reported for nc materials. Carsley et al (1997, 183-192) have studied nc Fe–10% Cu alloys with grain sizes ranging from 45 to 1,680 nm. In all cases, deformation in compression proceeds by intense localized shear banding. The stress-strain curves exhibited essentially elastic, perfectly plastic behavior; that is, no measurable strain hardening was observed. Shear banding is also the deformation mode observed in amorphous metallic alloys and amorphous polymers. The deformation shear banding in nc Fe–10% Cu was compared to that for metallic glasses, amorphous polymers, and coarsegrained polycrystalline metals after significant plasticity and work hardening had taken place. While this suggests a close similarity between deformation in nc materials and amorphous materials, not all tensile data on nc materials exhibit a lack of strain hardening. The Fe–10% Cu samples of Carsley et al. (1997, 183-192) showed shear bands even in their larger grained specimens (i.e., about 1,000 nm).

Theoretical Needs, Application, Opportunities and Challenges

Central to all of the above discussion is the lack of understanding of the microscopic deformation and fracture mechanisms in nc materials. Clearly, a stronger theoretical effort is needed to guide critical experiments and point the direction for optimizing properties. There has been limited work in this area, especially in Russia, in applying disclination theory to grain rotation (Romanov and Vladimirov 1992, 191), for example. However, a much larger theoretical effort is required. Another alternate deformation mechanism that may be important in nc materials is mechanical twinning (Huang et al. 1996). Little theoretical work has been carried out to address this possibility. Another important potential approach to understanding the deformation mechanisms in nc materials is to explore the rich older literature on shear banding—mechanical instabilities that have been observed in, for example, mild steels as grain size decreases; the approach to very low strain hardening (perfectly plastic behavior) is observed. Amorphous materials exhibit many of the phenomenological characteristics of deformation in nc materials, that is, shear banding, asymmetry between tensile and compressive behavior, and perfectly plastic behavior. However, deformation mechanisms are not well understood in amorphous materials either. Recent work on bulk metallic glasses may help clarify this.

Of the present or near-term applications for nc materials, the hard material WC/Co is an example of several important trends. Nanostructured WC/Co composites have been prepared that can have the following characteristics:

- nanoscale grain sizes after consolidation of powder
- hardness values about twice that of conventional micrograined WC/Co
- enhanced wear resistance and cutting performance

While this material has increased hardness and strength, preliminary reports from Stevens Institute of Technology in Hoboken, NJ, and the Royal Institute of Technology in Stockholm, Sweden, point to similar or increased fracture toughness values for nanostructured WC/Co. As noted earlier, single-phase nanostructured materials studied to date have exhibited high strength and hardness but brittle behavior at low homologous temperatures (< 0.5 TM). The results of these studies of WC/Co two-phase nanostructured materials suggest that the combination of high hardness/strength and toughness/ductility may be possible in multiphase nanostructured materials. Other examples that point to this possibility come from the work of Professor A. Inoue at Tohoku University in Japan, whose lab the WTEC panel visited. Inoue and coworkers have synthesized a variety of multiphase Al, Mg, and Ni-base alloys with nanoscale microstructures. Many of these alloys consist of nanocrystallites in an amorphous matrix. Some Al-rich alloys contain nanoscale quasi-crystalline particles surrounded by crystalline face-centered cubic Al. The fascinating properties of these multiphase nanostructured alloys include extremely high strength coupled with some ductility. Ductility is high in compression, but uniform elongation in tension is limited. Again, this behavior is analogous to that exhibited by ductile amorphous alloys. These results again suggest the possibility for development of nanostructured multiphase composites that combine extremely high hardness and strength with toughness and ductility. Such materials could have many applications as unique structural materials. Multiphase ceramic nanocomposites are the focus of several efforts. One significant effort is underway at the Research Center for Intermaterials of the Institute of Scientific and Industrial Research at Osaka University in Japan. Professor Koichi Niihara has a large effort studying micro-nano composites such as Al2O3/SiC that have enhanced toughness and also a variety of hard matrix/soft dispersion or soft matrix/hard dispersion nanocomposites. The enhanced toughness in some ceramic nanocomposites observed by Prof. Niihara has been verified by parallel studies of Dr. Steve Roberts at University of Oxford in England. While "low temperature" superplasticity has not been realized in nanostructured materials, enhanced superplastic behavior has been observed at elevated temperatures in terms of a somewhat lower temperature range for superplasticity and, perhaps more significant, a higher strain rate regime. While much needs to be understood about superplastic behavior of nanostructured materials, the possibility for using more conventional strain rates for forming has major industrial implications. This is of particular interest for near net shape forming of ceramics.

Perhaps the overriding issue for defining the mechanical behavior of nanostructured materials is the lack of understanding of the mechanisms for plastic deformation and fracture in these materials. An extensive experimental and theoretical research effort should be applied to solve these questions. Only with such understanding can the intelligent design of nanostructured materials with optimum mechanical properties be realized. Since many of the synthesis routes for nanostructured materials involve using powder or particulate products, compaction of such powders is needed to form bulk parts without coarsening the nanoscale microstructure. While there has been ongoing work on compaction and the thermal stability of nanocrystalline microstructures, more basic work is needed in this critical area. If some unique properties are limited to the finest grain sizes, methods must be found to stabilize the grain size while attaining theoretical density and complete particulate bonding. Furthermore, conventional processing methods are desirable if realistic scaleup and economy are to be realized. Near-term opportunities for structural applications of nanostructured materials are in the form of coatings prepared by, for example, thermal spray deposition or electrodeposition. There is a significant effort in the United States on thermal spray deposition of nc materials. Several groups in Canada lead in deposition of nc materials by electrodeposition methods. A spin-off company, Nanometals, formed under the auspices of Parteq Research and Development Innovations, has adopted a commercial electrodeposition method to produce nanostructured metals. Structural applications of nanostructured materials may be viewed as a focused approach to a long standing, well known design of materials with optimum mechanical properties. Use of nanoscale precipitates or dispersoids has been known for decades to improve mechanical behavior. The "new" emphasis on using nanostructured materials comes from 2 significant factors: 1. the special processing methods to push microstructures to the limits of the nanoscale 2. the unique properties and phenomena—as yet not well understood—when this limit is approached

Soft Magnetic Nanocrystalline Alloys

The discovery of nanocrystalline Fe-based soft magnetic materials is less than ten years old. The first class of such materials was the melt-spun Fe-Si-B alloys containing small amounts of Nb and Cu (Yoshizawa et al. 1988). The Fe-Si-B-Nb-Cu amorphous phase transforms to a body-centered cubic (bcc) Fe-Si solid solution with grain sizes of about 10 nm during annealing at temperatures above the crystallization temperature. The presence of small amounts of Cu helps increase the nucleation rate of the bcc phase while Nb retards the grain growth. These "Finemet" alloys provide low core losses (even lower than amorphous soft magnetic alloys such as Co-Fe-Si-B), exhibit saturation induction of about 1.2 T, and exhibit very good properties at high frequencies, comparable to the best Co-based amorphous alloys. These were first developed in Japan and have stimulated a large amount of research and development worldwide to optimize the magnetic properties. There has been relatively little work in the United States in this area, however. While many of the soft magnetic properties of Finemet-type nanocrystalline alloys are superior, they exhibit lower saturation inductions than Fe-metalloid amorphous alloys, mainly because of the lower Fe content to attain amorphization and because of the addition of Nb and Cu (or other elements to control the nucleation and growth kinetics). In order to remedy this problem, another class of Fe-based nanocrystalline alloys was developed by Inoue and coworkers at Tohoku University (Makino et al. 1997), which is commercialized by Alps Electric Co., Ltd., of Nagaoka, Japan. These "Nanoperm" alloys are based on the Fe-Zr-B system; they contain larger concentrations of Fe (83-89 at.%) compared to the Finemet alloys (~ 74 at.% Fe) and have higher values of saturation induction (~ 1.6-1.7 T). The Nanoperm nc alloys have very low energy losses at power frequencies (60 Hz), making them potentially interesting for electrical power distribution transformers. The issues of composition modification, processing, and the brittle mechanical behavior of these nanocrystalline/amorphous alloys are discussed by V.R. Ramanan in the first volume of this WTEC study, the proceedings of the May 8-9, 1997 panel workshop on the status of nanostructure science and technology in the United States (Ramanan 1998, 113-116).

Effective permeability, μe , vs. saturation magnetic flux density, Bs , for soft ferromagnetic materials (after A. Inoue 1997). While there has been extensive research on these alloys, particularly in Japan and Europe, most of the development has been carried out in Japan. The Finemet family of alloys is marketed by Hitachi Special Metals. Vacuumschmelze GmbH (Germany) and Impky (France) also market similar alloys. The Nanoperm alloys are being commercialized by Alps Electric Co. (Japan). No extensive research nor any commercialization of these materials has been carried out in the United States. The small single-domain nanocrystalline Fe particles in the amorphous matrix gives these alloys their unique magnetic behavior, the most dramatic being the lowest energy losses (narrowest B/H hysteresis loop) of any known materials, along with very high permeabilities. These alloys can also exhibit nearly or exactly zero magnetostriction. To date, these materials have been made by crystallization of rapidly solidified amorphous ribbons. Other methods that might provide geometrically desirable products should be explored or developed. Electrodeposition is one such method that requires further work. Electrodeposited nc Fe-Ni soft magnetic alloys are being developed in Canada. The brittle nature of these materials is a problem for scaleup and transformer manufacture. The brittleness problem must be solved by finding less brittle materials or applying the handling and processing knowledge that exists for embrittled (after annealing) metallic glasses.

Permanent Magnet Materials

The first attempts to produce nanoscale microstructures to enhance the magnetic properties of the Nb-Fe-B permanent magnetic materials used mechanical alloying of blended elemental powders followed by heat treatment (Schultz et al. 1987). Since the grain structure so obtained does not exhibit any crystallographic texture—and limits the energy product— special processing methods such as die-upsetting were used by Schultz and coworkers (1989) to provide the crystallographic anisotropy. While the coercivities of these nanocrystalline alloys are high, the remanent magnetization is decreased. Recent approaches to increasing the magnetic induction have utilized exchange coupling in magnetically hard and soft phases. The Fe-rich compositions (e.g., Fe90Nd7B3) result in a mixture of the hard Fe14Nd2B phase and soft aFe phase. The nanoscale two-phase mixtures of a hard magnetic phase and a soft magnetic phase can exhibit values of remanent magnetization, Mr, significantly greater than the isotropic value of 0.5 Ms. This "remanence enhancement" is associated with exchange coupling between the hard and soft phases, which forces the magnetization vector of the soft phase to be rotated to that of the hard phase (Smith et al. 1996). Two important requirements for alloys to exhibit remanence enhancement are a nanocrystalline grain size and a degree of coherence across interphase boundaries sufficient to enable adjacent phases to be exchange-coupled. The significant feature of the exchange coupling is that it allows crystallographically isotropic materials to exhibit remanence values approaching those achieved after full alignment. Such two-phase nanoscale ferromagnetic alloys have been prepared by nonequilibrium methods such as melt-spinning, mechanical alloying, and sputter deposition. Besides the high reduced remanence, the material cost is reduced by reduction in the content of the expensive hard rare earth-containing magnetic phase. The theoretical understanding of remanence enhancement appears to be developed to a degree enabling prediction of magnet performance; however, this performance, while a significant improvement over single-phase isotropic magnets, does not reach predicted values. Work is required on optimizing the orientation relationships between the hard and soft phases and the interphase properties (coherency) between them. Research on nanocrystalline hard magnetic alloys has received attention worldwide. The U.S. efforts are summarized in the article by G.C. Hadjipanayis (1998, 107-112). While less research seems to be carried out in the world on these materials compared to the nanocrystalline soft magnetic alloys, some efforts exist in most countries. Notable programs are those of L. Schultz and coworkers at the Institut für Festköper und Werkstofforschung (IFW) in Dresden and P.G. McCormick and coworkers at the University of Western Australia.

While the very low losses of the nc soft magnetic materials (Finemet or Nanoperm) are dependent on grain size for their properties, the hard magnetic nc alloys with remanence enhancement provide flexibility in processing, especially with powder materials. These remanence-enhanced nc hard magnetic alloys may find many applications as permanent magnet components.

Giant Magnetoresistance (GMR)

The phenomenon of giant magnetoresistance (GMR)—the decrease of electrical resistance of materials when exposed to a magnetic field—was first reported in a number of multilayer ferromagnetic/nonferromagnetic thin film systems (Baibich et al. 1988). More recently, GMR was observed in equiaxed granular nanocrystalline materials (Berkowitz et al. 1992). In particular, GMR systems with low saturation fields offer a wide area for application in magnetoresistive devices. GMR sensors have a higher output than conventional anisotropic magnetoresistive sensors or Hall effect sensors. They can operate at higher magnetic fields than conventional magnetoresistive sensors. In multilayer systems the antiferromagnetic alignment of the ferromagnetic layers in zero field becomes ferromagnetic as the field is applied and causes a decrease in resistance. Granular materials that show GMR consist of small ferromagnetic single-domain particles with randomly oriented magnetic axes in a nonmagnetic matrix. An external field rotates the magnetic axes of all magnetic particles. The rotation towards complete alignment of all magnetic axes again reduces the resistance in a similar way as for multilayers. The GMR in granular systems is isotropic. The explanation for the GMR is spin-dependent scattering of the conduction electrons at the ferromagnetic/nonmagnetic interfaces and, to a lesser extent, within the magnetic grains. The GMR scales inversely with the average particle diameter. There is worldwide research on the GMR effect. U.S. programs are reviewed by R. Shull and G.C. Hadjipanayis in the proceedings of the WTEC U.S. nanotechnologies workshop (Shull 1998, 43-58; Hadjipanayis 1998, 107-112). The NIST work described by Shull has provided material with the largest GMR values for the smallest switching fields. Japanese research on GMR includes studies in Prof. Fujimori's group at Tohoku University. While the theory for GMR of spin-dependent scattering referred to above has been used as an explanation, other explanations taking into account interaction between magnetic regions have been proposed (El-Hilo et al. 1994). Combined theoretical and experimental studies should help to clarify the mechanism for this effect

Other Ferromagnetic Nanocrystalline Materials

Magnetic nanocomposite refrigerants, which have four times the magnetocaloric effects of the best low temperature magnetic refrigerant, were developed by NIST and described by R. Shull (1998, 43-58). The entropy change at a given (low) temperature for a system of magnetic spins is enhanced when the isolated spins are clustered. Shull et al. (1993) have shown that the nanocomposite $Gd_3Ga_{5-x}Fe_xO_{12}$ gives superior magnetocaloric effects, which increase with x up to x = 2.5 and can be extended to higher temperatures than conventional materials. Magnetostrictive materials such as Terfonol-D ($Tb_{0.3}Dy_{0.7}Fe_2$) have been of scientific and technological interest in recent years. It is suggested by G.C. Hadjipanayis (1998, 107-112) that nanostructured magnetostrictive materials can have improved properties, such as lower saturation fields, with reduced anisotropy and in multilayers with alternate layers of magnetostrictive and soft magnetic materials that are exchange-coupled. Hadjipanayis states that most of the research in this area is carried out in Japan and Europe. Nanocrystalline magnetic materials offer perhaps the nearest-term prospect of significant applications of bulk nanostructured materials. The remanence enhancement in two-phase hard/soft magnetic materials can result in excellent energy products—comparable to those from rapid solidification processing routes—in powder composites. The flexibility in manufacturing this allows should provide many possible permanent magnet applications. Nanocrystalline soft magnetic materials have the lowest energy losses of any material. While problems such as mechanical brittleness remain, these materials promise to replace existing transformer core materials for power applications. The U.S. research and industrial effort in these materials lags the efforts in Japan.

Nanocrystalline Hydrogen Storage Materials

R.B. Schwarz (1998, 93-95) has pointed out that nanostructured materials offer several potential advantages for hydrogen storage materials. Rapid kinetics of absorption/desorption can be aided by refining the microstructure to the nanoscale. For example, nanoscale inclusions of Mg_2Ni in Mg catalyze the decomposition of the molecular hydrogen, increasing the hydrogen absorption/desorption kinetics. Another advantage of the nanoscale microstructure is that the alloy powder does not comminute on repeated charging/discharging with hydrogen. This is not strictly a bulk material, since powder agglomerates or green compacts can be used, thus obviating the need for compaction to theoretical density.

Nanocrystalline Corrosion-Resistant Materials

The limited work to date on corrosion resistance of nanocrystalline materials indicates that no generalizations can be made. Superior localized corrosion resistance in HCl was observed for nanocrystalline 304 stainless steel (Fe–18%Cr–8%Ni) prepared by sputter deposition (Inturi and Szklavska-Smialowska 1991 and 1992). This was attributed to the fine grain size and homogeneity of the nc material. However, the average dissolution rate of nc Ni was found to be higher than that for coarse-grained material (Rofagha et al. 1991).

Polymers, Polyelefin and Polymer-Silicate

There has been enormous interest in the commercialization of nanocomposites for a variety of applications, and several of these applications will be successful in the near future. Mineral fillers, metals, and fibers have been added to thermoplastics and thermosets for decades to form composites. Compared to neat resins, these composites have a number of improved properties including tensile strength, heat distortion temperature, and modulus. Thus, for structural applications, composites have become very popular and are sold in billion pound quantities. These filled thermoplastics are sold in even larger volumes than neat thermoplastics. Furthermore, the volume of fillers sold is roughly equal to the volume of thermoplastic resin sold. Clearly, the idea of adding fillers to thermoplastics and thermosets to improve properties, and in some cases decrease costs, has been very successful for many years. More recently, with advances in synthetic techniques and the ability to readily characterize materials on an atomic scale has lead to interest in nanometer-size materials. Since nanometer-size grains, fibers and plates have dramatically increased surface area compared to their conventional-size materials, the chemistry of these nanosized materials is altered compared to conventional materials. Polymer nanocomposites combine these two concepts, i.e., composites and nanometer-size materials. Thermoplastics filled with nanometer-size materials have different properties than thermoplastics filled with conventional materials. Some of the properties of nanocomposites, such as increased tensile strength, may be achieved by using higher conventional filler loading at the expense of increased weight and decreased gloss. Other properties of nanocomposites such as clarity or improved barrier properties cannot be duplicated by filled resins at any loading. Polymer nanocomposites were developed in the late 1980s in both commercial research organizations and academic laboratories. The first company to commercialize these nanocomposites was Toyota, which used nanocomposite parts in one of its popular car models for several years. Following Toyota's lead, a number of other companies also began investigating nanocomposites. Most commercial interest in nanocomposites has focused on thermoplastics. Thermoplastics can be broken into two groups: less expensive commodity resins and more expensive (and higher performance) engineering resins. One of the goals of nanocomposites was to allow substitution of more expensive engineering resins with a less expensive commodity resin nanocomposite. Substituting a nanocomposite commodity resin with equivalent performance as a more expensive engineering resin should yield overall cost savings. By a strict definition of nanocomposites, i.e., any filler submicron in size, there already are significant volumes of nanocomposites being produced. These amount to more than 20 million pounds. However, since these fillers are on the upper end of the nanocomposite size range, most sources have excluded them from consideration. This report covers these products but does not add their volumes into aggregate numbers of nanocomposite markets or production volumes to comply with the established convention. At this point in time, there has been much less commercial interest in thermoset nanocomposites compared to thermoplastics. This neglect may not continue much longer since thermoset nanocomposites have some distinct advantages over neat thermoset resins. This report covers some possible markets for thermoset nanocomposites. There are two types of recognized polymer nanocomposites:

- Thermoplastic nanocomposites: these materials are broken into two major categories, i.e., commodity resins and engineering resins. The potential of polymer nanocomposite commodity resin is covered for each resin. Engineering resin nanocomposites are restricted to resins that have been under development.
- Thermoset nanocomposites: thermoset nanocomposites have received less commercial interest in their development than thermoplastic nanocomposites, but these materials may be relatively straightforward to bring into production. Furthermore, thermoset nanocomposites can offer some significant advantages over conventional thermosets.

Polymer nanocomposite (PNC) is a polymer or copolymer having dispersed in it nanoparticles. These may be of different shape (e.g., platelets, fibers, spheroids), but at least one dimension must be ca. 1 to 50 nm (diameter of pencil lead is about 1 mm, or 1,000,000 nm). PNC with all three types of nanoparticles have been prepared (e.g., polycarbonate with carbon nanotubes, polyamide with iron oxide spheres), but only PNC with clay platelets are on the market for the structural (high volume) applications. These PNC's belong to the category of multi-phase systems (MPS, viz. blends, composites, and foams) that consume ca. 95 per cent of plastics production. The key to industrial success of MPS is the desired and stable morphology. Thus, these systems require controlled mixing/compounding, stabilization of the achieved dispersion, orientation of the dispersed phase, and optimization of interactions in the finished product. In spite of differences in the nature of the dispersed phase, the compounding strategies for all MPS, including PNC, are similar.

Mixing is the most important operation in polymer processing. Uniformity of the molecular weight and its distribution, entanglement, composition, as well as the stress history and temperature, is the key to good performance. Dispersive (or intensive) and distributive (or extensive) mixing of molten polymers is carried out within the laminar flow regime. The dispersive action is responsible for the breaking of original domains of the minor component into the desired size, while the distributive provides for their homogeneous distribution within the matrix. Microrheology may be used to describe the dispersive process in the shear or extensional flow field, while distributive mixing may be treated in terms of the laminar flow theory. Both models - microrheology and laminar mixing - indicate superiority of the extensional flow field. The presence of the vorticity component in the shear tensor is responsible for rotation of the flow elements that cause excessive energy consumption, polymer degradation and attrition of anisometric solid particles - shear mixing is inefficient. Mixing in the extensional flow field may be carried out by means of the convergent-divergent flow geometry. A motionless extensional flow mixer (EFM) and its newer, dynamic version (DEFM) are fully adjustable, general-purpose mixers used for homogenization of reaction products, removal of "fish eyes" from films, alloying, blending or dispersing of nanoparticles[4]. In the search for synergism the industry moves toward progressively complex MPS, e.g., foaming of nanoclay reinforced polymer blends. Here, the difficulty is the presence of several phases with potentially different thermodynamic and rheological character as well as reinforcing and foaming the appropriate blend components. However, once the difficulties are over, the advantages translate into superior performance. The clay-containing polymeric nanocomposites (CPNC) constitute new type of polymeric systems, with a great variety of molecular structures and processing difficulties. The only type of nanoparticles used in commercial production of CPNC is the lamellar, expandable, smectite clay. For the best performance of CPNC the clay platelets (ca. 1 nm thick and 300 nm diameter) must be homogeneously dispersed as individual crystalline lamellae (known as exfoliation). To accomplish this task, a balance between the thermodynamics and mechanical mixing must be achieved. Owing to the large aspect ratio, at a concentration of clay exceeding ca. 1 wt per cent the platelets start to interact, creating 3D structures. In commercial CPNC the clay content is about twice as large, hence there is local ordering. Generation of the most suitable orientation is the key for the optimization of the performance. Three recent developments in the CPNC science and technology are worth noting.

The use of an elongational flow field greatly accelerates dispersion and orientation, facilitating the processing. Several CPNC types (e.g., with polyethylene terephthalate, PET, polystyrene, PS, or polypropylene, PP, matrix) have been prepared using commercial EFM attached either to single-screw extruder (SSE) or twin-screw extruder (TSE). The results demonstrated significant beneficial role even of the standard EFM (developed for melt homogenization and polymer blending). For example, PET was extruded with organoclay using a TSE and the same machine, but with attached to it EFM. The presence of EFM increased the PNC properties over that from TSE by (data from EFM Inc.):

(i) Tensile modulus by ca. 67 per cent.

(ii) Tensile strength by ca. 60 per cent.

(iii) Impact strength 80 per cent.

In Union Chemical Laboratory, Industrial Technology Research Institute (UCL/ITRI) in Taiwan, PP was compounded with organoclay and a compatibilizer (maleated polypropylene, PP-MA)[5]. Three processing methods were used: #1: TSE; #2: TSE + EFM; and #3: SSE + EFM. The method #1 resulted in mixtures with clay dispersed as micron-sized aggregates. By contrast, process #2 produced well-intercalated CPNC with short stacks of organoclay uniformly dispersed in the molten PP. However, the best dispersion was obtained from #3, i.e., using SSE + EFM. This unexpected result was most likely due to higher thermo-mechanical degradation in TSE.

The ammonium-ion intercalated organoclay is the main type used for the production of CPNC. However, these materials start to degrade at temperatures above 180°C by the Hofmann elimination mechanism, leaving the clay surface bare. Numerical simulations were carried out to evaluate the effects of independent parameters on the degree of CPNC dispersion. The most dramatic worsening of the thermodynamic conditions for exfoliation was obtained in the presence of bare clay platelets - even as small amount as 1 or 2 per cent of clay surface precludes exfoliation.

Kato et al. from the Toyota Central R&D Lab proposed a revolutionary method of CPNC production. The method uses un-treated clay, e.g., sodium montmorillonite (Na-MMT), which becomes exfoliated and the morphology stabilized during the mixing/compounding step. The process uses a long (L/D = 77) TSE, divided by a pair of seal rings into three zones. In the first zone PP + PP-MA + Na-MMT were melt mixed together. In zone #2 ca. 10 per cent water was injected into the melt and the pressure of the water vapor was well controlled. Adsorption of water by clay platelets, followed by reaction with compatibilizer (PP-MA), resulted in exfoliation and stabilization of the dispersion. Finally, in zone #3 devolatilization took place. The resulting CPNC had almost the same properties as the conventionally prepared one, but at a fraction of cost. The method bypasses the need for the costly intercalation of Na-MMT with ammonium ions, avoids the introduction of readily degradable organoclay to molten polymer (which in turn gets degraded by combination of organoclay degradation and air) and at the same time eliminates the need of costly in time and energy pre-drying step. Evidently, the method is in an early stage of development. Its extension to other polymers (e.g., polyesters or polyamides), optimization of the compatibilizer, processing equipment and process parameters are required.

The main advantages of CPNC stem from the improved stiffness (by a factor of 2 at 5 wt per cent loading), reduction of gas and vapor permeability (by a factor of 100 at 10 wt per cent loading), and reduced flammability. Since these enhancements are achieved with only a slight increase of density, the transport, and packaging industries are expect to be the first to profit from this new technology. Up to now two principal obstacles hindered progress of CPNC into the market: the cost and reproducibility of performance. The demonstrated by Kato et al. suitability of Na-MMT reduces the cost of the nano-filler from ca. \$7 to \$2 per kilogram, and at the same time reduces the probability of thermo-mechanical degradation during processing and forming. Many questions remain the foremost being whether the achieved improvement of technology is sufficient to lower the cost and improve stability to satisfy the industry standards. Evidently, the future of CPNC technology hinges on the intelligent application of the available compounding and processing technologies, adopted for the specific needs of the nano-scale mixing.

The subject of hybrids based on layered inorganic compounds such as clays has been studied for a considerable time, but the area is enjoying a resurgence of interest and activity as a result of the exceptional properties which can be realized from such nanocomposites. Materials variables which can be controlled and which can have a profound influence on the nature and properties of the final nanocomposite include the type of clay, the choice of clay pre-treatment, the selection of polymer component and the way in which the polymer is incorporated into the nanocomposite. The last of these may be dictated by the available processing methods and whether the prospective user is an integrated polymer manufacturer or a specialist processor.

Common clays are naturally occurring minerals and are thus subject to natural variability in their constitution. The purity of the clay can affect final nanocomposite properties. Many types of clay are aluminosilicates, which have a sheet-like (layered) structure, and consist of silica SiO_4 tetrahedra bonded to alumina AlO_6 octahedra in a variety of ways. A 2:1 ratio of the tetrahedra to the octahedra results in smectite clays, the most common of which is montmorillonite. Other metals such as magnesium may replace the aluminium in the crystal structure. Depending on the precise chemical composition of the clay, the sheets bear a charge on the surface and edges, this charge being balanced by counter-ions, which reside in part in the inter-layer spacing of the clay. The thickness of the layers (platelets) is of the order of 1 nm and aspect ratios are high, typically 100-1500. The clay platelets are truly nanoparticulate. In the context of nanocomposites, it is important to note that the molecular weight of the platelets (ca. 1.3×10^8) is considerably greater than that of typical commercial polymers, a feature which is often misrepresented in schematic diagrams of clay-based nanocomposites. In addition, platelets are not totally rigid, but have a degree of flexibility. The clays often have very high surface areas, up to hundreds of m^2 per gram. The clays are also characterized by their ion (e.g. cation) exchange capacities, which can vary widely. One important consequence of the charged nature of the clays is that they are generally highly hydrophilic species and therefore naturally incompatible with a wide range of polymer types. A necessary prerequisite for successful formation of polymer-clay nanocomposites is therefore alteration of the clay polarity to make the clay 'organophilic'. An organophilic clay can be produced from a normally hydrophilic clay by ion exchange with an organic cation such as an alkylammonium ion. For example, in montmorillonite, the sodium ions in the clay can be exchanged for an amino acid such as 12-aminododecanoic acid (ADA):

$$Na^{+}\text{-CLAY} + HO_2C\text{-R-}NH_3^{+}Cl^{-} \text{ à.}HO_2C\text{-R-}NH_3^{+}\text{-CLAY} + NaCl$$

The way in which this is done has a major effect on the formation of particular nanocomposite product forms and this is discussed further below. Although the organic pre-treatment adds to the cost of the clay, the clays are nonetheless relatively cheap feedstocks with minimal limitation on supply. Montmorillonite is the most common type of clay used for nanocomposite formation; however, other types of clay can also be used depending on the precise properties required from the product. These clays include hectorites (magnesiosilicates), which contain very small platelets, and synthetic clays (e.g. hydrotalcite), which can be produced in a very pure form and can carry a positive charge on the platelets, in contrast to the negative charge found in montmorillonites.

The synthetic route of choice for making a nanocomposite depends on whether the final material is required in the form of an intercalated or exfoliated hybrid. In the case of an intercalate, the organic component is inserted between the layers of the clay such that the inter-layer spacing is expanded, but the layers still bear a well-defined spatial relationship to each other. In an exfoliated structure, the layers of the clay have been completely separated and the individual layers are distributed throughout the organic matrix. A third alternative is dispersion of complete clay particles (tactoids) within the polymer matrix, but this simply represents use of the clay as conventional filler.

In recent years, there has been extensive study of the factors which control whether a particular organo-clay hybrid can be synthesised as an intercalated or exfoliated structure. Since clay nanocomposites can produce dramatic improvements in a variety of properties, it is important to understand the factors which affect delamination of the clay. These factors include the exchange capacity of the clay, the polarity of the reaction medium and the chemical nature of the interlayer cations (e.g. onium ions). By modifying the surface polarity of the clay, onium ions allow thermodynamically favourable penetration of polymer precursors into the interlayer region. The ability of the onium ion to assist in delamination of the clay depends on its chemical nature such as its polarity. The loading of the onium ion on the clay is also crucial for success and it should be borne in mind that a commercial organo-clay may not have the optimum loading for a particular application. For positively charged clays such as hydrotalcite, the onium salt modification is replaced by use of a cheaper anionic surfactant. Other types of clay modification can be used depending on the choice of polymer, including ion-dipole interactions, use of silane coupling agents and use of block copolymers. An example of ion-dipole interactions is the intercalation of a small molecule such as dodecylpyrrolidone into the clay. Entropically-driven displacement of the small molecules then provides a route to introducing polymer molecules.

Unfavourable interactions of clay edges with polymers can be overcome by use of silane coupling agents to modify the edges. These can be used in conjunction with the onium ion treated organo-clay. An alternative approach to compatibilising clays with polymers has been introduced by TNO (Netherlands), based on use of block or graft copolymers where one component of the copolymer is compatible with the clay and the other with the polymer matrix. This is similar in concept to compatibilisation of polymer blends. A typical block copolymer would consist of a clay-compatible hydrophilic block and a polymer-compatible hydrophobic block. The block length must be controlled and must not be too long. High degrees of exfoliation are claimed using this approach.

The correct selection of modified clay is essential to ensure effective penetration of the polymer or its precursor into the interlayer spacing of the clay and result in the desired exfoliated or intercalated product. Indeed, further development of compatibiliser chemistry is undoubtedly the key to expansion of this nanocomposite technology beyond the systems where success has been achieved to date. Polymer can be incorporated either as the polymeric species itself or via the monomer, which is polymerized in situ to give the corresponding polymer-clay nanocomposite. The second of these is the most successful approach to date, although it probably limits the ultimate applicability of these systems. Polymers can be introduced either by melt blending, for example extrusion, or by solution blending. Melt blending (compounding) depends on shear to help delaminate the clay and can be less effective than in situ polymerization in producing an exfoliated nanocomposite. Both thermo sets and thermoplastics have been incorporated into nanocomposites, including:

- Nylons
- Polyolefins, e.g. polypropylene
- Polystyrene
- Ethylene-vinyl acetate (EVA) copolymer
- Epoxy resins
- Polyurethanes
- Polyimides
- Poly(ethylene terephthalate) (PET)

Polyolefin nanocomposites are useful for a wide variety of applications in packaging, engineering and fire retardant markets. These nanocomposites are near molecular blends of montmorillonite nanoclay and a polyolefin matrix. Concentrate trade name. Here performance properties are outlined for typical polyolefins with explanations of options for creating nanocomposites.

Polyolefins represent a wide range of homo and copolymers, principally containing propylene and ethylene. Also included are polyolefin/rubber alloys usually called TPO's and polyolefin/EVA blends. Because propylene and ethylene contents as well as polydispersities vary widely, some Nanomer Concentrate screening is required to determine the optimum match between the matrix and nanoclay surface treatment.

Polyolefin nanocomposites are produced via extrusion compounding using Concentrates. Refer to data for concentrate specifications and grade selection. Film or bottle grade products require twin screw compounders or intensive mixers. Most engineering and fire retardant applications can be accommodated using single screw technology. A specially designed screw, called a NanoMixer™, must be used. Compounding guidelines are available on Tech Data Sheets P-803 through P-806.

Nanocomposite performance depends on the degree of nanoclay de-agglomeration and dispersion within the polyolefin matrix. The process of de-agglomeration/dispersion is commonly referred to as exfoliation. Three factors influence exfoliation: the nanoclay surface treatment, nanocomposite loading, and the amount of dispersive energy used. The optimum nanoclay surface treatment is readily determined through screening, as described earlier. Nanomer loading will not hinder exfoliation within the 2-8 per cent wt/wt range and is optimized at 6 per cent wt/wt. The proper choice of compounding equipment and use of the guidelines assists in achieving the dispersive energy requirement. However, another factor impacting dispersive energy is matrix resin melt flow. Resin melt flow influences the level of shear resulting from dispersive energy input. The lower the melt flow, the greater the shear on Nanomer particles. It follows that exfoliation is maximized in lower melt flow resins, where shear is the highest.

Polyolefin nanocomposites exhibit mechanical properties similar to highly filled systems, but with the added advantage of low density. The density (and resulting part weight) is just slightly higher than unfilled resin. Nanocompositing also provides a high degree of reinforcement without significant loss of impact. In most cases notched Izod impact remains the same as neat (unfilled) resin.

Polyolefin nanocomposites have specific gravities and melt flows nearly the same as neat resin, so they process like neat resin. Part output rates will be significantly higher than conventionally filled composites and machine wear will also improve noticeably. Nanocomposites can be used in conjunction with conventional fillers, i.e. glass fiber or talc, to create synergistic materials with customized property sets. For example a 6 per cent wt/wt nanocomposite containing 10-12 per cent wt/wt glass fiber will deliver a property set similar to 30 per cent wt/wt straight glass fiber, but lower specific gravity and part weight. Polyolefin nanocomposites are potent char formers. In fire retardant (FR) applications they function by insulating the resin substrate from the air-surface interface, starving the flame by reducing the fuel needed to continue the burning process. The result is a dramatically lower peak heat release rate (HRR). Nanocomposites alone do not conform to a UL V0 rating. But in conjunction with reduced levels of conventional FR additives they act synergistically to achieve V0. Often traditional FR additives can be reduced by 50 per cent wt/wt. With less FR additive dilution and the enhanced reinforcing effect of nanocompositing, mechanical properties are restored, usually at lower cost than would be the case without nanocompositing. Another benefit of nanocomposite/FR additive synergy is lower density and part weight, since a lot of high density additive has been removed from the formulation.

As polyolefins take an increased role in packaging applications, gas barrier becomes an important requirement. Nanocomposites improve oxygen transmission rates (OTR) between 1.2-2X. Carbon dioxide transmission rates (CO2TR) improve 1.4-1.8X. There is also some improvement in water vapor transmission (WVTR), around 10-15 per cent. Improved gas barrier follows the tortuous path model and is valid for cast, blown and molded products. Present performance data for a variety of polyolefins. Predictably, nanocomposite films possess greater tensile strength and modulus, although elongation-atbreak is reduced. Simultaneous improvement in tensile properties and gas barrier provides the formulator with opportunities to downguage films and containers. Tear strengths in both the machine and transverse directions remain the same as neat polyolefin films of the same mil thickness. Tear strength of downguaged nanocomposite film drops in proportion to the reduction in film thickness.

The earliest example of the in situ polymerization method was work by Toyota on synthesis of clay-nylon nanocomposites and this remains probably the most studied system, including work by Bayer and Ube. In a typical synthesis ADA-modified clay is dispersed in the monomer caprolactam, which is polymerized to form the nylon-6-clay hybrid as an exfoliated composite. Complete exfoliation may be preceded by intercalation of the monomer in the clay. Generally, low concentration of clay (a few per cent) are incorporated in these nanocomposites, partly because this is often sufficient to modify the desired properties significantly, but also because higher levels of clay can adversely increase the system viscosity leading to poor processability, although the viscosity increase is shear rate dependent. Other nylons and copolyamides (e.g. nylon-6/6,6) have also been incorporated in clay nanocomposites. Functionality such as hydroxyl groups can be introduced into the onium salt modifiers to improve compatibility with the nylon via hydrogen bonding – this can lead to an enhancement of desirable nanocomposite properties. Similar modification of the chemistry may be required for successful exfoliation with other types of polymer system. In the case of ethylene-vinyl alcohol (EVOH) copolymers, use of hydroxylated quaternary ammonium ions improves compatibility between the clay and the EVOH by introducing favourable hydroxyl group interactions. In polypropylene (PP) nanocomposites, maleic anhydride grafted PP is used as a compatibiliser. Polymerization initiators can be anchored to the clay platelet surface and this approach has been extended to living free radical polymerization of styrene, where an initiating species bonded to the TEMPO free radical is attached to the surface of the clay platelets via ion exchange.

For preparation of nanocomposites from high temperature engineering thermoplastics, a major limitation in the use of conventional onium ion modified clays is the thermal instability of the alkylammonium species during processing. The block copolymer route developed by TNO (vide supra) offers one potential solution to this problem. Imidazolium salts such as that depicted. A further alternative is the use of phosphonium salts in place of ammonium salts—this can lead to an increase in the degradation temperature of the organo-clay from 200-300°C to >300°C. Wholly synthetic organo-clays clays, on the other hand, can exhibit thermal stability to beyond 400°C. By using these last two approaches, Triton Systems has succeeded in producing nanocomposites from high temperature resins such as polyetherimide (PEI). Forming the nanocomposite provides a potential route to producing 'speciality' engineering resin performance from cheaper 'standard' engineering resins.

Future goals include the extension of this nanotechnology to additional types of polymer system, where the development of new compatibilisation strategies is likely to be a prerequisite. Production of PVC-based systems is still some way off and challenges remain to be solved in PET nanocomposites. Additional reinforcement of clay nanocomposites by glass fibre is currently being investigated. There is also interest in the development of electrically conducting clay nanocomposites.

Polymer nanocomposites, particularly polymer-silicate nanocomposites, represent an emerging class of specialty plastic materials. The melt-processing route would be a powerful approach in the manufacture of these compounds. Techniques such as transmission electron microscopy and x-ray diffraction were used to determine that nylon-6 nanocomposites could be produced via melt compounding. As a result of the "nano"-sized dispersion, reinforcement efficiency is realized at low loading levels (3-5 per cent) which is a small cost in terms of specific gravity. Dramatic enhancements are observed in heat distortion temperature and stiffness. Besides mechanical properties, nanocomposites also have much lower gas permeability than the neat resin. Other potential benefits include good surface appearance, dimensional stability, and lower heat release. Nanocomposites, particularly polymer-silicate nanocomposites, are materials that have been given much attention recently and represent an emerging class of specialty plastic compounds. One of the reasons for this attention is that due to the "nano"-scale dispersion, which results in high aspect ratio and surface area, the reinforcement efficiency of polymer-silicate nanocomposites can be significantly better than conventional mineral fillers. The potential benefits of polymersilicate nanocomposites include higher heat resistance, increased flexural modulus, lower gas permeation rate, easy flow, dimensional stability, good surface appearance, and lower heat release rate. Typically, the type of silicate used in polymersilicate nanocomposites is montmorillonite which is a 2:1 layered smectite clay. This clay has a platy structure with individual platelets having thicknesses of one nanometer and surface lengths on the order of 100 to 1000 nanometers. Montmorillonite is hydrophillic which makes proper exfoliation and dispersion into conventional polymers difficult. Thus, montmorillonite is usually modified through substitution of sodium ions with organic onium ions. The result is an organo-clay complex.

The intercalation procedure expands the gallery and improves the compatibility of the complex with the polymer so that individual platelets can be more easily separated. Possible morphologies of polymers and organo-clay complexes.There are several potential approaches for the production of the polymer-silicate nanocomposite. One procedure involves the synthesis of the compound in the polymerization stage with the complex *in-situ* with the liquid monomer. Also, nanocomposites could be prepared from solvents in which both polymer and complex are dissolved. Another method involves the preparation of polymer-silicate nanocomposites via melt processing. In this process, the application of shear during compounding assists exfoliation and dispersion. The melt-compounding route is a powerful approach in the production of polymersilicate nanocomposites since existing technologies and equipment could be utilized and easily scaled to commercial quantities. Additionally, this approach allows specialty materials such as those with custom colours, static dissipation, or wear resistance to be readily manufactured.

Polymer-silicate compounds were prepared by compounding various organo-clay complexes at different loading levels with commercially available nylon-6 using a twin-screw extruder. Neat nylon-6 and the compounds were dried at 80°C under vacuum prior to compounding, molding, and rheometry. The extent of intercalation, exfoliation, and dispersion of the complexes in the compounds was determined. This was accomplished through wideangle x-ray diffraction (XRD) for preliminary determinations. Transmission electron microscopy (TEM) was used to confirm the morphology development established by XRD. Test specimens were prepared by injection molding. The mechanical and electrical properties of the specimens were evaluated according to ASTM test methods: D-638 for tensile, D-790 flexural, D-256 for Izod impact, D-648 for deflection temperature, D-955 for shrinkage, and D-570 for water absorption. The barrier capability of these materials was explored by measuring oxygen transmission rates (OTR) with a Mocan apparatus. OTR measurements were performed at constant temperature and humidity for 200 hours on specimens thermoformed from injection molded plaques. Compound melt flow behavior was analyzed with a capillary rheometer. The effect of shear rate at 235°C and melt flow stability at 235°C and 1000 sec^{-1} were determined.

Suitable morphology development in a nanocomposite via the compounding route consists of wetout, intercalation, exfoliation, and dispersion. Therefore, the selection of an appropriate complex for the matrix polymer plays a critical role in determining the potential degree of exfoliation and dispersion in a polymer-silicate compound. One procedure that can be used to assess the extent of intercalation and exfoliation is to analyze x-ray diffraction patterns. XRD patterns for nylon-6 compounds containing two different types of complexes. Finally, not only does this show that XRD patterns can be used to evaluate nanocomposite development but also that nylon-6 nanocomposites are indeed possible via melt compounding. The properties of a typical nylon-6 nanocomposite are given in Table and compared to the unfilled matrix nylon-6. At almost no cost in specific gravity, significant improvements in certain properties are realized. For instance, flexural modulus is increased by almost 50 per cent and deflection temperature raised by over 35°C. Also, tensile strength is increased by over a third. For comparison purposes, the properties of typical mineral and glass fiber filled nylon-6 compounds are given. Reinforcement efficiencies of nanocomposites compared to conventional mineral fillers on a weight basis.

Besides mechanical and heat resistance property improvements, nanocomposite compounds can exhibit dramatically enhanced barrier properties as represented by lower oxygen transmission rates. Nylon-6 nanocomposites can achieve an OTR almost four times lower than the unfilled nylon-6. The reduction in OTR is typically attributed to the increase in effective diffusion length as solutes must travel a tortuous path around well-dispersed platelets of high aspect ratio. It is also important to point out that as a result of the nanometer length scale that excellent transparency is retained in sheets and films formed from nylon-6 nanocomposites. There has been some concern about the ability to process nanocomposites due to the high surface area of the nanoclay upon exfoliation. For instance, the effect of other high surface area additives such as high structure carbon black is to dramatically increase the viscosity. The viscosity of a typical nylon-6 nanocomposite is compared to the viscosity of the neat polymer as a function of shear rate. Contrary to the concerns, the viscosity is observed to be lower than the neat resin at the shear rates of most interest to potential processors. As a result, nylon-6 nanocomposites may be processed as easily as the neat resin. There are several potential explanations for the shear thinning at higher shear rates. One possible reason could be due to orientation of the platelets upon shear. Another reason could be that the polymer has been degraded during compounding with the nanoclay. However, degradation was not obvious in the nylon-6 nanocomposite samples. The melt flow stability of a nylon-6 nanocomposite was determined. The viscosity was found to not significantly decrease with time. Additionally, the viscosity is equivalent to the viscosity measured under similar conditions during prior testing of the effect of shear rate. Therefore, the degradation if present, would not be expected to continue on further processing. The dimensional stability of nylon-6 nanocomposites is compared. The unfilled resin and other typical reinforced compounds. As seen, the water absorption of the nanocomposite is dramatically reduced by as much as 55 per cent. This is comparable to a filled compound which contains much higher content. This is due to the barrier properties since resin replacement cannot entirely account for all of the change as is the case for conventional mineral and glass compounds. Also observed in table is that the shrinkage of nylon-6 nanocomposites is more similar to the base resin due to the lower filler content. Additionally, the shrinkage in the transverse direction does not differ significantly from the flow direction. This gives a warpage factor much closer to one as shown.

Polymer-clay nanocomposites have been developed via the compounding route in nylon-6. XRD and TEM analysis confirmed that good exfoliation and dispersion was attained. Notable property improvements were observed in tensile strength, flexural modulus, heat deflection temperature, and oxygen transmission rate. Proposed application areas include those requiring structural properties. For structural applications, nylon-6 nanocomposites exhibit equivalent or better properties than typical 20-30 wt per cent mineral filled compounds depending on the property being investigated. It is important to note that these gains come at almost no cost in specific gravity from the unfilled nylon-6. However, the properties of nylon-6 nanocomposites have yet to approach those of glass fiber reinforced compounds. Even so nylon-6 nanocomposites are excellent candidates when improved heat deflection temperature, stiffness, or strength is required especially when glass fiber reinforcement cannot be considered either due to specific gravity, mold shrinkage/warpage, or wear. Another potential application area is barrier. A significant four-fold improvement was observed in the oxygen transmission rate. This may allow nylon-6 nanocomposites to be utilized in certain packaging and film applications.

Dendrimer Nanocomposites in Medicine

Dendrimer based nanocomposites 1-5 are novel organic-inorganic hybrid nanoparticles synthesized by dispersing very small inorganic domains in nanoscopic size polymeric networks. Due to the molecular level mixing of their components they display unique material properties in addition to properties represented by their individual components. Dendrimers are often used as templates in making these particles because they can be made uniform, and using these well-controlled and uniform templates leads to well controlled and uniform composite particles. Apart from obvious materials science applications, such as catalysis and photonic materials, dendrimer nanocomposites (DNC) have a great potential in biomedical applications due to their controlled composition, predetermined size, shape and optional surface functionalities. They may be used as drug delivery vehicles to deliver bioactive guests such as silver or for an incorporated radioactive isotope. Radioactive dendrimer nanocomposites can be delivered directly to the tumor by injecting nanoparticle solutions directly into the tumor microvasculature.6 If a short-lived isotope is used, a procedure like that may serve as an alternate method to direct irradiation. It has also been envisioned that if the surface of such a nanoparticle had been equipped by appropriate targeting moieties the nanoscopic device could find its way selectively to cancerous cells anywhere in the body.

A relatively new type of hybrid is based on the silsesquioxane cage structure, which can be regarded as one of the smallest forms of silica or even as 'molecular silica'. This approach has been developed by Hybrid Plastics (USA) and is called polyhedral oligomeric silsesquioxane (POSS) nanotechnology. The particle diameter of the POSS molecules is in the range 0.7-30 Å, on average much smaller that the diameter of, for example, colloidal silica particles. Polymers containing the POSS molecules are therefore true nanocomposites. The physical form of the materials varies from liquids through waxes to crystalline solids. The cost of the POSS chemicals is projected to fall between that of fumed silica and the silicones. The POSS structures can be functionalised with a wide variety of groups, leading to a range of POSS monomers. The nature of the functional group determines compatibility with the polymer matrix. The R group can be simple alkyl, cycloalkyl or aryl, or reactive/ polymerisable groups such as acrylic, a-olefin, styrene, epoxide, carboxylic acid, isocyanate, amine, alcohol and silane.

Using such functionalisation, the POSS structures can be either copolymerised with a range of monomers, or grafted onto polymer chains. A variety of phase morphologies is possible, ranging from phase separated to phase inverted to single phase, truly molecular level reinforcement. The POSS compounds can be readily incorporated into polymers using standard processing methods. Interestingly, and in contrast to clay nanocomposites, incorporation of POSS can lead to a reduction in the polymer melt viscosity, which should represent a distinct advantage for many applications. A major effect of POSS incorporation, e.g. in polypropylene, is retention of modulus above the glass transition temperature (Tg) of the unreinforced polymer. Increasing the concentration of the POSS structures in copolymers can have a dramatic effect in increasing the Tg of the polymer. POSS nanocomposites occupy the performance region indicated as 'Hybrids'. Flammability behaviour of the polymer is also claimed to be improved by incorporating POSS.

POSS silanols can be used as coupling agents for fillers and offer a number of advantages over conventional silane coupling agents, including moisture stability, low volatility, no requirement for water addition and no formation of volatile by-products. Surface coverage by POSS cages is also very high. Use of POSS surface treatment can enhance the reinforcing effect of the filler. POSS products are attracting interest in a wide range of materials markets, including aerospace, sporting goods, medical applications and electronics.

Nanoparticles, Nanoclusters & Nanocrystals

Nanoscience has taken scientists around the world by storm. It claims to revolutionize the world we live in with radical breakthroughs in areas such as materials and manufacturing, electronics, medicine and healthcare, environment and energy, chemical and pharmaceutical, biotechnology and agriculture, computation and information technology. Nanoparticles are small clusters of atoms about 1 to 100 nanometers long. 'Nano' derives from the Greek word "*nanos*", which means dwarf or extremely small. It can be used as a prefix for any unit like a second or a liter to mean a billionth of that unit. A nanosecond is a billionth of a second. A nanoliter is a billionth of a liter. And therefore a nanometer is a billionth of a meter or 10^{-9} m.

- If a baseball is the size of Earth, a nanoparticle would be the size of an apple.

- We can also compare it with things in the natural world.
- The smallest synthetic nanotechnology made so far are the individual components of an Intel Pentium III microprocessor and they span about 200 nanometers.

Nanoparticles are larger than individual atoms and molecules but are smaller than bulk solid. Hence, they obey neither absolute quantum chemistry nor laws of classical physics and have properties that differ markedly from those expected. There are two major phenomenons that are responsible for these differences. First is the high dispersity of nanocrystalline systems. As the size of a crystal is reduced, the number of atoms at the surface of the crystal compared to the number of atoms in the crystal itself, increases. For example, a 4 nm diameter CdS nanoparticle has about 1500 atoms, of which about a third are on the surface. Properties, which are usually determined by the molecular structure of the bulk lattice, now become increasingly dominated by the defect structure of the surface. The second phenomenon occurs noticeably only in metals and semiconductors. It is called size quantisation and arises because the size of a nanoparticle is comparable to the de Broglie wavelength of its charge carriers (*i.e.* electrons and holes). Due to the spatial confinement of the charge carriers, the edge of the valance and conduction bands split into discrete, quantized, electronic levels. These electronic levels are similar to those in atoms and molecules. The spacing of the electronic levels and the bandgap increases with decreasing particle size. This is because the electron hole pairs are now much closer together and the Coulombic interaction between them can no longer be neglected giving an overall higher kinetic energy. This increase in bandgap can be observed experimentally by the blue-shift in the absorption spectrum or sometimes even visually by the colour of the samples. A larger bandgap means that more energy is required to excite an electron from the valance band to the conduction band and hence light of a higher frequency and lower wavelength would be absorbed. The increase in molar absorption coefficient shows that the oscillator strength also increases as particle size decreases. This is due to the strong overlapping of the wave functions of confined charge carriers. Experiments carried out in the past have found the absorption spectrum to be potential dependent, with decreasing wavelengths as applied potential decreases. This change is fully reversible before the threshold potential.

The first generation of nanofillers is emerging from R&D into commercial application in composites. Federal funding for nanotechnology research and development in USA has increased six-fold, from $116 million (USD) in 1997 to an estimated $961 million in 2004. In the U.S. alone, the FY 2005 budget request that President George W. Bush has sent to Congress calls for a total National Nanotechnology Initiative (NNI) budget of $982 million, which represents a 2 per cent increase. With all this research, what has come out of the R&D pipeline ready for commercial application? As it turns out, plenty. The following are new and, surprisingly, not-so-new applications of two types of nanotechnology: ceramic nanoparticles and carbon nanotubes used as highly functional fillers/additives in polymer products of interest to composites manufacturers. Introduction of these materials can result in the improvement of the polymer's mechanical properties in molded parts and coatings on a wide variety of fronts, including increased part strength, ductility and dimensional stability as well as resistance to damage by abrasion and exposure to caustic chemicals and/or extreme heat. Nanotechnology involves the control and manipulation of materials at the nanoscale—particle sizes from 1 to 100 nanometers (nm), where 1 nm equals 1 *billionth* of a meter— to create new materials and structures that have novel properties due to their small size. Conventional multifunctional materials, such as fiber-reinforced or filled polymers, require a trade-off between various mechanical and physical properties, cost and processability. However, combinations of properties derived at the nanoscale provide opportunities to circumvent these traditional trade-offs. The transition from microparticles to nanoparticles leads to a number of changes in physical properties. One of the major factors is the increase in the ratio of surface area to volume. As the surface area of a particle increases, the portion of its constituent atoms at or near the surface increases exponentially, creating more sites for bonding, catalysis or reaction with surrounding materials, resulting in improved properties such as increased strength or chemical or heat resistance. Additionally, the fact that nanoparticles have dimensions below the critical wavelength of light can render them transparent, which has implications for part aesthetics. One of the most common groups of the new generation of commercial nanotechnologies is ceramics. Ceramics can be categorized into two subgroups: silicates (or silicon oxide), generally in the form of nanoscale flakes of clay, and metal oxide ceramics such as titanium, zinc, aluminum and iron oxides.

Nanoclays—The most common type is montmorillonite (MMT), a layered aluminosilicate in the smectite family of clays. MMT is a component of bentonite clay, formed when volcanic ash settled into ancient brine seas, and now mined in Wyoming, western Europe, China, Australia, Iran and elsewhere worldwide. Unlike clay minerals such as talc and mica that have been used as fillers for years, MMT can be delaminated and dispersed into individual layers only one nanometer thick by about 70 nm to 150 nm across. The result is a radical increase in the surface area-to-volume ratio: With a surface area of 750 m2/g, 20g of MMT platelets could cover a football field! Southern Clay Products Inc. (Austin, Texas), a manufacturer of functional clay-based additives, began MMT product development in the early 1990s. The company's Cloisite additives, commercially available since 1999, consist of organically modified, nanometer-sized MMT platelets. Unseparated MMT particles are often referred to as tactoids. When tactoids are partially separated in a polymer, they are referred to as being intercalated. When thoroughly separated, individual platelets are said to be exfoliated. To form a nanocomposite, the clay must first be modified with an appropriate organic surface treatment to enable complete dispersion into and miscibility with the polymer matrix. Then the tactoids must be exfoliated into platelets either through the shear stresses exerted during melt compounding or via chemical processes during polymerization of the matrix. As Karl Kamena, Southern Clay's commercial manager of Cloisite nanoclays, describes it, "The incorporation of single-digit percentages of high surface-area nanoclays into host polymer systems improves the performance of the specific polymer matrix without significant deterioration of properties." The numerous benefits that Cloisite can confer on the host polymer include increased mechanical properties, such as modulus, melt strength and scratch resistance, as well as increased heat resistance, dimensional stability and flame retardation.

The extent of enhancement is related directly to the degree of exfoliation. Currently, about 40 per cent of the MMT platelets can be fully exfoliated, with the remainder forming tactoids of 2 to 10 platelets. Southern Clay continues to improve exfoliation methods, anticipating that, eventually, remaining tactoids will contain as few as one to three platelets. In 1998, Southern Clay began working with General Motors (GM, Detroit, Mich.) and resin manufacturer Basell (Hoofddorp, The Netherlands) to develop a nanoclay-reinforced thermoplastic olefin (TPO) that would outperform the talc-filled TPO then in use. An 8-micron Cloisite clay, each particle of which contains one million platelets, was selected as the filler. Work carried out jointly by the companies allowed the new material to be formulated to different shrinkage rates and to match colour with conventionally filled molded-in-colour parts. Early on, the 8-micron particles had a tendency during the extrusion stage (when the clay was mixed into the TPO) to aggregate into 200-micron clumps rather than exfoliate, causing surface defects and painting problems in the injection molded parts. The team then proved that this issue could be resolved by making adjustments to the extrusion temperature, the extruder screw rotation speed, the material feed rate and other variables during the nanocomposite manufacturing process. GM first introduced a 3 per cent nanoclay-filled TPO on the step assists for the 2002 GMC *Safari* and Chevrolet *Astro* vans, which had been using a 20 per cent talc-filled TPO. The same material moved into a higher volume application in the body side trim moldings for the 2004 and 2005 Chevrolet *Impala*, molded at Blackhawk Automotive Plastics (Troy, Mich.). Will Rodgers, GM staff research scientist, says GM saw weight savings of 3 to 21 per cent on these redesigned parts. With increased confidence in the material, GM now is using 7 lb of molded-in-colour nanocomposite parts to make this year's *Hummer H2 SUT* cargo bed trim (the center bridge, the sail panels and the box rail protectors), all molded by SportRack Automotive (Sterling Heights, Mich.). These are the first parts that GM has designed as nanocomposites from conception, choosing the material primarily because of its low coefficient of thermal expansion (CTE) and improved appearance. For example, by varying the surface finish of the tool, three levels of gloss were molded into the box rail protectors to match adjoining parts, something that was never possible with talc-filled TPO. GM also reports better scratch/mar performance, increased cohesive strength between the paint and the part, greater ductility at low temperatures, lower flammability, a larger injection molding process window, sharper feature lines, improved knit line appearance and the elimination of tiger striping, (Tiger striping is caused when a filled material groups itself into lines of higher and lower fill weight or viscosity as it flows into a mold, giving the part a striped appearance.) Although the MMT-filled TPO is about an order of magnitude more expensive than the talc-filled material (about $3/lb vs. $0.35/lb), the improvements in part performance and appearance apparently justify the cost.

Metal oxides—Scratch resistance is a highly sought quality in coatings, such as automotive finishes. However, use of silica or clay fillers in quantities sufficient to improve scratch resistance tends to make the coating hazy and reduce its viscosity. Alumina particles are harder than silica but also lead to haziness. This is particularly troublesome in clear coats. Nanophase Technologies Corp. (Romeoville, Ill.) and BYK-Chemie (Wesel, Germany) have teamed to address this problem with two new additives marketed under the tradenames Nanobyk-3600 and Nanobyk-3601. Both additives contain Nanophase's 37-nm diameter Nanodur aluminum oxide particles in concentrations of 30 to 50 per cent by weight, and are intended for use with aqueous (Nanobyk-3600) and nonaqueous UV-curable coatings (Nanobyk 3601). Nanophase Technologies produces its aluminum oxide nanoparticles via physical vapor synthesis: Arc energy is applied to a metallic precursor to generate a vapor at a high temperature. A reactant gas is introduced to the vapor and then the mixture is cooled at a controlled rate so that it condenses and forms nanoparticles, which then are coated with a thin polymeric shell that ensures their compatibility with various polymers and fluids. When the additives are blended into resins and coatings at very low 1.5 to 6 per cent concentrations, scratch resistance reportedly goes up dramatically. In Nanophase tests, particle loadings of 1 per cent resulted in five- to ten-fold increases in scratch resistance, depending on the resin formulation tested, with less than 0.5 percent haze. (At greater than 1 per cent haze, or about a 2 per cent loading of Nanodur alumina particles, a resin is no longer considered "clear.")

Exploration of carbon nanotubes is relatively new. Single wall nanotubes (SWNTs) are long, thin cylinders made of carbon atoms arranged in a graphitic lattice structure that may or may not have closed ends. Multiwall nanotubes (MWNTs) are concentric cylinders of carbon in a similar graphite structure, held together by weak intermolecular-level forces. SWNTs and MWNTs have length-to-diameter ratios that can range from 100:1 to 10,000:1. Their large surface-area-to-volume ratios mean that carbon nanotubes (CNTs) affect matrix properties far beyond what would be predicted from rule-of-mixtures calculations. Nanotubes are currently being commercialized in applications where electrical conductivity and increased mechanical performance are primary design drivers.

Electrical properties. SWNTs can possess electrical conductivity of 1 billion amps/cm2, Young's modulus of more than 1 TeraPascal and a maximum strain of 10 per cent. MWNTs have lesser properties but are easier to produce, though they also are less well understood, due to their greater complexity and variety. Orders of magnitude less costly than SWNTs, MWNTs are more attractive for commercial applications. While they have been used for the last 10 years in thermoplastics as a direct replacement for carbon black in electrical applications, recent developments are enabling their use in thermosets, and significant efforts are underway to exploit their mechanical properties as well. Carbon black and carbon nanotubes are both pure graphitic carbon and so have the same electrical properties. Carbon black particles, however, are spherical, and the rod-like CNTs have a higher aspect ratio, which means far fewer CNT particles are needed to form a conductive "bridge" through a nonconductive polymer. Typical carbon black loading in polymers is 7 to 18 percent (depending on the polymer and the level of conductivity needed), which will make the polymer brittle. CNT's achieve the same conductivity in 1 to 3 percent loadings. Moreover, the CNT's conductivity-versus-loading curve is more gradual and more easily tailored to a specific level of conductivity than carbon black's curve.

In 1984, Hyperion Catalysis (Cambridge, Mass.) was the first company to file a patent on a nanotube manufacturing method. Hyperion's MWNTs have 8 to 12 walls and are about 10 nm wide by 10 micrometers long, an aspect ratio of 1,000:1. They are made through a process which Hyperion's sales and marketing manager Pat Collins describes as "continuous, catalyzed vapor phase decomposition and recombination, using light hydrocarbon feedstock." When an atomized metallic catalyst comes in contact with a hydrocarbon gas in a reactor under appropriate time, pressure, and temperature conditions, nanotubes will grow on the catalyst. Each nanotube has a tiny remnant of catalyst inside it. The MWNTs, as formed, tend to clump together in aggregates, attracted to each other by intermolecular Van der Waals forces. The high shear forces which are necessary to disperse the CNTs into the resin are obtained by using a high-viscosity resin and a twin-screw extruder typically used to compound thermoplastics. The disentangled and dispersed nanotube/resin mixture is formulated in concentrations of up to 20 per cent. A wide variety of nanotube-filled thermoplastics can be purchased, either ready to use with low CNT loadings or as a master batch with higher nanotube concentrations, which then can be let down (diluted) to the appropriate usage level. The ready-to-use resins are offered in antistatic, electrostatic dissipative and conductive grades, with volume resistivities ranging from 1010 ohm/cm for antistatic grades to 1 ohm/cm for conductive grades (as compared to 1.68 x 10-6 ohm/cm for copper). Because the nanotubes are in such low concentrations and are so small, their addition has little effect on the base resin's properties, allowing the material to be processed on the injection molder's existing equipment under conditions similar to those used to process the base resin. Hyperion's continuous process runs 24 hours a day, seven days a week, producing tons of pure (no byproducts, such as amorphous carbon) nanotubes per week. Automotive fuel systems have to be designed for electrostatic discharge. The turbulent flow of fast-moving fuel creates a static charge which must be dissipated. If charge is allowed to build up it can cause pinholes to develop in the fuel system and may damage electronic sensors. On rare occasions, sparks may even discharge during refueling. The fuel system, which includes the filler neck, inlet check valves, fuel module (pump/sender unit) inside the tank, the fuel lines and the fuel filter, must all, therefore, be conductive. (The Society of Automotive Engineers (SAE) will soon issue its newly revised *Recommended Practice, SAE J1645*, which will outline how all materials that touch fast-moving fuel must be grounded.) Fuel systems static dissipation is a rapidly growing segment of Hyperion's business, according to Collins. BASF (Ludwigshafen, Germany) has already commercialized a conductive polyoxymethylene (POM) with Hyperion nanotubes, specifically for automotive fuel systems. CNT-reinforced nylon 6, manufactured by Degussa (Parsippany, N.J.), is used for the tubing in the flexible portion of the fuel line in 70 percent of cars made in North America. O-rings in the system also must be conductive or they will build unwanted capacitance. Precix (New Bedford, Mass.) recently introduced its F98 series of O-rings, made by incorporating Hyperion's multiwall nanotubes into Precix's own proprietary flouroelastomer resin system. Compared to the company's carbon black-filled O-rings, the F98 compound is said to have three to five times greater resistance to fuel vapor permeation, with significantly reduced fuel swell (the expansion of part dimensions due to absorption of the fuel) as well as a durometer of 79 versus 88, allowing for much lower insertion pressure when assembling parts. Hyperion also supplies various conductive thermoplastics to manufacturers such as GE Plastics (Pittsfield, Mass) for use as injection molding material for electrostatically painted automotive body parts, such as mirror housings, door handles and fenders.

According to a *CT* web search, the Renault *Clio* and *Mégane* currently sport fenders made from a nanotube-filled polyamide/polyphenylene ether (PA/PPE) blend. Since CNT loadings are typically 2 to 3 percent rather than the 12 to 15 per cent necessary with carbon black, the plastics do not become brittle, thus negating the need for impact modifiers which would produce the undesirable side effect of reducing the plastic's glass transition temperature (Tg). The CNT-reinforced parts need no primer and can be painted at the same temperature and on the same electrostatic line used with metal parts, reducing cost and improving colour matching. Because the CNT concentrations are low, the surface finish also is smoother than plastics with other types of conductive fillers. Collins adds that Hyperion now has branched out beyond thermoplastics into thermosets, with nanotube-filled versions of Resolution Performance Products' (Ennis, Texas) Epon 1009 and Epon 828 resins. Epon 1009 is a high molecular weight, solid pellet form of bisphenol-A, which can be compounded into liquid epoxies or vinyl esters. Hyperion sells the Epon 828 nanocomposite in paste form, which is made by grinding the Epon 1009 nanocomposite pellets into a powder and blending the powder into the 828 with a multishaft double planetary mixer in a liquid dispersion process. Collins also notes that Hyperion will introduce easier-to-use liquid master batches of other nanotube-filled epoxies in the near future, primarily for electrical applications, such as conductive epoxies, potting compounds and connectors. Although the nanotubes may add \$5/lb to \$6/lb to the final polymer mix, Collins says there are many applications where the cost is justified either through cost savings in the final product manufacturing process or because the necessary properties cannot be obtained any other way.

Mechanical properties. Zyvex Corp. (Richardson, Texas) approaches the market from a different angle, seeking to use specially modified carbon nanotubes to increase the mechanical performance in sporting goods, aerospace and defense applications. Zyvex does not make its own nanotubes, but instead uses SWNTs and MWNTs from other suppliers, and certifies them for quality and batch-to-batch consistency using a series of standard analytical tests. Zyvex uses a patented process to functionalize the surface of the nanomaterials with a non-covalently bonded "Kentera" molecule which non-covalently bonds the nanomaterial surface via pi-pi bonding and also adheres well to the host matrix. Pi-pi bonding, which involves a slightly stronger molecular force than the intermolecular forces that cause the nanotubes to clump together, enables the Kentera molecule to slip between the nanomaterials and disperse them. Zyvex offers its Nanosolve functionalized nanotubes and nanofibers as dispersions in solvents or water additives, or in solvent-free concentrates, which can be used in thermoplastics and a subset of thermosets comprising polyurethanes and epoxies. This includes concentrates in Resolution Performance Products' Epon 826, 828 and 862 at customized weight loadings specific to an application, whether it is thermal, electrical or mechanical. Marni Rutkofsky, Nanosolve chemist and technical sales representative, is overseeing third-party ASTM physical and mechanical testing on several materials. She says preliminary results at less than 5 per cent loading of MWNTs in thermoplastic polycarbonate show increases of 80, 50 and 150 per cent, respectively, in tensile strength, elongation and Young's modulus. Results at similar loading in polyurethane show a four-fold increase in thermal conductivity. Data on epoxy is expected later this year.

One beneficiary of Zyvex research is Easton Sports (Van Nuys, Calif.) a leading manufacturer of bicycling, hockey, baseball, softball and archery equipment. Three years ago, Easton embarked on an ambitious project to explore the advantages that nanotubes could bring to bicycle components bladder molded at its production facility in Dongguan City, Guangdong Province, China. Easton drum winds its own prepreg and planned to incorporate CNTs into the resin before introducing the resin to the fiber. The challenge was to find a nanotube chemistry that would remain well dispersed in the resin during the course of Easton's manufacturing process and yield improved properties at a reasonable cost. Zyvex and Easton tried several iterations of the resin filled with the CNT/Kentera chemistry before developing a formulation that met the challenge. "We were looking more for strength increase than weight decrease," says John Harrington, VP of Easton's Bicycle Div. "Our products are already some of the lightest on the market, and we want to be conservative with weight reductions when we are using a new material." However, Easton has shaved 10g to 20g from most of the CNT-filled prepreg components, while achieving an average 15 per cent increase in product flexural strength. Road, mountain, and triathlon handlebars, forks, seatposts and OEM seatstays and chainstays are already on the market. The CNT Technology logo distinguishes components which have incorporated nanotubes, and most of the rest of Easton's bicycle components will follow suit. Retail prices on the logoed components have gone up an average of only 10 per cent—an increase that Harrington says consumers are willing to pay for the increase in performance. While nanosized fillers aren't an overnight revolution, it is clear that there is significant activity on many fronts, with increasing use on all. Beyond the applications highlighted here, nanofillers have been incorporated into in golf clubs and golf balls, tennis rackets and a number of other products, with many more to come. The main barriers to nanomaterials use now seem to be development cost and the end-user's ability to obtain proper dispersion in and compatibility with resins. However, all of the nanosuppliers interviewed above are focused on making materials less expensive and more accessible by scaling up production, standardizing products, issuing test data and offering them in user-friendly forms.

You must be wondering by now after all the 'over-the-head' theory about how different nanoparticles are from bulk and atomic material, what this really mean to us average everyday individuals. A lot!! Due to the two unique phenomenon which occurs in nanoparticles, their properties (electrical, optical, chemical, mechanical, magnetic, etc.) can be selectively controlled by engineering the size, morphology, and composition of the particles. These new substances will have enhanced or entirely different properties from their parent materials. Nanoparticles can make metals stronger and harder, give ceramics enhanced ductility and formability, enable normally insulating materials to conduct heat or electricity, and make protective coatings transparent. Hence, industries can re-engineer many existing products and design novel new products/processes that function at unprecedented levels. Use the menu bar on the top to read about more about some examples of the possible applications.

Solar cells are photovoltaic devices, which are devices that can convert light into electricity. They are much in demand as they offer an inexhaustible and environmentally benign energy source. Most solar cells are made of amorphous silicon. The problem with this is that the silicon must be of a very high purity and have a near perfect crystal structure. This makes it very expensive to produce. The efficiency of such a cell is also very small, typically converting only 13-18 per cent of sunlight to electricity. However, low efficiency wouldn't matter if huge arrays of cells could be produced cheaply. After all, nature's solar cells, chloroplasts in plants are less than 1 per cent efficient. Most solar powered devices rely on the same principle: a photon of sunlight boosts an electron in the material into a mobile state so that it can be used to generate electricity. The problem with this simple mechanism is that the electrons are negatively charged and will leave a positive charge. These opposite charges attract one another and therefore will tend to recombine, squandering the absorbed energy as heat or as re-emitted light. Silicon solar cells use an electric field to push the negatively charged electrons and positive charges apart. While chloroplasts adopt a more subtle approach of separate charges by making a distinction between the units that generate the electron and those that transport it away. Deciding to copy nature's trick, Michael Gratzel and Brian O'Regan at the Swiss Federal Institute of Technology began research and produced the Grätzel cell in 1991. The Grätzel cell uses intensely coloured organic dye molecules to capture light energy to inject an electron from the dye into a semiconductor such as titanium dioxide (TiO_2). This remarkably efficient charge separation reaction initiates current flow and the output of electrical energy by the cell.

The ideal material used in the cell ´must have a high surface area for light absorption and charge separation. Nanoparticles, having a comparable surface area to volume ratio, provides for just that. Titanium dioxide nanoparticles are used to make nanoporous thin film supported upon a glass substrate. The material obtained has optical transparency, excellent stability and good electrical conductivity. The benefit of these novel photoelectrical solar cells is that they can be fabricated from cheap, low purity materials by simple and low cost procedures. Contrary to expectation, some of the new devices also have strikingly high conversion efficiency. The size-tunable bandgaps of the semiconductor nanoparticles, due to size quantisation, also means more efficient solar cells can be produced for photovoltaics (electricity production) and water splitting (hydrogen production) processes.

Electrochromism is defined as a reversible colour change of a material caused by the application of an electrical current or potential. It has potential applications in making smart windows for the modulation of incoming light, as well as displays, a field in which competition with liquid crystal displays (LCD) is possible.

There used to be only two types of electrochromic devices. In the first type, the electrodes are made of conducting glass and covered with an organic or inorganic polymer. The two materials would usually display complementary electrochromism and hence produce the same colour change when one is oxidized while the other is reduced. This type of construction is used in electrochromic windows. It is bistable, which means that once the colour change has occurred, the state of the device remains even in absence of applied voltage. The limitation of this typ e of electrochromic devicesis the slowness of the colour change, due to thelow migration rate of the counterions in the bulk polymer. It is also difficult to obtain strong colour changes or bright colours. In the second type, two complementary electrochromic molecules are dissolved in a solvent. This type of system is very simple to build, reacts very fast and can produce dark or bright colours. The drawback of this type is that an electrical current is needed to maintain the coloured state because the two types of coloured molecules diffuse through the system and react with each other to restore the bleached states. It therefore cannot be used for large area devices or for battery-powered displays.

We therefore need a system which is both bistable and changes colour rapidly. This can be done by attaching a suitable molecule that is colourless in the oxidized state and coloured in the reduced state onto the surface of a colourless semiconductor on conducting glass. When a sufficiently negative potential is applied, electrons are injected from the conducting glass into the conduction band of the semiconductor and reduce the adsorbed molecules. The reverse process occurs when a positive potential is applied and the molecules get bleached. The advantage of this system is that combines immobility of the electrochromic material with the rapidity and colouration efficiency of molecular systems. However, a single molecular monolayer does not absorb a perceptible amount of light. This is where nanoparticles come in!

By using nanocrystalline semiconductor films, it was possible to amplify the light absorption of the molecular monolayer to obtain visible colour changes. The nanocrystalline layer are highly porous and usually the real surface of a 5 μm thick layer is about 500 times the projected area. Therefore, as light passes through the layer, it crosses several hundreds of monolayers of coloured molecules giving a strong absorption.

Some commercial products have already been produced! The most successful is the automatic dimming rear-view mirror sense ambient light and glare from behind, and reduce annoying glare from bright lights. While the company NTera has produced paper quality displays. These displays are bi-stable, not requiring much power keep to a pixel coloured, and reflective, not needing power for backlighting, and therefore can operate with very little power. Coupled with the fast switching, (*i.e.* changing of colour) due to amplification with nanoparticles, this technology could in most flat panel displays.

The invention of the transistor in 1948 by John Bardeen and William Shockley, triggered a new era in electronics. The trend since then has been to create smaller and smaller product using fewer chips of greater complexity and smaller 'feature' sizes. The development of integrated circuits and storage devices have continued to progress at an exponential rate. At present it takes two or three years for each successive halving of component size. However, the technologies currently in use have fundamental limits below, which the devices no longer function in a predictable manner. For instance, oxide layers used in Complimentary Metal Oxide Semiconductors (CMOS) devices are becoming so thin that leakage currents are conducted quantum mechanically by electron tunneling. Therefore, replacement technology to advance miniaturisation even further down, eventually to the dimensions of single atoms and molecules, is keenly sought. In 1985, Dmitri Averin and Konstantin Likharev, proposed the idea of a new three-terminal device called a single electron tunneling (SET) transistor. Two years later Theodore Fulton and Gerald Dolan at Bell Labs in the US fabricated such a device and demonstrated how it operates. Since then much research has gone into this area and better single electron devices has been fabricated. These devices usual have a molecule or atom as a functional unit and operate based on the quantum effect. Due to the small dimensions of the device (a couple of nanometers and hence is a nanoparticle), there are repulsive Coulombic interactions between the electrons and energy is required to add an extra electron. This energy cost can be tuned to zero by applying a voltage to the gate electrode. The electrostatic potential at this gate will be such that an extra electron can hop from the source onto the molecule. However, due to Coulomb repulsion, a second, extra, electron cannot hop on at the same time. The first electron must leave the molecule, moving into the drain before the next electron can enter. This one-by-one electron motion, is known as the single electron quantum tunneling. The creation an ultrafast, single electron transistor would lead to the development of "quantum" computers with supercomputer powers and the size of a thumbtack. However, there is still a long road ahead before atomic or molecular transistors can be assembled into viable, dense, fast logic-circuits. Currently these single electron devices are still no competition for silicon transistors. But they will serve both scientifically, for studying electron motion through nanoscale objects, and technologically, for developing chemical techniques with which to fabricate electronic devices on single molecules.

Nanoart comes in two flavors:

- Works created by artists envisioning our nanotechnological future
- Creations and designs produced by nanotechnologists

Let's look at some intriguing examples of nanoart...

In 1997, researchers at Cornell University created the world's smallest guitar. Carved out of crystalline silicon it measures only 10 micrometers long, about the size of a single cell, and has six strings, each about 50 nanometers wide. Using an atomic force microscope (AFM) the nanoguitar can be played, but it cannot be heard as its strings resonate at inaudible frequencies.Its creation demonstrates a new technology that could have a variety of uses in fiber optics, displays, sensors and electronics leading to a new generation of microelectromechanical devices. In 1998, writing and printing techniques were miniaturized by a research team led by Chad Mirkin and Charles and Emma Morrison of Northwestern's Center for Nanotechnology when they created the world's smallest pen, utilized by an automated four-colour nanoplotter. The nanolithography had previously allowed us to draw tiny lines with only a single ink or type of molecule. Now, with the nanoplotter, multiple inks, or different kinds of molecules can be placed side by side with such accuracy that the chemical purity of each line can be retained. By 2001, the Northwestern University researchers Seunghun Hong and Mirkinin had invented a nanoplotter that can draw multiple copies of a nanometre-sized pattern: at assorted places on a surface, simultaneously. Their eight-pen nanoplotter could be adapted to use a thousand pens!

The examples so far has focused only on small dimensions of nanoparticles. There are other properties that can be exploited.

Nanoclusters

A nanocluster is a fragment of solid comprising somewhere between a few atoms to a few tens of thousands of atoms. Nanoclusters are therefore a novel state of matter with properties that are neither those of a bulk crystal nor those of individual atoms and molecules. Over the past 10 years huge advances have been made both in the synthesis of size-tunable, monodisperse (i.e. size-selected) nanoclusters of various chemical compositions and in the development of techniques for their assembly into nanostructured solids (facilitating the synthesis of what have been termed "designer materials"). Clusters with sizes in the order of several nanometers can have physical properties with are very different from the bulk material. In just the same way are the (electronic) properties of a cluster very different from those of the individual atoms which make up the cluster. This makes the nanoclusters a very attractive research area in the solid state physics. One of the most characterizing properties of these clusters is the very large fraction of surface atoms. This makes the cluster very sensitive to changes at the border of the clusters. Influences from the outside of the clusters are therefore having a great impact on the properties of these clusters. The study of these influences at the border is the main topic of our study. In our research we try to investigate in which way the vibrational state of a cluster is influenced by the surrounding matrix. At the same time we want to get a detailed picture of how these clusters are being formed from a (over)saturated solution, formed by ion implantation. We focus on two different cluster systems. On one hand we study Sn clusters in SiO_2, which is a soft metal in a hard matrix and on the other hand we study Co clusters in Ag which is a hard metal in a soft matrix. These two systems are so to say each others counterpart in the dynamical behavior we expect. Nanoclusters have a very high zeta potential (negative electrical charge) and are small enough to easily penetrate any cell membrane in the body. Zeta potential represents a basic law of nature. It plays a vital role in the health of all forms of plant and animal life. It is the force that maintains the discreteness of the trillions of circulating cells that nourish an organism. If zeta potential is low, toxins can not be suspended for elimination and nutrients can not be suspended for transportation to the cells. The whole system becomes clogged. Nanoclusters, with their high zeta potential, may increase the absorption of nutrients and the elimination of toxins by helping to maintain the proper polar attraction of negatively charged nutrients to the positively charged intracellular fluid. When these two electrical charges are in the appropriate balance, nutritional absorption and cellular metabolism are at optimum levels for maximum energy and an overall sense of well being.

The properties of silica clusters at temperatures from 1500 to 2800 K are investigated using classical molecular dynamics simulations for particles containing up to 1152 atoms. It was found that atoms in the cluster were arranged in a shell-like structure at low temperatures, and that the radial density profile peaked near the outer edge of the particle. Smaller clusters have much higher pressures, with the magnitudes corresponding quite well to those obtained from the Laplace–Young equation, when evaluated in a self-consistent manner using our derived surface tension. The computed surface tension did not show any significant size-dependent behavior in contrast to the decreasing surface tension observed for Lennard–Jones liquid clusters. Finally the computed diffusion coefficients in the liquid state are seen to be larger than bulk computed diffusivities. One of the most import methods of synthesizing nanoparticles is through vapor phase nucleation and growth. Several methods are used, including combustion, plasmas, thermal reactors, and evaporation-condensation. The evolution of an aerosol undergoing a gas-to-particle growth process is described by a master equation, the "aerosol general dynamic equation" (GDE). The GDE is a population balance equation for the behavior in time and space of the particle size distribution function, which can include all the driving forces for particle growth (nucleation, surface growth, coagulation/ coalescence, transport, etc.). Less conventionally, but of interest for nanoparticle synthesis research, the GDE can be extended to track the degree of coalescence of particle aggregates. One of the biggest challenges in the implementation of the GDE is knowledge of the rates of these individual processes. For nanoparticle synthesis, one of the most fundamental and important kinetic properties of interest in nanoparticle growth is the sintering or coalescence rate between particles during vapor-phase growth. A knowledge of these rates and their dependence on process parameters ultimately impacts the ability to control primary particle and agglomerate growth, which are of critical importance to nanoscale particles whose properties depend strongly on size, morphology, and crystal structure. Previous experimental and theoretical studies of sintering in flame reactors demonstrated good agreement for the case of titania particle growth, 7–9 based on a characteristic coalescence time obtained from a solid-state diffusion model, in which the characteristic time is written as

$$t_f = 3k_B T n_p / 64 p D s n_0,$$

where T is the gas temperature, vp is the particle volume, D is the diffusion coefficient usually reported as an Arrhenious s_i is the surface tension, and v_0 is the molecular volume for diffusion. In the case of silicon dioxide (silica), however, the primary particle size is typically underpredicted based on a viscous flow coalescence time written as

$$t_f = 2hR_p/s,$$

where h is the viscosity, and R_p is the radius of the particle. Ehrman *et al.* discussed the possible causes of the enhanced rate of silica particle sintering observed in an experiment, including the presence of impurities such as alkali metals which could result in a lower viscosity of silica. On the other hand, in theoretical considerations kinetic coefficients such as the diffusion coefficient and the viscosity, as well as the surface tension, were taken from the properties of bulk material. However, we know that what makes the nanoparticles so interesting is that their fundamental properties are intimately related to the physical size of the component. For an example, a 10-nm-diameter iron particle has almost 20 per cent of its atoms on the surface. This high fraction of surface atoms alters properties such as the melting point and presumably the surface tension14 and accounts for the high reactivity of nanoparticles both toward each other and also toward other materials and gases. One of the interesting aspects of nanoparticles properties is the variation of the internal pressure with particle size. Laplace's equation

$$DP = 2s/R_p,$$

tells us that the pressure difference between a particle and the surrounding environment DP goes as the surface tension s, divided by the particle radius. This implies that as the particle becomes smaller, the pressure will approach infinity as the radius becomes vanishingly small. Of course, this assumes that the surface tension is size independent, but in fact, at some point, the surface tension will tend toward zero. Nevertheless, we should expect that the pressure of a particle can be many hundreds of atmospheres, and may impact the chemical, phase and crystalline structure of particles. For an example, the pressure inside a 10-nm-diameter TiO_2 particle s = 0.5 J/m^2 is on the order of 2000 atmospheres. Such elevated pressures may have a significant impact on properties such as diffusion coefficients and the viscosity, which in turn affect the rates of particle sintering. Given these constraints, data obtained at large grain sizes when extrapolated to smaller dimensions may lead to considerable errors. Experimental studies are certainly desirable for investigating these characteristics; however, experiments aimed at determining the fundamental transport and thermodynamic properties of fine aerosols are not easily obtained, because of the ensemble nature of most studies.

Here we study the nanoscale silica clusters of different sizes (N=72, 288, 576, and 1152) and at different temperatures (1500–2800 K) using a classical moleculardynamics (MD) simulation to determine the structural properties, internal pressure, diffusion coefficients, and surface tension. We will particularly focus on the particle size dependence of these properties, and its influence on coalescence/sintering modeling.

The clusters consist of N classical atoms which interact through a pair model potential developed by Tsuneyuki *et al.*,which takes into account the steric repulsion due to atomic sizes, screened Coulomb interaction from charge transfer, charge-dipole interactions due to the large electronic polarizability of the anions, and dipole-dipole interactions to mimic the short- and intermediate-order interaction. The interaction potential has an analytical form of the Born-Huggins-Mayer type

$$V_{ij} = q_i q_j / r_{ij} + a_{ij} \exp(-b_{ij} r_{ij}) - c_{ij} / r_{ij}^6,$$

where q_i is the charge of the i-type atom, r_{ij} is the interatomic distance, and a_{ij}, b_{ij}, and c_{ij} are the parameters taken. For higher temperature the potential was modified by adding a Lennard-Jones term, in order to avoid the occasional overriding of the O-O repulsion barrier at a shorter interatomic distance at a higher temperature.

The modified interatomic interaction is referred to as the TTAMm potential with the form $V_{ij}^* = V_{ij} + 4e_{ij}\,[(s_{ij}/r_{ij})^{18} - (s_{ij}/r_{ij})^6]$, where e_{ij} and s_{ij} are the usual Lennard–

Jones parameters. The equations of motion for each atom was solved using the standard velocity version of the Verlet algorithm. In our calculations we constructed a silica cluster with an appropriate number of atoms, and placed it in the center of a spherical cavity of radius R^*, which comprised the calculational cell. The center of the calculation cell ($r50$) coincides with the center of mass of the cluster. The radius of the cell was kept at 2 nm for all clusters. In order to conserve mass at higher temperatures, we force any cluster fragment which escapes the cluster due to evaporation, and reaches the boundary of the cell, to be elastically reflected. All clusters were initially heated to a high temperature liquid state at T=4000 K. Several clusters were then prepared at a variety of temperatures between 1500 and 2800 K by slow numerical cooling. Clusters were allowed to anneal for 35000 time steps (with a time step dt = 1 – 4 fs), followed by a test at constant energy to evaluate if the system had reached an equilibrium state. Production runs for cluster properties were performed at a fixed temperature, and typically required about 1.5×10^5 time steps to obtain a reasonable statistical averaging of cluster characteristics. In order to verify the equilibration procedure for each prepared cluster, we gradually decreased the temperature from 4000 to 1500 K, followed by a reversal of this procedure to the desired target temperature. A comparison of cluster properties such as the radial density distribution obtained by cooling and then reverse heating, shows good agreement. The total angular momentum of cluster motion is always reset to zero by the transformation of velocities in order to evaluate transport properties such as diffusion coefficients. It is not conducive to analyze the cluster behavior at higher temperatures T>3000 K, where a significant dissociation of fragments takes place during the time interval of the calculation (t µ 0.15 ns). This is because this interatomic potential cannot realistically describe the vaporization processes of silica. For instance, in the calculations we observe the vaporization of SiO_2 molecules, while it is well known that silica vaporizes by decomposition to SiO and O_2.

At lower temperature T = 1500 K in a solid "glass" state, all clusters exhibit an oscillating density distribution. With increasing temperature there seems to be a tendency to dampen out the radial density gradient, though some peaks still remain. The radial dependence of the mass densities of different-sized clusters at a temperature T=2080 K. Note that the interpeak space is about the mean Si-Si interatom distance which is 0.3175 nm. To explore the effect of density, we show the radial mass density distribution for a 576-atom cluster at three temperatures. It is clear that the radius of the SiO_2 nanoparticles and, consequently, the averaged density, are effectively insensitive to temperature over quite a large temperature range (1500–2800 K). This is in accordance with the fact that bulk silica has an extremely low thermal expansion coefficient. Over this temperature interval a cluster with N=288 atoms has a radius R=0.92 nm and an average density of p=53.2 g/cm^3. The 576-atom cluster has a radius R=1.16 nm and a density p=52.9 g/cu cm and the cluster of N=1152 atoms has R=1.5 nm and p=2.8 g/cu cm. The radius of a cluster is defined as the distance where the density drops to 0.5 of its interior value. Note that smaller clusters have a higher density, but the values are in close agreement with that calculated for bulk silica in the coesite phase, with a density of 2.9 g/cu cm. The density profiles presented in Figs. 20.3 and 20.4 show that near the surface the clusters have a maximum in the density, just before the rapid decrease at the cluster edge. A similar behavior was observed by Roder *el al* using a BKS interaction potential. They obtained a density of 2.3 g/cu cm. This implies that clusters of different sizes have similar shell structures. Moreover, the cluster surface has the same structure and width regardless of size. The rapid density decrease extends over a distance which is more than a Si-O bond length equal to 0.162 nm. A careful temporal observation makes it clearer that the cluster can be separated into "core"and surface components. On the surface both silicon and oxygen have dangling bonds, with the oxygen atom tending toward the surface. These fragments residing on the surface tend to be highly mobile, so that when averages of the density are taken we see a smooth monotonic decrease in the density profile at the edge of the cluster corresponding to about 0.2–0.3 nm. In general, then, we observe an enhanced density for smaller clusters, small thermal expansion, and an oscillating density profile.

In order to obtain further insight into this oscillating radial density behavior, we consider the structure of the clusters more closely. A cross-section slice for two different clusters (N=288 and 576 atoms) at T=500 K. The slice is taken through the center of a cluster of thickness 0.328 nm (which is slightly larger than two of nominal Si-O bond lengths) What is being shown are Si and O atom positions following averaging over 75 ps. The slice map of smaller cluster looks like a network of atoms organized in rings. The larger cluster has apparently a spherically symmetric shell structure, and the external shell incorporates both O and Si atoms. The defects in the smaller cluster are denoted by atom trajectories that wander. For an instance, in Figs. 20.5 (a) and 20.5 (c) the arrow labeled by 1 shows a vacancy for a silicon atom. The arrow labeled by 2 is another example of a defect and the enhanced mobility of atoms containing dangling bonds. The distribution of oxygen and silicon atoms as a function of radial position is presented for the cluster prepared with 576 atoms (192 Si atoms and 384 O atoms). It clearly indicates that oxygen is preferentially found on the surface layer of the cluster under all conditions. Note that the change of temperature has a pronounced effect on the internal arrangement of the particles. In particular, at T=1600 K the peak in the Si concentration corresponds to a minimum in the O atom density. With increasing temperature to T=1680 K, we find that a thermally induced structural transition takes place. The second Si peak labeled by 1 in Figs. 20.6 (a) and 20.6 (b) shifts, and the second peak in O atoms distribution labeled by 2 disappears and indicates an altogether different shell structure. In order to understand the change in the cluster structure we analyze the angular distribution, varying the temperature. The O-Si-O and Si-O-Si angular distributions are shown for a cluster with 576 atoms. At a temperature of T=1600 K, the Si-O-Si angle distribution has two maxima which are at 97.7° and 142°. The former denotes the presence of four member rings, and the latter refers to six member rings. As the temperature increases to

T=1680 K, the Si-O-Si angular distribution has only one maximum at 142°. The observation of the angular distribution change clarifies the structural rearrangement in the shell structure presented in Figs. 20.6 (a) and 20.6 (b). Note that the O-Si-O angular distribution only slightly broadens with increasing temperature as expected, and retains a monomodal character. The results agree well with the angular distributions in the amorphous and molten bulk SiO_2 studied by Vashishta *et al.* who used a three-body potential. For example, at T=2500 K the O-Si-O and Si-O-Si angular distributions have maxima at 110.1° and 141° for the cluster with 576 atoms, and at 109.5° and at 142.5°, respectively, for bulk material. The angular distributions are, however, much broader for the clusters compared to the bulk because of the spherical surface and shell structure. A similar broadening was observed by Roder *et al.* The partial pair correlation functions were calculated for the O-O, Si-O, and Si-Si bonds to understand the enhanced density of smaller particles and mechanism of cluster expansion at higher temperature. It turns out that the first and second peaks of the O-O and Si-O pair correlation functions have the same positions for all clusters. In contrast to this fact the different clusters have various Si-Si pair correlation functions. One can see that the smaller cluster (N=288 atoms) has a closer Si-Si packing than the largest one (N=576 atoms at T=1500 K. With increasing temperature the peaks of the O-O and Si-O pair correlation function become broader, and their positions remain independent of temperature, whereas the Si-Si pair correlation function is sensitive to the temperature. Its first peak at r = 0.317 nm does not respond to temperature, but a second peak shifts to the larger coordinate. Thus, we can conclude that the variation of the density with the cluster size as well as the expansion of a cluster with temperature are explained by the change in the Si-Si atom packing due to the strong repulsive nature of the Si-Si interaction.

Nanocrystals

Semiconductor nanocrystals (also called nanocrystals or quantum dots) can be described as a state of matter intermediate between molecular and bulk crystalline. Their small size and high optical activity make them a very interesting material for study in disciplines ranging from opto-electronics (integrating optical and electronic signals on a single substrate, i.e. crystalline Si) to fluorescence microscopy (as a bright, photostable and minimally perturbative fluorescent marker). This work will concentrate on experiments designed to optically characterize silicon nanocrystals as fluorescent beacons in solution since in biology fluorescent labels are invariably used to image structures or track/probe particles/biomolecules in the liquid state. Images of single immobilized Si nanocrystals on quartz substrates will also be presented. Just as bulk semiconductors are characterized by the structure of their conduction and valence bands as being direct or indirect, so are semiconductor nanocrystals. Direct semiconductor nanocrystals (e.g. CdSe, CdS) are characterized by having the minimum transition energy to promote an electron from the valence band to the conduction band without a change in the electron momentum (the energy of this energy separation is known as the band-gap). For indirect semiconductors however excitation at the band gap energy must be accompanied by a change in the electron's momentum (supplied by a phonon). While both types of semiconductors are efficient absorbers of light, it has only been the direct semiconductors which have been emissive enough to be used as an optical source (e.g. in semiconductor GaAs quantum well lasers, Saleh and Teich, 1991) or as fluorescent labels (CdSe) (Chan and Nie, 1998, Bruchez et al, 1998). The discovery of porous silicon as an efficient luminescent material (Canham, 1990) has created great excitement as a prospective candidate for opto-electronic circuit integration. [Integrating direct semiconductor structures on silicon has been unsuccessful due to lattice mismatch]. Though the waited for advances have not materialized, it has been the starting point for experimental and theoretical investigations on nanocrystalline silicon.

Nanocrystalline materials are build from nanoparticles. These particles are also called quantum dots or artificial atoms. These nanoparticles are usually spherical colloids of a semi-conducting material with a diameter of nanometers. Because these colloids are so small new material properties exist which are not present in bulk materials.

An important property is that the energy difference between the energy bands, the valence and conducting bands in bulk materials, increases and separate energy levels can be formed. These changes in the energy levels and the small size of the particles result in a new optical and electrical properties which can be used in the devices which were mentioned before.

Due to their novel properties and varied potential applications, nanocrystalline materials, with typical grain sizes <100 nm, are attracting increasing attention from researchers all over the world. These materials are traditionally classified as one-dimensional (1-D), or layered structures, two-dimensional (2-D), or rod- or wire-like, and three-dimensional (3-D), or equiaxed, with the 1-D layered and 2-D wire-like structures being extensively investigated for coating and electronic applications and the 3-D equiaxed structures exploited for potential structural or bulk applications. Because of the small grain size of these materials, and consequently the large volume fraction of atoms in or near the grain boundaries, these *materials* exhibit properties that are often superior, and sometimes completely new, in comparison to those of conventional coarse-grained *materials*. However, it should be realized that a full understanding of the structure, thermal stability, and properties of these novel materials is not yet available. A number of issues and concerns still need to be solved.

Nanocrystalline materials can be synthesized by a number of techniques with the starting material existing either in the vapor, liquid, or solid state. Inert-gas condensation, in which the evaporated material is quenched onto a cold substrate, was the first technique used to synthesize nanocrystalline metals and alloys. Subsequently, plasma processing and other physical and chemical vapor deposition techniques have also been used. Electrodeposition and rapid solidification processing use the liquid state as the starting material. Mechanical alloying, friction stir welding, severe plastic deformation, spark erosion, sliding wear, and repeated cold rolling produce nanocrystalline materials in the solid state. While some of these techniques are commercially in use for large-scale production purposes (e.g., inert-gas condensation, electrode-position, and mechanical alloying), many of the other techniques are essentially used for laboratory investigations. The choice of technique for the synthesis of nanocrystalline materials is determined by the ease of the process, its economic viability, its scalability, and the purity of the end product desired. Most of the techniques referred to previously produce the nanocrystalline material in the form of powder. The application of nanocrystalline materials requires that these powders are consolidated to full density with minimal or no porosity and this is not a trivial matter. Full densification with complete inter-particle bonding requires exposure of the powder to high temperatures and pressures for extended periods of time, which is likely to coarsen the micro structural features. Retention of the material in the nanocrystalline state, however, warrants that the powder is not taken to high temperatures for long periods of time; this may not result in complete densification. Thus, successful consolidation to full density requires innovative methods of consolidation. However, since nanocrystalline materials have a large fraction of the atoms in the grain boundaries, the (effective) diffusivity of nanocrystalline materials is much higher than that of a coarse-grained material of the same composition. This will facilitate achieving full consolidation of nanocrystalline materials at temperatures 300°C to 400°C lower than those required for the coarse-grained materials. Successful consolidation of nanocrystalline powders has been achieved by electro-discharge compaction, plasma-activated sintering, shock (explosive) consolidation, hot isostatic pressing (HIP), Ceracon processing, hydrostatic extrusion, strained powder rolling, and sinter forging. Consolidation is not required if the powder could be used as-is (e.g., in slurries, as catalysts, or for coating purposes).

Nanomaterials: Commercial Applications

This quote from "The Next Big Thing Is Really Small" by Jack Uldrich and Deb Newberry sums it up nicely: This is not to say that nanotechnology is a far-off, fuzzy, futuristic technology. It is not. It has already established a beachhead in the economy. The clothing industry is starting to feel the effects of nanotech. Eddie Bauer, for example, is currently using embedded nanoparticles to create stain-repellent khakis. This seemingly simple innovation will impact not only khaki-wearers, but dry cleaners, who will find their business declining; detergent makers, who will find less of their product moving off the shelf; and stain-removal makers, who will experience a sharp decrease in customers. This modest, fairly low-tech application of nanotechnology is just the small tip of a vast iceberg—an iceberg that threatens to sink even the "unsinkable" companies." Researchers at Pacific Northwest National Laboratory have developed a coating process to make sponge-like silica latch onto toxic metals in water. Self-Assembled Monolayers on Mesoporous Supports easily captures such metals as lead and mercury, which are then recovered for reuse or contained in-place forever.

PNNL One example of a SAMMS nanocomposite (Self-Assembled Monolayers on Mesoporous Supports). An hexagonally close-packed cluster of tubular pores (end view) is shown in the foreground. A single pore, in this case coated with a mercaptopropylsiloxy monolayer, is shown in the background. A model of one surfactant molcule is also shown. A plastic nanocomposite is being used for "step assists" in the GM Safari and Astro Vans. It is scratch-resistant, light-weight, and rust-proof, and generates improvements in strength and reductions in weight, which lead to fuel savings and increased longevity. And in 2001, Toyota started using nanocomposites in a bumper that makes it 60 per cent lighter and twice as resistant to denting and scratching.

Impact: Will likely be used on other GM and Toyota models soon, and in other areas of their vehicles, as well as the other auto manufactures, lowering weight, increasing milage, and creating longer-lasting autos. Likely to impact repair shops (fewer repairs needed) and auto insurance companies (fewer claims). Will also likely soon be seen everywhere weight, weather-proofing, durability, and strength are important factors. Expect NASA, the ESA, and other space-faring organizations to take a serious look, soon, which will eventually result in lower lift costs, which will result in more material being lifted into space.

Electron micrograph of typical silicon nanocomposite cross section showing uniform distribution of conductive carbon nanotube network. Photo courtesy of U.S. Air Force.

Nanocrystals are an ideal light harvester in photovoltaic devices. (They) absorb sunlight more strongly than dye molecules or bulk semiconductor material, therefore high optical densities can be achieved while maintaining the requirement of thin films. Perfectly crystalline CdSe nanocrystals are also an artificial reaction center, separating the electron hole pair on a femtosecond timescale. Fluorescent nanocrystals have several advantages over organic dye molecules as fluorescent markers in biology. They are incredibly bright and do not photodegrade. Drug-conjugated nanocrystals attach to the protein in an extracellular fashion, enabling movies of protein trafficking. (They) also form the basis of a high-throughput fluorescence assay for drug discovery.

Stain-repellent Eddie Bauer Nano-Care khakis, with surface fibers of 10 to 100 nanometers, uses a process that coats each fiber of fabric with "nano-whiskers. Developed by Nano-Tex, a Burlington Industries subsidiary. Dockers also makes khakis, a dress shirt and even a tie treated with what they call "Stain Defender", another example of the same nanoscale cloth treatment.

Impact: Dry cleaners, detergent and stain-removal makers, carpet and furniture makers, window covering makers.

BASF's annual sales of aqueous polymer dispersion products amount to around $1.65 billion. All of them contain polymer particles ranging from ten to several hundred nanometers in size. Polymer dispersions are found in exterior paints, coatings and adhesives, or are used in the finishing of paper, textiles and leather. Nanotechnology also has applications in the food sector. Many vitamins and their precursors, such as carotinoids, are insoluble in water. However, when skillfully produced and formulated as nanoparticles, these substances can easily be mixed with cold water, and their bioavailability in the human body also increases. Many lemonades and fruit juices contain these specially formulated additives, which often also provide an attractive colour. In the cosmetics sector, BASF has for several years been among the leading suppliers of UV absorbers based on nanoparticulate zinc oxide. Incorporated in sun creams, the small particles filter the high-energy radiation out of sunlight. Because of their tiny size, they remain invisible to the naked eye and so the cream is transparent on the skin. Sunscreens are utilizing nanoparticles that are extremely effective at absorbing light, especially in the ultra-violet (UV) range. Due to the particle size, they spread more easily, cover better, and save money since you use less. And they are transparent, unlike traditional screens which are white. These sunscreens are so successful that by 2001 they had captured 60 per cent of the Australian sunscreen market.

Impact: Makers of sunscreen have to convert to using nanoparticles. And other product manufactures, like packaging makers, will find ways to incorporate them into packages to reduces UV exposure and subsequent spoilage. The $480B packaging and $300B plastics industries will be directly effected.

Using aluminum nanoparticles, Argonide has created rocket propellants that burn at double the rate. They also produce copper nanoparticles that are incorporated into automotive lubricant to reduce engine wear. AngstroMedica has produced a nanoparticulate-based synthetic bone. Human bone is made of a calcium and phosphate composite called Hydroxyapatite. By manipulation calcium and phosphate at the molecular level, we have created a patented material that is identical in structure and composition to natural bone. This novel synthetic bone can be used in areas where natural bone is damaged or removed, such as in the in the treatment of fractures and soft tissue injuries.

Biologically Related Aspects

Biological molecules and systems have a number of attributes that make them highly suitable for nanotechnology applications. For example, proteins fold into precisely defined three-dimensional shapes, and nucleic acids assemble according to well-understood rules. Antibodies are highly specific in recognizing and binding their ligands, and biological assemblies such as molecular motors can perform transport operations. Because of these and other favorable properties, biomolecules, biophysics, and biology are themes that run through all of the topics of this report. Although very promising, the bio-related aspects of nanoparticles, nanostructured materials, and nanodevices, are, for the most part, not as well developed as the nonbiological ones. However, a number of recent workshops (e.g., the U.S./EC Workshop on Nanobiotechnology), symposia (e.g., the Conference on Molecular Nanotechnology, and the Symposium on Bio-Nano Electronics), and books attest to the fact that many of the novel developments in this field are poised for rapid expansion.

This study is organized along the lines of the main report. It first puts into perspective current research directed toward biological synthesis and assembly as it pertains to the building blocks of nanotechnology. It then focuses on the current state of the art in biological aspects of dispersions and coatings, high surface area materials, and functional "building blocks" atoms nanoparticles layers nanostructures assembly dispersions and coatings high surface area materials functional nanostructures consolidated materials synthesis large biological component less of a biological component. Organization of the WTEC study; sections with large biological content are indicated. nanostructures. Finally, it ends with a section on the challenges and opportunities for the biological aspects of nanotechnology.

Molecular Manufacturing (MM)

Definition Manufacturing today works by cutting or deforming large chunks of matter, then fastening together the remaining pieces into products. Molecular manufacturing plans to be more efficient and make better products by assembling products directly from the smallest pieces: atoms and molecules. The basic idea is to develop a small set of chemical reactions that can be applied repeatedly to build large molecules, then control the sequence and/or position of the reactions by computer in order to build engineered molecular systems. Molecular manufacturing is different from biology in that biological systems are not engineered. The functional properties of a cell, or even a protein, are complex and hard to predict. However, the process of building protein molecules from small molecular fragments is quite programmable, and scientists are developing the ability to design and synthesize proteins with desired properties. This will allow protein chemistry to be used in an engineering, rather than a biological, context. This would be one approach to molecular manufacturing. Other approaches using different kinds of chemistry may also work, producing better materials. Eric Drexler has proposed two kinds of chemistry that might work. The first, in 1981, was protein chemistry as described above. The second, in 1992, was three-dimensional carbon lattice, similar to diamond. This would be synthesized by direct robotic operation in vacuum. Although the required chemical reactions have not yet been demonstrated, some of them have been simulated in detail. Despite this, some scientists assert that this kind of chemistry can't possibly work as planned.

Discussion of molecular manufacturing has been distorted by several factors. From the beginning, it has been associated with "gray goo" (runaway biosphere-eating self-replicators), leading to excessive fear and attempts to counteract that fear. It has also been associated with with extreme science-fictional projections-though it can be hard to tell fantasy from sober calculation, because the calculations predict some pretty amazing capabilities. In reaction to these factors, some mainstream scientists have attempted to shut down discussion entirely by declaring that molecular manufacturing is impossible or that its major proponents are not credible. Discussion has been further distorted by a variety of widespread conceptual confusions.

The position that molecular manufacturing is impossible is not supportable. Living organisms are not an example of molecular manufacturing, because they are not based on engineering but rather on interlocking complex systems. However, the biochemistry of life could be adapted almost unchanged to a molecular manufacturing system.

A few prominent scientists have nonetheless claimed that molecular manufacturing is impossible, and others have echoed them. However, study of their objections shows that the arguments are weak, based on intuition rather than calculation. Some of the arguments elevate engineering difficulties to the status of fundamental limitations. Others are built on basic misunderstandings of the proposals. See *"Molecular Manufacturing Myths"* for examples of arguments that have been advanced against the theory, and why they are incorrect. Although there are some practical questions remaining to be answered, there is no scientific study demonstrating a limitation or problem with molecular manufacturing theory. Indeed, the chemical mechanisms of life demonstrate that machines of a sort can be constructed out of molecules and can build more machines. Thus the scientists who dismiss the entire field out of hand have not bothered to study it in any depth. An intermediate position is that some form of molecular manufacturing will be possible, but will be limited. This is usually based on the assumption that engineering cannot improve on biology. This may be true for complex problems, but not for the relatively simple tasks that are tackled by engineering.

Perhaps because of the opposition of these scientists, not many people are working on molecular manufacturing in the lab; most of the work has been theoretical study, which is easier to do without funding. For now, it remains a controversial theory. But after more than a decade, no one has managed to find a real problem with the theory, and evidence is accumulating in favor of it. Some predicted phenomena, such as exceptionally low friction in certain cases, have now been observed. A single atom has been mechanically removed from a crystal and put back in the same place. Although these were not intended as demonstrations of molecular manufacturing theory, they indicate that at least some of the predictions are valid.

Separating science from science fiction, and alarmism from serious policy concerns, will not be an easy task. Most nanotechnology researchers have not paid enough attention to molecular manufacturing to be able to give informed commentary. The section on *"Molecular Manufacturing Myths"* lists some of the common misconceptions about molecular manufacturing.

According to its proponents, molecular manufacturing could be revolutionary. Because chemistry is very precise and repeatable, the manufacturing operations should be reliable enough to allow complete automation. Because the chemical steps are simple, a single fabricator could build a wide variety of products by changing the control program. Furthermore, machines built this way could have very precise features. This is a good thing, because building with molecules would be quite slow, and small machines can be built more quickly. In fact, if the calculations are right, a complete nanoscale manufacturing system could build a complete copy of itself in a few hours-and this remains true if many small manufacturing systems are combined into one large one, so that even human-scale factories and products can be built quite rapidly. Finally, it appears that the materials produced by the manufacturing system may be extremely high-performance, allowing products to be far more compact and powerful. The section on *"Risks and Benefits of Molecular Manufacturing"* discusses some implications of the technology.

Exponential general-purpose molecular manufacturing — that's a mouthful, but what does it mean? Let's take the phrase apart to see why it is so important.

Manufacturing: The ability to make products, in this case ranging from clothing, to electronics, to medical devices, to books, to building materials, and much more.

Molecular manufacturing: The automated building of products from the bottom up, molecule by molecule, with atomic precision. This will make products that are extremely lightweight, flexible, durable, and potentially very 'smart'.

General-purpose molccular manufacturing: A manufacturing technology that will find many applications across many segments of society. Its extreme flexibility, precision, high capacity, and low cost will cause rapid adoption almost everywhere, and therefore will have disruptive effects in many industries.

EXPONENTIAL general-purpose molecular manufacturing: The word exponential refers to the rapid pace — probably unprecedented — at which this technology may be deployed. A compact automated molecular manufacturing system will be able to make more manufacturing systems. We're talking about factories that can build duplicate factories — and probably do it in less than a single day. The math is simple: if one factory makes two, and two factories make four, then within ten days you could have one thousand factories, in ten more days a million factories, and ten days after that a billion factories. Within the span of just a few weeks, in theory, every household in the world could have one of their own, to make most of the products they need, at just the cost of raw materials.

Exponential general-purpose molecular manufacturing means a manufacturing system capable of making a wide range of technologically advanced products, far superior to what we have today, much cheaper, much faster, and able to multiply its own source of production exponentially.

From *Responsible Nanotechnology*. See *Unanswered Questions, Part 1* for details and comments.

Myths

Perhaps the biggest myth is that is impossible. Before we get into the alleged reasons, it helps to remember how mundane it is: it's just very small machines, making chemistry happen according to plan. The explanations given here are accurate, but simplified. If you want more detail, please contact the author at cphoenix [at] CRNano.org. He will be happy to defend and explain these statements in as much detail as necessary.

1. "*The chemistry won't work.*"

Myth: You can't build machines with chemistry because you can't pick up individual atoms.

Fact: This is a *straw man*. No one has proposed picking up individual atoms. What can be done-it happens all the time in nature-is to transfer groups of atoms from one molecule to another.

Myth: Atoms won't stay where they're put, because of *quantum uncertainty*, *Heisenberg uncertainty*, thermal noise, etc.

Fact: That depends on how the atoms are connected to other atoms. For example, diamonds don't evaporate. Not at all-not even one atom per year. This objection is like arguing that airplanes are impossible because they can't be built out of mud.

Myth: You can't make chemical reactions happen where you want them because it's too hard to control the surrounding atoms. (Sometimes stated as: It would take too many "fingers.")

Fact: Again, this depends on how the atoms are connected. Underwater chemistry is complex, but biology manages to do it. Chemistry in other environments may be a lot simpler. There's quite a bit of evidence that points to several different ways to build programmable shapes. If someone says it's impossible, ask for their evidence, and ask them about DNA folding, organic synthesis of graphite and fullerenes, and removal and replacement of single silicon atoms.

2. "*The machine won't work.*"

Myth: Chemistry is too complex, and it'll be difficult or impossible to build machines with it.

Fact: The point of molecular manufacturing is to use just a few reactions over and over again to build aggregated molecules. Building a brick wall is easy because you only need one kind of brick. Similarly, building a shape out of diamond is expected to be easy if you can add the atoms a few at a time, using the same reaction in multiple programmed locations. This plan shifts the complexity from chemistry to positional control-which can be controlled by software, and software is easy to write.

Myth: There's no way to control such small machines.

Fact: There are lots of ways, including simply running wires to them. Whoever says this is obviously unfamiliar with the scientific literature.

Related Myth: There's no way to power such small machines.

Fact: Again, there are lots of ways, with chemicals, electricity, light, and even sound pressure. If someone says this, ask them about the "nano-elevator" which might be controlled by chemicals, electricity, or light, and which may allow "tight control over the timing of a reaction."

Myth: Brownian motion (or thermal noise) will prevent the machines from working.

Fact: Thermal noise will make things wiggle a bit. But this has been taken into account in detailed engineering calculations. Large machines also have to deal with vibration. And cellular machinery, which is built a lot more loosely, manages to work just fine.

Related Myth: Nanomachines can't work because they don't use Brownian motion like biology does.

Fact: Several nanomachine designs do use Brownian motion. Others simply work around it. Biology doesn't always use it either. There's nothing magical about Brownian motion, and no rule that it has to be used, any more than larger machines must all use electricity.

Myth: Friction will prevent the machines from working.

Fact: With precise molecular construction techniques, surfaces can be made hard and smooth enough and positioned so that the friction drops to near-zero. This is called "superlubricity" and has recently been demonstrated.

Myth: Things are weird ("quantum") at that small scale, and familiar engineering just won't work.

Fact: Cells are chock-full of molecular machines like turbines, trusses, ratchets, and so on. It's possible to build things that depend on quantum effects, but we already know of lots of molecular machines that don't.

Myth: Entropy will keep molecular machines from working.

Fact: Entropy can be held at bay by putting energy into a system. Your computer does trillions of calculations without a single error. Your body is constantly rebuilding itself. Entropy doesn't magically get worse at the nanoscale-in fact, the more precisely a machine is built, the more efficiently it can run. Entropy has been accounted for in molecular machine designs, and it turns out not to be a big deal. (On average, entropy would require less than one error per billion tons of diamond - one defect per 1.6x10^25 tons.)

Unraveling the Big Debate over Small Machines:

> *Nanobot naysayers argue that molecular manufacturing is impossible, but the evidence goes against them*
>
> Bailey thoroughly and systematically rebuts the arguments that have been used to retard the growth of our understanding of *molecular manufacturing*. This article sets the record straight, once and for all, and is well worth reading.
>
> —Rocky Rawstern, Editor Nanotechnology Now. August 2004

In the spirit of "myth busting," and presented in an article for BetterHumans, writer Patrick Bailey sets forth on a journey of explanation, into the nanorealm. In *Unraveling the Big Debate over Small Machines,* Bailey carefully lays out the history of several complex and key issues; issues that have been used to slow the pace of our understanding of molecular manufacturing.

In just over 2,500 words, Bailey tells an intriguing tale about truths, concepts, and competing points of view, set in the last gasps of the old Millenium, fast forward to the present.

In easy to understand language, from a trained observer's point-of-view, read here a story that is sure to steer the debate.

In *Italics*, here are a few of his well-considered observations. "Arguments" here indicates a combination of myth and straw man.

Behind the public face of nanotechnology—the press releases, science fiction novels and environmental doomsday reports—a debate has been evolving and swirling for the past decade around a fundamental question: Can tiny machines build things useful to humans by moving molecules or even individual atoms?

(Dr K. Eric) *Drexler's proposal for molecular manufacturing is to use a process called mechanosynthesis or "machine-phase" chemistry. "By guiding and positioning molecules, molecular manufacturing will control which molecules react, and where, building up complex structures with atomically precise control," he says.*

What quickly became apparent is that the debate is larger than any individual scientist and the consequences more significant than simply settling who's right and who's wrong. This is a debate over conceptions, misconceptions, theories and misrepresentations of physics and chemistry, not men.

This isn't a debate of concern only to those in the ivory tower of academia. A major goal of nanotechnologists is to create nanoscale robots—nanobots—that can perform various functions at the nanoscale. These functions include molecular manufacturing: Using nanobot designs, such as fabricators and assemblers, to build products with atomically precise control. Considering the possible applications of nanobots and molecular manufacturing in medicine, computing, industrial production and more, the debate's importance becomes apparent.

One of the arguments is *"what we know about chemistry doesn't support Drexler's nanobot and molecular manufacturing proposals. Two key arguments here are known as the "sticky fingers" and "fat fingers" problems. Both state that the fingers used by a nanobot are not able to move individual atoms with the precision needed to make molecular manufacturing feasible."*

Sticky fingers refers to the inherent chemical bonding properties of atoms, which, says (Dr. Richard E.) Smalley, will prevent accurate atomic positioning because the atoms will stick to the end of a nanobot finger. Similarly, fat fingers refer to nanobot fingers being too large to move individual atoms.

Another case for the argument: *"When you have a tiny robot, the fingers on the robot have to be comparable to the size of the object that's being picked up or smaller, I mean you can't pick up a pea using a boxing glove very well and when you get down to the scale of atoms there's nothing smaller than an atom, so that it's not quite clear how that works. And atoms have a tendency to be very reactive so they stick to the fingers and you can't pull them out in the first place...it absolutely wouldn't work."* (Professor George Whitesides)

And Bailey's counter: *"Statements about sticky and fat nanofingers, however, overlook a very important point: Grabbing individual atoms is not part of Drexler's proposal for nanobot and molecular manufacturing. "No one has seriously proposed picking up individual atoms with pincers or fingers," says Chris Phoenix, director of research at the Center for Responsible Nanotechnology (CRN) in Brooklyn, New York. "What can be done—it happens all the time in nature—is to transfer atoms from one molecule to another.""*

If fat and sticky fingers were the only opposition arguments, chances are the debate would have ended long ago. Clearly, molecular manufacturing doesn't require fingers at all. There are other arguments, however, that, if valid, cause considerable problems for molecular manufacturing.

Another of the arguments is based on *"a chemistry-based objection claiming that enzyme catalysis only works in water."*

And the counter: *"Discussing enzyme catalysis is important, not because enzyme catalysis is necessary or proposed by Drexler, but because Smalley has used it to reject mechanosynthesis—Drexler's proposal for molecular manufacturing." ... Nearly two decades ago, (Alexander) Klibanov and his then graduate student Alex Zaks demonstrated that enzyme catalysis in a nonaqueous (non-water-based) environment is possible. Their findings were first published in 1984.*

At the end of the day, it's not just understanding the technical issues, but the enormous *human implications* that matters most.

Bailey has taken a nanoScalpel to the straw men, and they are no more. Now maybe we can forge ahead and fully explore the many paths that lead to molecular manufacturing.

From : *Nanobot naysayers argue that molecular manufacturing is impossible, but the evidence goes against them.* 8/16/2004

Bailey holds a masters degree in philosophy from the University of South Florida. He specializes in the philosophy of mind, neuroscience, and theory of knowledge, and is a technical writer for the IT industry.

BetterHumans is an editorial production company that's dedicated to having the best information, analysis and opinion on the impact of advancing science and technology.

3. "There's no point"

Myth: Mechanical chemistry will be far too slow to build anything useful.

Fact: A mechanical chemistry machine won't be too slow to build additional tiny machines. In fact, a nanoscale robot arm doing chemistry should be able to build its own mass in a few hours-maybe even minutes. Detailed plans exist for combining lots of these tiny machines into an integrated factory that could make large products. So there's no problem with productivity.

Myth: Biology is better than anything we can build.

Fact: Biology never invented the digital computer, internal combustion, or even the wheel. Try putting three hundred actual horses under the hood of your sports car! Biology is very good at surviving on its own in ecosystems. But engineering, whether large-scale or nanoscale, is better at simple powerful things that modern humans tend to want-like compact motors and fast computers.

Related Myth: Nanomachine designs are unnatural and therefore inefficient. (Sometimes stated as: nanomachines are too rigid and have to spend lots of energy to do anything.)

Fact: Biology includes some great tricks-and nanomachines can use them too. But sometimes, rigid is better (see the "friction" myth). And by the way, if someone gives diffusion as an example of biology being better, they're wrong: making something diffuse takes just as much energy as dragging it (the Einstein, Stokes, and Shannon equations cancel out).

Related Myth: Other technologies will do just as well.

Fact: Precise nanoscale machines have several special advantages. Being small, they can operate more quickly-and thus more efficiently. Being precise, they can have very low friction. And being built with atoms, which are discrete, pliant, and identical, each copy of a machine can be very predictable and reliable. All of this is basic physical law, not speculation.

Myth: It's too hard to do-it'll take many decades if it can be done at all.

Fact: We won't know for sure until it's done. But it doesn't look that hard to design a few chemical reactions, and a few molecular machines, and some larger systems, and some software to pull it together, and a nano-manipulation or enzymatic system to build the first one. Each step requires only engineering, not complex system design. The military payoff alone will probably tempt nations to spend billions on developing it. A lot of people are now saying it could take ten to fifteen years; I think it could be as little as five, especially if one or more nations are already working on it.

Myth: It'll be nearly impossible to design products for a molecular manufacturing system, so it won't have much impact.

Fact: Once a few basic molecular machines have been designed, they can be recombined in many ways to make almost any product you can think of. Software engineers do this kind of thing all the time, compiling a few instructions into large programs-and never having to think about the physical chips. And most of today's products are a lot less complex than today's software programs.

From *Leading nanotech experts puts 'grey goo' in perspective*.

"Contrary to previous understanding, self-replication is unnecessary for building an efficient and effective molecular manufacturing system. Instead of building lots of tiny, complex, free-floating robots to manufacture products, it will be more practical to use simple robot arms inside desktop-size factories. A robot arm removed from such a factory would be as inert as a light bulb pulled from its socket. The factory (*) as a whole would be no more mobile than a desktop printer and would require a supply of purified raw materials to build anything."

"An obsession with obsolete science-fiction images of swarms of replicating nanobugs has diverted attention from the real issues raised by the coming revolution in molecular nanotechnologies," said Drexler. "We need to focus on the issues that matter - how to deal with these powerful new capabilities in a competitive world."

—Mike Treder, Executive Director of CRN, said, "We hope that this article will advance the discussion of the actual implications of molecular manufacturing. There is no need for panic, but there are urgent concerns that must be addressed before the technology arrives." This article is well worth a read - *link*.

4. "It's all just science fiction"

Myth: Molecular manufacturing hasn't appeared in peer-reviewed scientific literature.

Fact: This is incorrect. For example, there's Eric Drexler's 1981 PNAS paper, Ralph Merkle's publications in *Nanotechnology* in the mid-1990's, and a wide range of theory and experiment papers on mechanosynthesis

Myth: Nanobots just aren't well thought out: no power, control, heat dissipation, etc.

Fact: There's a lot of fiction and "artist's conception" floating around. But anyone who tars all nanomachine work with that brush is seriously ignorant. There's a lot of serious engineering work, with nanomachines, robots, and even desktop integrated manufacturing systems worked out in detail.

5. "It's all about dangerous self-replicating nanobots"

Myth: Products of molecular manufacturing must be small.

Fact: It won't be hard to integrate many nanoscale production systems into a factory that joins their output to make large products. This will be a lot more convenient and efficient than trying to keep track of zillions of tiny assemblers and nanobots.

Myth: Every product of molecular manufacturing will self-replicate.

Fact: This would be as inefficient as building an oil refinery and a machine shop into every car. In fact, even the manufacturing systems won't self-replicate, since they'll require preprocessed chemicals.

Myth: A laboratory accident could unleash a "gray goo" of nanobots on the world.

Fact: During the development as well as the use of mechanical molecular manufacturing systems, there will be no reason to build anything even close to a self-replicator. Gray goo would take a lot of hard, pointless design work. Even for military purposes, it would be quite inefficient (think of the car with the built-in oil refinery). A bigger concern appears to be deliberate production of non-replicating weapons.

We encourage debate. Please use this form to tell us what you think.

Risks and benefits of Molecular Manufacturing

The risks and benefits of molecular manufacturing depend heavily on several factors: what it is capable of, when it will be developed, and how it is used. There is substantial disagreement over the first two questions, and almost no information about the third.

As explained above, the idea that molecular manufacturing is impossible is not credible. A separate question is how difficult it will be to develop and whether it will be worth doing. Other technologies are advancing so rapidly that if molecular manufacturing is not developed by 2050 or so, it may not have a very large impact. If it is limited to biochemical materials such as protein, its products may be limited in strength (though still extremely compact compared to today's computers and MEMS).

Despite the weight of rhetoric indicating that molecular manufacturing will have little or no impact, the weight of calculation indicates that it will. In the absence of demonstrated problems with the theory, this section is written assuming product performance consistent with carbon-lattice construction-diamond, graphite, and carbon nanotubes, also called buckytubes or fullerenes.

The high speed of small systems implies that manufacturing systems based on molecular machines will probably be able to produce their own mass in an hour or so. (Bacteria can do this, and bacteria spend only a fraction of their energy on reproduction.) Building complicated shapes from simple chemistry (under direct computer control), there is no apparent reason why manufacturing systems cannot duplicate themselves. This implies that the systems would become extremely cheap, and manufacturing capacity could increase to any desired level almost immediately.

The products of molecular manufacturing would be very high performance. Even with biochemistry, computers would be millions of times more powerful than today's. With carbon lattice, motors would also be millions of times more powerful. Computation and actuation would be a miniscule (and virtually free) fraction of any human-scale product. And carbon-based materials may be hundreds of times as strong as plastics or even metals. With molecular-scale features, very low friction, and very strong materials, the main limitation on products would be the skill of the designers. The benefits are obvious. Cheap manufacturing of advanced products would allow inefficient infrastructures to be replaced-including our entire energy infrastructure, and much that is wasteful about our agriculture. This would greatly reduce our ecological footprint. Medical research and treatment would be greatly accelerated by the ability to design medical equipment, including massive sensor arrays and tiny robotics, and have them built instantly at negligible cost. Medical research would also be significantly helped by advanced computers. Cheap local self-contained manufacturing offers the possibility of eliminating material poverty. Advanced materials, extremely compact avionics, and rapid, inexpensive manufacturing would allow remarkably cheap access to space and its resources.

However, these capabilities also have downsides. Advanced weapons could also be developed quickly. An arms race based on this technology would be far less stable than the nuclear arms race: weapons would be easier to develop and build in secret, easier to use and control, and harder to stockpile (since they would be almost immediately obsolete) than nuclear bombs. An unstable arms race leading to a devastating war is probably the largest as well as the earliest danger posed by this technology.

Criminal technologies, from weapons to spy systems to communication to smuggling, would also get a boost from the ability to fabricate advanced products as needed. It remains to be seen whether defensive or offensive technologies will be stronger. Even if defensive technologies are adequate, they will probably be used unevenly. This implies that at least some criminal and terrorist acts using the technology will be successful.

And this brings us to the next risk: oppression. For a variety of reasons (including a desire to maintain artificial scarcity), established government and business interests will want to retain control of molecular manufacturing and its effects. A technology that lets each individual design and build whatever they want does not appear compatible with centralized control. But if the technology is too restricted, a thriving black market will arise to supply citizens with restricted and valuable products. Unfortunately, this would also supply the criminals and perhaps even lead to loss of military control, leading to a spiral of further repression.

Nearly free manufacturing could lead to an orgy of conspicuous consumption that could damage the ecosystem. And advanced products, up to and including brain and body modification, could shake up society.

Quantum Dots

Quantum dots are objects so small that adding or removing a single electron represents a significant change. While the integration of quantum dots into computers threatens to make Pentium processors as obsolete as vacuum tubes, such objects are nothing new to biological systems. Take for instance the *electron transport chain* used by nearly every living organism for *energy conversion*. Numerous proteins in this chain, as well as some smaller organic electron carriers all act as quantum dots. The image above is a model of part of the Iron Sulfur Protein (ISP), which uses its Iron Sulfur cluster to shuttle electrons from one quantum dot (quinone) to the next (Cytochrome C_1). As it does so, *conformation* changes throughout the multi-enzyme complex that it is a part of pump protons across a membrane resulting in a molecular battery. *Single electron transfers* abound in biological systems. Catalases are enzymes that use quantum dots to convert the damaging H_2O_2 into water. The photosynthetic machinery of plants use numerous quantum dots to absorb light and pass their energy and electrons to the complex "photosystems" that convert this energy into nanomechanical fuel, otherwise known as sugar.

In the world of computer science, quantum dots are viewed from a different perspective than in biochemistry. While in biochemistry they are generally referred to as *redox groups*, chip designers refer to them as *qubits* [Collins]. Researchers such as Yasunobu Nakamura at the NEC Fundamental Research Laboratories in Tsukuba, Japan, have combined a *superconducting* phase with a quantum dot in an attempt to build a *quantum computer*. Such a computer would use qubits much as a modern computer uses bits. However, qubits may have an advantage over macroscale bits due to the *quantum mechanical* principle of superposition, or the ability to be in two states at once. This principle may allow a single qubit to be used in more than one computation simultaneously. Numerous quirks of quantum mechanics are being found to be advantageous to *quantum computing*. For instance, the "quantum mirage" discovered at IBM provides evidence that physical wires may not be necessary in some nanoelectronic circuits. A promising tool for the organization of quantum dots into nanoelectronic circuits is *self-assembly*. Living systems have mastered the art of using self-assembly to generate useful structures. For instance, Prof. Dr. Dietmar Pum at the Center for Ultrastructure Research has demonstrated that *S-layer proteins* (Crystalline bacterial cell surface layers) can be effectively used to organize quantum dots (5 to 15 nm in diameter), regularly spaced up to 30 nm with oblique, square or hexagonal lattice symmetries. Such well ordered surfaces organize themselves, and thus, require a minimal amount of effort to prepare. The S-layers are able to self-assemble on a variety of solid supports, including *semiconductors*. Thus, it may not be too long before these nanoscopic arrays are integrated into memory chips, nanoprocessors, and optical devices.

Seeing Things Smaller Than Light

The primary sense that scientists use for observation is vision. Of course, vision depends on light (electromagnetic waves) bouncing off of the object and into the scientist's eye. Microscopes have been used for centuries to gather light so that scientists can see smaller and smaller objects. For some time, it seemed that improvement on this method (termed "optical microscopy") would allow us to see smaller and smaller objects without limit. For instance, this concept was expressed in 1733 by the poet Jonathan Swift when he wrote:

So, naturalists observe, a flea
Has smaller fleas that on him prey;
And these have smaller still to bite 'em;
And so proceed *ad infinitum*.

That concept may still be true; however, due to the delocalized nature of light, traditional optical microscopy encounters a *far-field diffraction limit*, which limits the resolution to half of the wavelength of light being used. Since the smallest frequencies of light that humans can see is around 400 nm, 200nm is the theoretical limit of simply focusing light into a human eye or detector (the practical limit is even larger). Thus, in order to observe and manipulate nanoscale objects, it is necessary to cross the diffraction barrier.

Scanning Probe Microscopy (SPM) is a recently developed family of techniques that can be used to produce images of nanoscale surfaces with resolution reaching down to the *sub-angstrom level*. Each of these methods has its own unique strengths and weaknesses.

- Scanning Tunneling Microscopy (STM)
- Scanning Force Microscopy (SFM)
- Scanning Nearfield Optical Microscopy (SNOM)
- Magnetic Force Microscopy (MFM)
- Scanning Capacitance Microscopy (SCM)

Mechanical VLSI

The miniaturization of mechanical devices (MicroElectroMechanical Systems—MEMS) is gaining increasing attention due to numerous commercial applications. Clocks, accelerometers, ink jet printer heads, color projection displays, *scanning probe microscopy*, pressure, temperature, chemical and vibration sensors, light reflectors, switches, vehicle control, pacemakers and data storage are only a few examples. Each of these applications uses MEMS in a unique way, but they are all based on fundamental micromechanical components.

Two fundamental components in MEMS devices are actuators and cantilevers. A mechanical actuator is a device that can convert an electrical signal into mechanical motion. For instance, if one applies a voltage to a quartz crystal, it will change its size in a very precise and predictable way. Such an actuator (known as a *piezoelectric* material) is the heart of a digital watch. A cantilever is basically a small diving board. Cantilevers (and diving boards) bend when pressure is applied to them and oscillate in a way similar to a spring when given the chance (e.g. if they are "driven" at their resonance frequency by an actuator). Such a micromechanical component is what makes atomic force microscopy sensitive enough to see single molecules.

The miniaturization of electronic components is more established than for mechanical components. Fabricating smaller diodes and transistors allows more circuits to be integrated onto a single chip. Combinations of basic components such as diodes and transistors results in new, larger scale components such as a logic gate, which can carry out a single step of logical computation. Over the years, the integration of these logic gates has increased from "small-scale integration" (SSI, less than 10 gates) to medium-scale integration (MSI, 10-100 gates) to large-scale integration (LSI, 100-5000 gates) to today's very-large scale integration (VLSI, with more than 5000 gates).

The first steps towards integrating mechanical systems into a single chip have been taken by researchers at IBM. The device known as the *'Millipede'* is an array of cantilevers integrated on a single chip. Each cantilever is sensitive enough to read and write data as nanometer scale indentations in a polymer film, giving it a capacity of 400-500 Gb/in^2. While a single cantilever would be to slow to read all of the pits, integrating them significantly increases the speed, making the "Millipede" a realistic data storage device. Numerous variations on this theme can be conceived. For instance, researchers at Northwestern University have demonstrated that coating the cantilevers with ink allows one to write nanoscale patterns, and have named the method *Dip Pen Nanolithography*. Since this method can be operated in a parallel fashion, it may not be long before arrays of AFM cantilevers coated with ink begin to lay down nanoelectronic circuits.

Since these mechanical systems are controlled electronically, they could conceptually be added as "drop-in" components in current chips. Indeed, most micro-machines are fabricated by the same semiconductor fabrication methods used to make computer chips. Since these methods are approaching the nanoscale, it is reasonable to expect that the basic units of both electronic and mechanical integration will eventually reach atomic and molecular dimensions. Since atoms and molecules are the fundamental building blocks of all matter, we can expect revolutionary capabilities from such NanoElectroMechanical Systems (*NEMS*), also known as *nanomachines*.

Electrons

The electron is a most convenient particle (or wave) for microscopy and lithography, not to mention electronic devices such as your computer. The mass of this subatomic particle is so small that only a negative exponent can do it justice: about 10^{-27} g. That is about one billionth of a billionth of a billionth of a gram. See what I mean? Too small for words. Compared to this super-small mass, it has a quantized negative charge of (elementary charge = $1.6*10^{-19}$ Coulomb). As this charge is exactly the same for all electrons, it is possible to precisely predict and control the movement of electrons. This tendency for very small objects such as electrons to be identical to one another is one of the motivating factors behind nanotechnology. That is, unlike components used in larger scale manufacturing, there need not be an inspector to check the mass and charge of each electron in a nanodevice. The small and uniform *mass to charge ratio* of an electron makes its momentum extremely sensitive to electric fields, which exert a force on charged particles. Such fields can be used to accelerate, direct, and focus electrons in the form of an *electron beam*. Scanning an electron beam over a surface, and measuring its interactions with the surface is a method known as *Scanning Electron Microscopy (SEM)*, which allows nanoscale resolution of conducting surfaces. Furthermore, if a surface is chosen that reacts in response to exposure to electrons then an electron beam can be used as a writing tool for the rapid patterning of surfaces with ultra-high resolution, a method known as *Electron Beam Lithography (EBL)*.

Now, consider the wave properties of this beam. The *De Broglie wavelength* of a particle can be related to its momentum by the relatively simple formula l = h/p, where l is the wavelength, h is Plank's constant and p is the momentum of the particle. Electrons differ from photons in that their mass is much larger, which results in a larger momentum and smaller wavelength. While a photon has a wavelength much larger than the size of one atom, the wavelength of an electron is far smaller than an atom. Thus, unlike optical microscopy or lithography, the *far-field diffraction limit* is not a barrier to the resolution of electron microscopy or lithography. Rather, other difficulties must be (and are being) overcome, such as our ability to focus the beam on a spot so small as a single atom. In Feynman's founding speech of 1959, he explained how electron microscopy might one day see individual atoms:

> "I would like to try and impress upon you while I am talking about all of these things on a small scale, the importance of improving the electron microscope by a hundred times. It is not impossible; it is not against the laws of diffraction of the electron. The wave length of the electron in such a microscope is only 1/20 of an angstrom. So it should be possible to see the individual atoms."

It is remarkable how prophetic Feynman was in his speech, seeing as this exact feat was was accomplished 22 years later with the invention of *Scanning Tunneling Microscopy (STM)*. STM is indeed invaluable for determining atomic precision within conducting samples and even selectively positioning individual atoms. However, even with its sub-atomic resolution, STM relies on slow mechanical actuators and is not currently useful for industrial *nanolithography*. Thus, it is still entirely possible that continued improvements in the focusing of electron beams (or other particle beams) might result in more rapid atomic resolution.

Alchemy

Nanotechnology can be qualitatively understood as an approach to alchemy that is based on the current model of matter. For more than 2000 years, the Aristotelian model of matter went relatively unquestioned in the western world. In this model, matter consisted of four elements: earth, air, fire and water. The variety of materials in existence was explained by the idea that the ratios of these elements within a material determined the form that the matter took. Based on this theory, the idea that one might rearrange the elements to create a desired material naturally followed. The arts of alchemy were early attempts to achieve this goal. In the last centuries, alchemy became more of a philosophy while its techniques were divided between chemistry, nuclear physics and various other disciplines.

In the 18th and 19th century, a far more accurate and productive model of the elements was finally developed. Lavoisier started the change with the discovery of a material that he called "oxygen" and began to arrange the other known elements into four groups. Dalton then proposed the importance of atomic masses, after which

Mendeleyev noticed a pattern that resulted in the periodic table of the elements. Early in the 20th century, Pauling provided an accurate model of how these elements *bond* together to form molecules. These molecules can in turn interact by means of *weak interactions*. The chemical bonds and molecular interactions seem to be flawlessly explained by *quantum mechanics*, making our model of matter relatively complete. With the new model being far more complex than the original one, the alchemist's goal of arbitrarily rearranging the elements became too confusing to pursue directly, until recently. During the last century, physicists and mathematicians have been ironing out the kinks in quantum mechanics. Chemists have found numerous ways to synthesize and characterize a plethora of molecules. Biologists have made enormous progress in understanding the molecular mechanisms and weak interactions that allow life to function. Meanwhile, the industrial revolution has provided us with machines, computers and other tools enabling further progress.

Atomic Precision

Our technology has now reached a point where the idea to more or less arbitrarily arrange the elements to create materials ranging from *nanomachines* to food is beginning to seem achievable. In 1959, Feynman translated the goals of alchemy into more modern terms, proposing that we miniaturize our tools for observing and manipulating matter until atomic precision is reached. In 1991, Drexler coined the term *molecular nanotechnology* referring to thorough, inexpensive control of the structure of matter. In the last decade, the concept of *nanotechnology* has been accepted by the scientific community and numerous organizations have been formed. Unlike alchemy, nanotechnology is based on a more modern model of matter. Nanotechnology does not try to transmute elements as it is far cheaper to extract and purify elements or molecules from existing materials. Arranging the atoms in a material to get the desired elements where we want them is a far more economical approach but still requires significant effort and cooperation. Thus, nanotechnology is an *interdisciplinary* subject drawing from biology, chemistry, physics, engineering and computer science, to name a few. Each discipline contains an important piece of the puzzle, and as the puzzle is assembled we are witnessing remarkable advancements in our ability to economically control the structure of matter.

The latest examples include *molecular imprinting*, *single molecule detection*, *DNA arrays* and *better sunscreens*. However, with nanotechnology just now beginning, what might result is still a matter of imagination. For instance, the Drexlerian nanotech community is looking forward to a future where a nanorobotic *assembler* could be programmed to build just about any stable arrangement of atoms, including more assemblers. Others imagine *quantum computers* or long term *life extension*. Only time will tell whether these dreams will be realized or face the same fate as the philosopher's stone and the elixir of life. Regardless of whether the ultimate goal of atomic precision is reached, novel *materials*, *methods* and *models* are resulting from this second attempt at alchemy.

REFERENCES

Attard, G.S., et al. 1997. *Science* 278: 838. Averback, R.S., J. Bernholc, and D.L. Nelson. 1991. *MRS Symposium Proceedings* Vol. 206.

Blumenthal, D.M., M. Gluck, K.S. Louis, M. Stoto, and D. Wise. 1986. "University-Industry Research Relationships in Biotechnology: Implications for the University," *Science* 232: pp. 1361-1366 (June 13).

BMBF. 1996. "Delphi-Bericht 1995 zur Entwicklung von Wissenschaft und Technik – Mini-Delphi." Bonn, author's translation.

Brown, J.S., and Paul Duguid. 2001. Don't count society out. *Societal Implications of Nanoscience and Nanotechnology.* National Science Foundation. Arlington, VA. March.

Drexler, E.; Smalley, R.E.: 2003, 'Point-Counterpoint: Nanotechnology', *Chemical and Engineering News*, 81(48), 37-42.

Drexler, K. Eric. 1986. *Engines of Creation.* Garden City, N.Y.: Anchor Press/Doubleday.

Drexler, K.E.: 1981, 'Molecular Engineering—An Approach to the Development of General Capabilities for Molecular Manipulation', *Proceedings of the National Academy of Sciences of the United States of America-Physical Sciences*, 78, 5275-5278.

Esant. 1999. "Economic and Social Aspects of Nanotechnology." First draft report of working group 6 of the Euroconferences on Nanoscience for Nanotechnology.

Estermann, M., L.B. McCusker, C. Baerlocher, A. Merrouche, and H. Kessler. 1991. *Nature* 331: 698.

Fedor N. Dzegilenko, Deepak Srivastava, Subhash Saini, "Simulations of carbon nanotube tip assisted mechano-chemical reactions on a diamond surface," *Nanotechnology* 9(December 1998): 325-330.

Galison, P.: 1997, *Image and Logic: A Material Culture of Microphysics*, University of Chicago Press, Chicago.

George M. Whitesides, "The Once and Future Nanomachine," *Scientific American* 285(September 2001): 78-83.

Gleiter, H. 1990. *Progress in Materials Science* 33: 4. Guinier, A. 1938. *Nature* 142: 569; Preston, G.D. 1938. *Nature* 142: 570.

Gutte, B., Dannigen, M. & Wittschieber, E. (1979) *Nature (London)* 281, 650-655.

Hadjiconstantinou, N.G. 1999. Combining atomistic and continuum simulations of contact-line motion. *Physical Review E.* Vol. 59(2).2, 2475-2478.

Hadjipanayis, C.G. 1998. Nanostructured magnetic materials. In *R&D Status and Trends*, ed. Siegel et al.

Kline, R.; Pinch, T.: 1996, 'Users as Agents of Technological Change: The Social Construction of the Automobile in the Rural United States', *Technology and Culture*, 37, 763-95.

Kroes, P. & Meijers, A. "The Dual Nature of Technical Artifacts—Presentation of a New Research Programme." Techné 6, 2 (2002): 4-8.

Kuusi, O. and M. Meyer. 2001. "Technology Generalizations and Leitbilds – The Anticipation of Technological Opportunities." Manuscript submitted to *Technological Forecasting and Social Change.*

Martin, T.P., U. Naher, H. Schaber, U. Zimmerman. 1993. *Phys. Rev. Lett.* 70: 3079.

Mizrach, S., *Should there be a limit placed on the integration of humans and computers and electronic technology?*

Philipse, A.P. 1998. "Colloidal Dispersions," section 3.4, A. ten Wolde (ed.), *Nanotechnology: Towards a molecular construction kit*. STT Report #60, The Hague, 171-178.

Pierce, N.A. and M.B. Giles. 2000. Adjoint reconstruction of superconvergent functionals from PDE approximations. *SIAM Rev.* Vol. 42(2). 247-264.

Rietman, E.A.: 2001, 'Drexler hypothesis of a universal assembler is supported not by theoretical arguments alone but by existence proof in the form of biological life', in: *Molecular Engineering of Nanosystems*, Springer, New York & Berlin.

Rinzler, A.G.; Liu, J.; Dai, H.; Nikolaev, P.; Huffman, C.B.; Rodriguez-Macias, F.J.; Boul, P.J.; Lu, A.H.; Heymann, D.; Colbert, D.T.; Lee, R.S.; Fischer. J.E.; Rao, A.M.; Eklund, P.C.; Smalley, R.E. 1998.

Roco, M.&Bainbridge, W. (eds.). Converging Technologies for Improving Human Performances, National Science Foundation report, 2002.

Smalley, R.E. 2000. Hearing of the Senate Science and Technology Caucus on Nanotechnology. April 5, (unpublished).

Smalley, R.E.: 1999, 'Prepared written statement and supplemental material', Rice University, 22 June.

Smalley, R.E.: 2001, 'Of Chemistry, Love and Nanobots', *Scientific American*, (Sept.), 76-77.

Smith, A., *The Wealth of Nations: An Inquiry into the Nature and Causes*, Random House, New York, 1994.

Smith, D., H. Babcock, and S. Chu. 1999. Single-polymer dynamics in steady shear flow. *Science*. Vol. 283. 1724-1727. March.

Snider, G.L.; Orlov, A.O.; Amlani, I.; Zuo, X.; Bernstein, G.H.; Lent, C.S.; Merz, J.L.; Porod, W. 1999. "Quantum-dot cellular automata: Review and recent experiments (invited)", *Journal of Applied Physics*, 85, 4283-4285.

Snow, E.S., P.M. Campbell, and F.K. Perkins. 1997. Nanofabrication with proximal probes. *Proceedings of the IEEE* 85: 601-11.

Sterman, J. D. (1989b). Modeling Managerial Behavior: Misperceptions of Feedback in a Dynamic Decision Making Experiment. *Management Science*, 35(3), 321-339.

Steve, Jurvetson, "Transcending Moore's Law with Molecular Electronics," Nanotechnology Law & Business Journal, Vol. 1, No. 1, article 9, p. 9.

Stipe, B.C.; Rezaei, M.A.; Ho, W. 1998. "Single-molecule vibrational spectroscopy and microscopy", *Science*, 280, 1732-1735.

Tighe, T.S., J.M. Worlock, and M.L. Roukes. 1997. Direct thermal conductance measurements on suspended monocrystalline nanostructures. *Appl. Phys. Lett.* 70: 2687-9.

Tihamer Toth-Fejel, "Modeling kinematic cellular automata: An approach to self-replication," Environmental Research Institute of Michigan, Ann Arbor, MI, 4 October 2000.

Tushman, Michael and Philip Anderson (1996), Managing Strategic Innovation and Change: A Collection of Readings. London: Oxford University Press.

2

Carbon Nanotubes and Buckytubes: Major Applications

Carbon Nanotubes, Nanoparticles, Nano-material & Nano-structuring Techniques

One-third of the tubes are metallic or quasi, the others are semiconductors with moderate band-gaps. If solid particles become small enough, in the nanometer range, they become electronically comparable to atoms and molecules. Instead of the laws of classical physics, they become quantum systems with discrete energy states. This is the basis for a number of useful applications. Quantum dots are reminiscent of the rhetorical question raised by Greek philosophers in antiquity: consider a sand-heap. If one removes from it a single grain of sand, obviously it is still a sand-heap. One continues doing so. At which stage does the sand-heap become something else? One of the most exciting applications of quantum dots is the single electron tunneling transistor. An electrometer measures electric charge. A single electron transistor is, at present, the most sensitive electrometer. Small alterations of the charge distribution beneath a small isolated island of aluminum atoms markedly affect the conductivity across that island. When a gate voltage is applied to an electron donor material, such as tellurium, placed underneath the island, changes in conductivity can be monitored. Nanocrystals are indeed important in single electron devices operating at room temperature, such as supersensitive electrometers and memory storage. Capped nanocrystals of both metals and semiconductors by virtue of their size possess capacitance in the range of attofarads (1 aF = 10^{-18} F). Charging a nanocrystal with an extra electron perturbs it to such an extent that the next electron requires an appreciable change in the charging potential. This is often seen as a 'Coulomb staircase' in the current-voltage tunneling spectra. Indeed, the charging energy varies linearly with the inverse of the diameter of the nanocrystal. Such sensitivity to single electron charging makes a nanocrystal an ideal candidate for use in single electron transistors and memory devices. There are metallic and there are semiconducting nanoparticles. The former enjoy industrial and medical applications, for their magnetic properties in particular. The latter have uses in electronics and optoelectronics. One can also write with nanoparticles, dip-pen lithography is one of the techniques. More generally, the challenge is to organize nanoparticles in three dimensions. The natural and the artificial once again meet! Biosystems, such as enzymes, antibodies, and nucleic acids have sizes commensurate with those of man-made nanoparticles. Hence, it becomes tempting to marry and hybridize them into nanocomposites, which is one of the most recent developments. Two of the main thrusts of nanotechnology have been self-assembled ordered arrays of atoms with specific properties (electrical conductivity, magnetism, *etc.*) and well-defined, molecule-like clusters of atoms endowed with specific properties. The keyword here is specificity.

The standard account of the birth of nanotechnologies dates it to the lecture by Richard P. Feynman on December 29, 1959 at the annual meeting of the American Physical Society held at Caltech. The books under review rightfully salute Feynman as the Founding Father of the field. As is well-known, the brilliant Caltech physicist ended his talk with a dare, announcing two prizes financed with his own money for the miniaturization of the pages of a book and for the building of a miniscule electric motor, only 1/64 inch cube. Feynman's lecture makes interesting reading. Its perspective is identical to that in Jules Verne's books, not visionary in the sense of a prophecy, imagining revolutionary developments and devices. The outlook is much more down to earth. Jules Verne and Richard P. Feynman took heed of recent scientific and technological advances to extrapolate to what the future would almost certainly witness, in a prospective.

In his lecture Feynman took as his starting point the feasibility of scaling down the contents of the world's libraries. He guesstimated their holdings at about 24 million volumes. He figured that their total information could be contained on an area of about 35 pages of the *Encyclopaedia Britannica.* The discussions of potential economic, medical, and environmental benefits may have given the false impression that nanotechnology will create a wondrous utopia in which all human problems are solved and we all live happily ever after. This is even more mistaken than the idea that new technologies always cause more problems than they solve. Many of the main constraints and difficulties faced by people are based not on technology or its lack, but instead by the very nature of the world we live in and the essence of our humanness. Increasing affluence based on molecular manufacturing won't end economic problems any more than past increases in affluence have. Wilderness can still be destroyed; people can be oppressed; financial markets can be unstable; trade wars can be waged; inflation can soar; individuals, companies, and nations can go into debt; bureaucracy can stifle innovation; tax levels can become crippling; wars and terrorism can rage. None of these will automatically be stopped by advanced technology. What is more, the potential benefits of new technologies aren't automatic. Nanotechnology could be used to restore the environment, to spread wealth, and to cure most illness. But will it? This depends on human action, working within the limits set by the real world. This chapter first describes some of the limits to what nanotechnology can accomplish, and then some of the adverse side effects of its basically good applications. The next will discuss the problem of accidents, which seems manageable, and then the far greater problem of potential abuse of new capabilities.

The world imposes limits on what we can do. Technology in general (and nanotechnology in particular) can provide padding for us as we throw ourselves against these hard, sharp limitations, and can sometimes help us slip past old limits through previously unknown gaps. Eventually, though, we will encounter new limits. In the end, solid constraints will limit human action no matter how much we juggle atoms and molecules, or the bits and bytes of information. Let's look at some of these, starting with the most abstract and long term—the most definite and hardest to avoid—and moving toward the more personal and near term. Many problems differ fundamentally from the material problems of limited matter and energy: they involve information. Some of the most precious stores of information in the world today are the genetic codes of the biosphere. This information, different for virtually every individual organism, is the product of millions of events that we are incapable of modeling or recreating. When this information is lost, it is lost forever. When the atoms encoding this information are thoroughly scattered, there seems to be no way to retrieve it. With any species, most genetic information is shared in common, found in all members of that species. But the variations in genetic code between individuals are important, both to the individuals themselves and to the health and prospects of the species as a whole. Consider the northern white rhino, whose numbers have dropped to an estimated thirty-two animals, or the California condor, of which only forty remain, all in captivity. Even if biologists succeed in reestablishing these species—eight condors were hatched in 1989—much of the diversity of their genetic information has been lost. Worse yet are extinctions of species for which no tissue samples were saved. The future may see some amazing recoveries: dry skin and bones may yield a complete set of genes when sifted by molecular machinery, and even current techniques have been used to recover genes from an ancient leaf, almost 20 million years old. Our eyes and instruments cannot yet tell us how much information from the past remains, but we do know that genetic information is being lost every day, and once lost, it is irretrievable.

People have often been wrong about physical limits, confusing the limits of their technology with the limits of the possible. As a result, learned men first dismissed the idea of heavier-than-air flight, and then dismissed the idea of flying to the Moon. Yet physical limits are real, and all technology—past, present, and future—will stay within those limits. There is even reason to suspect that some of those limits are where the learned now believe them to be. Nanotechnology will make it possible to push closer to the real limits set by natural law, but it will not change those laws or the limits they set. It will not affect the law of gravity, the gravitational constant, the speed of light, the charge of the electron, the radius of the hydrogen atom, the value of

Planck's constant, the effects of the uncertainty principle, the principle of least action, the mass of the proton, the laws of thermodynamics, or the boiling point of water. Nanotechnology won't make energy or matter from nothing. It seems a good bet that no one will build a faster than light spacecraft, or an antigravity machine, or a cable twice as strong as diamond. There are limits. Science today may be wrong about some limits, but scientific knowledge is practically defined to be our best information about how the world works, so it isn't wise to bet against it. There will be claims that nanotechnology will be able to do things that it can't, or that capabilities are around the corner when they aren't. Sometimes these will be innocent errors, sometimes they will be culpably stupid errors, and sometimes they will be what amounts to fraud. Among the problems that nanotechnology cannot solve is that of misguided claims, by people calling themselves "scientists," "engineers," or "businesspeople," that they have a big technical breakthrough worth a fortune. Every interesting new technology, particularly in its early days, is a chaotic mix of competent workers and charlatans. For every Thomas Edison inventing useful products such as light bulbs or the precursor of movie projectors, there were people promoting electric hairbrushes to cure baldness, and electric shoes, electric belts, electric hats—the list goes on—that authoritatively claimed cures for infertility, overweight, underweight, and all the ills and discomforts of mankind. Today, we laugh at the credulity of our forefathers who bought these gadgets; we shouldn't, unless we laugh at our own times as well.

Natural law imposes limits, but so does the nature of human beings. These will continue as long as people do. Reproduction is a deeply ingrained instinct enforced by the march of time, which ruthlessly discards the genetic material of all who neglect it. Many would argue that the Earth is already overpopulated. While nanotechnology could enable the current population, and even a greatly increased one, to live more lightly on the Earth, there will still be limits to Earth's capacity. The norms of human life are shaped by ancient patterns: high rates of infant or childhood mortality have been facts of life for millennia, and having many, many children has been a way to ensure that one or two will survive to work on the farm, and to care for you in your old age. Large families naturally become traditional. When modern medicine and reliable food supplies change those conditions—as they have, in cultural terms, virtually overnight—behavior does not shift as quickly. The result is the Third World population boom. In Western countries, where there has been time for behavior to adapt, a huge family is the exception. It might seem that our problem is solved. Molecular manufacturing can make everyone wealthy, and wealthy populations today have stable or shrinking populations. The Earth can support more people with advanced technologies, and these will also open up the vast room and resources of the world beyond Earth. Would that this were true. If 99 per cent of the people in a population respond to wealth by reducing childbearing, the population will indeed stabilize or shrink, for a while. But populations are not uniform. What of the 1 per cent, say, who are members of a minority with different values? If that minority has a growth rate of 5 per cent per year. then in ninety-five years they will be the majority, and in one thousand years their population will have grown by a factor of 1,500,000,000,000,000,000,000, if resource limits or genocide haven't intervened. Note that the Hutterites of North America, a reasonably wealthy religious group viewing fertility control as a sin and high fertility as a blessing, have managed an average of ten children per woman. Given enough time, exponential growth of even the smallest population can consume all the resources in reach. The right to reproduce is often regarded as basic, as illustrated by the outrage at reports of forced abortion in the People's Republic of China. The Hutterites and many others regard it as part of their freedom of religion. But what happens when parents have more children than they can support—does redistribution solve the problem? If reproduction is not forcibly suppressed, and if resources are forcibly and repeatedly redistributed so that each human being has a roughly equal share, then each person's share will steadily shrink. Even given the most optimistic assumptions regarding available resources, with a policy of resource redistribution and unlimited reproduction, the amount per person would eventually be insufficient to sustain life. This policy must be avoided, because if it is followed, it will kill everyone. As soon as we grant that any entity is entitled to certain rights—whether that entity be a human child, an animal, or some future artificial intelligence—the question arises of who is responsible for providing resources to support it when it can't do so for itself. The above argument indicates that a policy of coercion by some central power to compel the entire population to support an exponentially exploding population of these individuals would lead directly to disaster. Ultimately, this responsibility must rest with the entities' initiator: the designer of the artificial intelligence, the owner of the pet, the parents of the child. No new technology can magically remove the limits imposed by natural law, and thereby lift the burden of human responsibility.

Every time a technology solves a problem, it creates new problems. This doesn't mean that the change is neutral, or for the worse, of course. The Salk and Sabin vaccines for polio virtually destroyed the iron-lung industry, and the pocket calculator virtually destroyed the slide-rule industry, but these advances were worth the price of some economic adjustment. Molecular manufacturing and nanotechnology will bring far greater changes, placing far greater strains on our ability to adapt. We shouldn't be surprised when basically beneficial applications make someone miserable. Our lives are largely centered around problems. If we can solve many of these problems, the centers of our lives will shift, creating fresh problems. This section sketches some of the issues of change and adaptation more to raise questions than to offer solutions. Molecular manufacturing offers the possibility of drastic change, a change in the means of production more fundamental than the introduction of industry, or of agriculture. Our economic and social structures have evolved around assumptions that will no longer be valid. How will we handle the changes in the way we work and live? Nanotechnology will have wide-ranging impact in many areas, including economic, industrial, and social patterns. What do historical patterns in similar circumstances tell us about the future? Any powerful technology with broad applications revolutionizes lives, and nanotechnology will be no exception. Depending on one's point of view, this may sound exciting or it may sound disturbing, but it most certainly does not sound comfortable. In comparison to many projections of the twenty-first century, though, nanotechnology may lead to comparatively comfortable change. The changes most often projected—for a future not including nanotechnology—have been ecological disaster, resource shortages, economic collapse, and a slide back into misery. The rise of nanotechnology will offer an alternative—green wealth—but that alternative will bring great changes from the patterns of recent decades. Times of rapid technological change are disconcerting. For most of humanity's existence, people lived in a stable pattern. They learned to live as their parents had lived—by hunting and gathering, later by farming—and changes were small and gradual. A knowledge of the past was a reliable guide to the future. Sudden changes, when they did occur, were apt to be ruinous: invasions or natural disasters. These sudden changes were fought or repaired or survived as best one could. Making major changes by choice was rare, and radical innovations were generally for the worse: the old ways at least ensured the ancestors' survival, the new might not. This made cultures conservative. It is only natural that there be efforts to resist change, but before undertaking such an effort, it makes sense to examine the record of what works and what doesn't. The only examples of successful change fighters have been communities that have created and maintained barricades to isolate themselves from the outside world socially, culturally, and technologically. For the two centuries before 1854, Japan turned its back on the outside world, following a deliberate policy of seclusion. The leaders of Albania restricted contacts for many years; only recently have they started to open up. Isolation attempts have worked better on a smaller scale, when participation is voluntary rather than decreed by government. Today, within the Hawaiian island chain, the tiny, privately owned island of Niihau, sixteen miles long and six miles wide, is deliberately kept as a preserve of the nineteenth-century Hawaiian lifestyle. Over two hundred full-blooded Hawaiians there speak the Hawaiian language and use no telephones, plumbing, television, and no electricity (except in the school). The Amish of Pennsylvania have no surrounding ocean to help maintain their isolation, but rely instead on tight social, religious, and technological rules aimed at keeping external technology and culture out, and themselves grouped in; those who leave the fold are excluded. On a national scale, attempts to take only one part of the package—whether social or technological—haven't done well at all. For decades, the Soviet Union and the Eastern bloc nations welcomed Western technology but attempted tight restrictions on the passage of people, ideas, and goods. Yet illegal music, thoughts, literature, and other knowledge still crept in—as they do into the Islamic countries. Fighting technological change in society at large has had little success, where that change gave some large group what it wanted. The most famous fighters of technological change—the Luddites—were unsuccessful. They smashed "automated" textile machinery that was replacing old hand looms during the early industrial revolution in England, but people wanted affordable clothing, and smashing equipment in one place just moved the business elsewhere. Change has sometimes been postponed, as when a later group, under the banner of "Captain Swing," smashed hundreds of threshing machines in a wide area of southern England in 1830. They succeeded in keeping the old, labor-intensive ways of harvesting for over a generation.

In previous centuries, when the world was less tightly connected by international trade, communications, and transportation, delays of years and even decades could be enforced through violence or legal maneuvers such as tariffs, trade barriers, regulations, or outright banning. Attempting to stop or postpone change is less successful today, when technology moves internationally almost as easily as people do—and human travel is so easy that 25 million people cross the Atlantic each year. Change fighters find that the problems they create mount with time. Products made using the old, high-cost techniques are uncompetitive. There is no way to bring back the "old jobs": they no longer make sense. But old habits die hard, and these same responses to the prospect of technological change continue today—ignoring it, denying it, and opposing it. Societies that have fought change, as Britain did, have fallen behind in a cloud of coal smoke. Why did the Luddites respond violently? Perhaps their response can be attributed to three factors: First, the change in their lives was sudden and radical; second, it affected a large group of people at one time, in one area; and third, in a world unprepared for rapid technological change, there was no safety net to catch the unemployed. While local economies might have been able to absorb a trickle of hungry laid-off workers, they lacked the size and diversity needed to offer other employment options quickly to large numbers of unemployed. In the twentieth century, however, societies have of necessity become somewhat better adapted to change. This has been a matter of necessity, because sluggish communities soon fall behind. In the ancient days of peasant stability, there was no need for institutions like Consumer Reports to study and rate new products, or regulators like the Environmental Protection Agency to watch over new hazards. We developed the needs, and we developed the institutions. These mechanisms represent important adaptions, not so much to the technologies of the twentieth century, but to the increasing change in technology during the twentieth century. There is great room for improvement, but they can perhaps provide a basis for adapting to the next century as well. Even with the best of institutions to cushion shocks and discourage abuse, there will be problems. The very act of solving problems of production—of increasing wealth—will create problems of economic change.

Over centuries, the trend has seemed to be toward centralization, beginning with the rise of factories and industrial towns. What drove these developments was the high cost of machinery and plant operations, the need to be near power sources, the impracticality of transportation among many small, dispersed sites, and the need for face-to-face communication. Beginning with the first industrial revolution, factories employed large numbers of people in one place, leading to overcrowding and making local economies dependent on one industry and sometimes on a single company. Costly equipment necessitated central locations for textile production, rather than the cottage industries where a lone woman could earn a livelihood carding wool and creating thread on a spinning wheel (providing the origin of the term *spinster*). By the 1930s, the belief in the virtues of centralization and central planning—the supposed efficiencies and economies of scale—led to nationwide or continentwide experiments in centralization. But over the last decade, these large-scale experiments have been dismantled, from Britain's privatization of nationalized utilities to the beginning of a return to the market system in Eastern European countries. Because the old limits on transportation, energy sources, and communication have fallen, business is now decentralizing. Between 1981 and 1986, the Forbes 500 companies cut their employees by 1.8 million. But during those same years, total civilian jobs went up by 9.2 million. Start-up companies created 14 million jobs; small companies created another 4.5 million. Telecommuting is booming, as are new businesses, independent professionals, and cottage industries. We've also seen the resurgence of small, but highly diverse stores: gourmet food shops, specialty ethnic shops, tea and coffee purveyors, organic and health food stores, bakeries, yogurt shops, gourmet ice-cream stores, convenience stores offering twenty-four-hour access, shops selling packaged food plus snacks. These stores epitomize something fundamental: At some point, what we want is not a standard good at an ever cheaper price, but special things customized to meet our own individual tastes or needs. The trend for advanced technologies seems to be leading away from centralization. Will nanotechnology counter or accelerate this trend? By reducing the cost of equipment, by reducing the need for large numbers of people to work on one product, and bringing greater ability to produce the customized goods that people want, nanotechnology will probably continue the twentieth-century trend toward decentralization. The results, though, will be disruptive to existing businesses. The computer industry perhaps provides a clue to what might happen as costs are lowered by nanotechnology. The computer-software industry is characterized by the garage-shop start-up. When your equipment is cheap—inexpensive PCs built around low-cost chips—and you can make a product by throwing in some ingenuity and human labor, it's possible to start a new industry on a shoestring. In 1900, when cars were simple, there were many car manufacturers. By the 1980s, if you weren't an industrial giant like General Motors or Ford, Honda or Nissan, you had to be John De Lorean to even get a shot at acquiring the capital to play in the business. If molecular manufacturing can slash the capital costs for producing cars or other plant-intensive equipment, we will see the equivalent of garage-shop businesses springing up to offer new products, and hiring workers away from the industrial giants of today just as the personal computer has destroyed the dominance of the mainframe.

The U.S.A. dream is to be an entrepreneur, and the technological trends of the twentieth century point in that direction. Nanotechnology probably continues it. In one area, however, the late twentieth-century trend has been toward uniformity. The nations of Western Europe are in the process of uniting under one set of economic rules, and parts of Eastern Europe are anxious to join them. More and more supranational and transnational organizations knit the world together. The growth of trade has motivated economic integration. Molecular manufacturing will work against this trend as well, permitting radical decentralization in economic terms. This will help groups that wish to step aside from the stream of change, enabling them to be more independent of the turbulent outside world, picking and choosing what technologies they use. But it will also help groups that wish to free themselves from the constraints of the international community. Economic sanctions will have little force against countries that need no imports or exports to maintain a high standard of living. And export restrictions will likewise do little to hamper a military buildup. By weakening the ties of trade, molecular manufacturing threatens to weaken the glue that holds nations together. We need that glue, though, to deal with the arms control issues raised by molecular manufacturing itself. This problem, caused by the potential for decentralization, may loom large in the coming years.

Lester Milbrath, professor of sociology and political science, observes, "Nanotechnologies will create the problem of how to meaningfully and sustainably occupy the time of people who need not perform much work in order to have a sufficiency of life's goods. Our society has never faced this problem before, and it is not clear what social restructuring will be required to have a good society in those circumstances. We face much deep social learning." The world has had little experience with what anthropologists call "abundance economies." The native American tribes of the Pacific Northwest were one of those rarities. Ruth Benedict, in her classic book *Patterns of Culture*, wrote, "Their civilization was built upon an ample supply of goods, inexhaustible, and obtained without excessive expenditure of labor." The Kwakiutls became famous for their "potlatches": contests in which they sought to shame their rivals by heaping more gifts upon them than they could ever return. The potlatches would often be a year in preparation, last for days, and occasionally involve destruction of entire buildings. It was certainly a colorful form of keeping up with the Joneses. What will motivate us, once we have achieved an abundance economy? What will we regard as worthwhile goals to pursue? Increased knowledge, new art, improved philosophy, eliminating human and planetary ills? Will we find ourselves creating a better, wiser world, or sunk in boredom and jaded now that we have all and want nothing? If boredom gets out of hand, the lively spectacle of wealthy donors seeking to outdo each other to endow the arts, aid the poor, and do other good deeds for the sake of prestige would be welcome. What will happen as life spans continue to lengthen and the time needed to make a living decreases? Even today, there are people who, when confronted with the prospect of a significantly longer life span, exclaim that they couldn't imagine what they would do with all that time. This response can be hard to understand, when it would take a thousand years to walk all the world's roads, more thousands of years to read all the world's books, and another ten thousand years to have a dinner conversation with each of the world's people—but tastes differ, and even a few decades of bad television might make anyone long for the peace of the grave.

A major concern, and certainly the single area of greatest upheaval, is employment (which may become hard to distinguish from leisure). Once, people had little choice of employment. To keep a full belly, most had to work at the only job available: peasant farming. Eventually, people will have a complete choice of employment: they will be able to keep a full belly and a wealthy lifestyle while doing whatever they please. Today, we are about halfway between those extremes. In advanced economies, many different jobs are deemed useful enough that other people will offer an adequate income in exchange for the result. Some people can make a living doing something they enjoy—is this work, or leisure? The impact of nanotechnology on patterns of employment will depend on when it arrives. Current demographics show a shrinking supply of young people entering the work force. Agriculture, the assembly line, and entry level service jobs are experiencing a labor shortage, and no relief is in sight. If these trends continue, nanotechnology may show up in the midst of a shortage of labor. If it arrives late enough, it may compete with industries that are already nearing full automation; "job displacement" may mean replacing an industrial robot with a nanomachine. Employment patterns have shifted radically in the past. One hundred and fifty years ago, the United States was an agricultural nation—69 per cent of all people worked the land and a growing percentage worked in industry doing things like building steam locomotives for Baldwin Locomotives Works or tanning leather for the giant Central Leather monopoly. By the early twentieth century, agriculture was waning in numbers but increasing in productivity; most people worked in industry, and the tiny information and service sector was beginning to grow. Today, the picture has reversed: 69 per cent of employed Americans work in information or service jobs, only 28 per cent work in industrial production, and 3 per cent in agriculture. This tiny fraction feeds the other 97 per cent of Americans, exports hugely to other countries, and receives subsidies and price support payments to stop them from growing even more food. Manufacturing, even without nanotechnology, seems to be heading toward a similar condition. With an ever-declining percentage of our population working in manufacturing, we have as everyday products things that were once available only to kings and the high nobility. Yet owning multiple suits of clothes, having personal portraits of ourselves and family members, having music upon our command, having a personal bedroom, and having a coach awaiting our need—these are now regarded as being among the bare necessities of life. It may be possible to adjust to even greater wealth with even less required labor, but the adjustment will surely cause problems. In a world in which nanotechnology reduces the need for workers in agriculture and manufacturing still further, the question will be asked, "What jobs are left for people to do once food, clothing, and shelter are very inexpensive?" Again, the twentieth century provides some guidelines. As technology has reduced costs by efficiently producing many units of an identical item, people have begun to demand customization to meet individual needs or preferences. As a result, there are ever more jobs in producing custom goods. Today, semi-custom goods that try to help us meet our needs or express our taste abound: designer linens, ready-to-wear fashions, cosmetics, cars, trucks, recreational vehicles, furniture, carpeting, shoes, televisions, toys, sports equipment, washing machines, microwave ovens, food processors, bread bakers, pasta makers, home computers, telephones, answering machines—are all available in large and ever-changing variety.

Just as varied is the fabulous wealth and diversity of information produced in the twentieth century. Information products are a large factor in the economy: Americans buy 2.5 billion books, 6 billion magazines, and 20 billion newspapers each year. In recent years, new magazines have been invented and launched at the rate of one every business day of the year. A visit to a well-stocked magazine rack shows only a hint of the wealth of highly specialized publications, each one focused on a specialized interest or attitude: hotdog skiing, low-fat gourmet cooking, travel in Arizona, a magazine for people with a home office and a computer, and finely tuned magazines on health, leisure, psychology, science, politics, movie stars and rock stars, music, hunting, fishing, games, art, fashion, beauty, antiques, computers, cars, guns, wrestling. Motion pictures, which started as a flock of independent production companies and then consolidated into the great studios of the 1930s, have since followed the decentralization and diversification trends of recent years. Now an expanding range of film entertainment comes via network TV, cable channels, private networks, videotapes, music videos. Independent producers are aided by the technology innovations of cable, direct broadcast satellites, videotape technology, laser disks, videocameras.

The arts have burgeoned, with the general public as the new patron of the arts. Any artist or art form that could find and satisfy a market boomed in the twentieth century. Not just the traditional arts of actors, writers, musicians, and painters, but all forms of "domestic" artistry have grown to unprecedented levels: landscape and interior design, fashion design, cosmetics, hairstyling, architecture, bridal consulting. Providing for these demands are some of the "service and information" jobs created in the late twentieth century. "Service" jobs include many ways of helping other people: from nursing to computer repairs to sales. In "information" jobs, projected to have the fastest percentage growth over the next decade, people find, evaluate, analyze, and create information. A magazine columnist or TV news producer obviously has an "information" job. But so do programmers, paralegals, lawyers, accountants, financial analysts, credit counselors, psychologists, librarians, managers, engineers, biologists, travel agents, and teachers. "Increasingly," states *Forbes* magazine, "people are no longer laborers; they are educated professionals who carry their most important work tools in their heads. Dismissing them from their jobs, cutting them off from their places of employment may hurt them emotionally and financially. But it doesn't separate them from their vocation in the same way that pushing a farmer off his freshly seeded land does. For centuries workers were more dependent on a particular physical setting than they are now. Modern occupations generally give their practitioners more independence—and greater mobility—than did those of yesteryear." These human skills that people carry with them will continue to be valued: managing complexity, providing creativity, customizing things for other people, helping people deal with problems, providing old services in new contexts, teaching, entertaining, and making decisions. A reasonable guess would be that many of the service and information industries of the twentieth century will continue to evolve and exist in a world with nanotechnology. What is harder to imagine would be what new industries will come into being once we have new capabilities and lower costs. Along with the old economic law of supply and demand is another governing factor: price elasticity effects. People's desire for something is "elastic": it expands or contracts when the cost of something valuable goes down or up. If the price of a flight to Europe is five hundred dollars, more people will take a European vacation than if the price is five thousand dollars. When you had to hire a highly trained mathematician to do equations, calculation was slow and expensive. People didn't do much of it unless they absolutely had to. Today, computers make calculation cheap and automatic. So now businesses do sophisticated financial modeling, chemists design protein molecules, students calculate orbital trajectories for spaceships, children play video games, moviemakers do ever more amazing special effects, and the cartoon—virtually extinct because of high labor costs—has returned to movie theaters, all because computers permit cheap calculation. Nanotechnology will offer new, affordable capabilities to these and other people. Today, it's as hard to predict what new industries will be invented as it would have been for the creators of the ENIAC computer to have predicted cheap, handheld game computers for children.

So rather than producing drastic unemployment, nanotechnology seems likely to continue the trend already seen today, away from jobs that can be automated and into jobs where the human perspective is vital. But the true possibilities are, as always in the modern world, beyond predicting. Major shifts in demographics always cause disruptions. Even when we know they are coming, we never prepare for them. Our plans are based on expectations of what will happen. If things don't go as expected, we find that we have "malinvested." Houston real estate was valuable and looked to become even more so when times were good for the oil business there; when the fortunes of the oil business changed, Houston real estate was found to have been overbuilt, overpriced, and many millions of dollars were lost. Lengthening life spans push people toward taking a longer-term perspective, but rapid rates of change force a shorter-term perspective in investments. Turbulence in technology and in governmental monetary policy have already shortened time horizons. Businesspeople once routinely built plants with a thirty-year useful life. Today, the rate of change is too fast, and uncertainty regarding inflation and potential changes in tax laws is too great for such investments to make sense. Faster change will shrink time horizons further. Governments have taken on themselves the burden of looking a lifetime ahead, and the Social Security Administration is in for some rough times. When Otto von Bismarck, Germany's Iron Chancellor, came up with the notion of a guaranteed old age pension, it was a cynically clever and low-cost way to gain popular goodwill. So few people lived to age sixty-five that the amounts paid out in pensions were a pittance. After watching the German experiment for a handful of years, other governments began following suit. None of them expected a world like ours where a baby girl born in the United States today has an average life expectancy of 78.4 years—double that of Bismarck's time—and even this estimate is based on the faulty assumption that her medical care will be no better than her great-grandmother's was. At present, the Social Security Administration has two models: one they call "positive" and one they call "negative." In the "positive" model, people work like dogs until old age, retire, and promptly die—presumably before they've had a chance to collect substantial social security or medical benefits. In the "negative" model, people retire early, develop illnesses that require medical intervention, and then live a long time making doctor visits and hospital stays during those years. Plans based on these models deserve to be disrupted. A better, more realistic scenario would have people living and able to support themselves for a long time, with illnesses that can be handled easily and inexpensively. Present social security benefits are enough to provide a certain standard of living—food, housing, transportation, and so forth. In a future of great material wealth, these benefits will be easy to provide, and present projections of economic woe resulting from an aging population will seem quaint.

Back in the seventies, author Alvin Toffler brought out a book called *Future Shock*, describing how disturbing rapid change is for people. The book was a best-seller, but how much actual future shock has been seen in the past decade? Most people seem to have come through the last two decades pretty much all right, not in a state of shock at all. Rather than being shocked by technology, they are instead annoyed about pollution and traffic. Does this mean Toffler was wrong in predicting future shock? It's true that technology has been advancing rapidly in many areas over the past twenty years. But consider the average person's home life: How much of this rapid technological advance has shown up there? A great deal, yet most of it is hidden, unlike the earlier part of the century, where obvious change was the norm. Electric lights and appliances, automobiles, telephones, airplanes, radio, and television affected almost everyone's private life. One person's life could span the time from horse-and-buggy travel to watching the Moon landings on television. In contrast, the past twenty years have seen new technologies move more quietly into the home. The VCR and microwave oven don't seem nearly as revolutionary as earlier inventions. Telephone answering machines are useful but haven't caused major changes in lifestyles. Fax machines are handy, but they're much like having very fast mail, and as this is written, fax machines aren't yet in most homes. So it's not surprising that the average person has felt little future shock lately. New medicines taken as pills—which may be radically improved—look just like the earlier pills. The computerized bills that come in the mail aren't any more exciting to pay than the old human-prepared bills. This situation is unlikely to last. How much longer can technology advance so rapidly in so many fields without major effects on our lifestyles? There's been a respite from future shock in the last three decades; people have had a chance to catch their breath. When nanotechnology arrives, will future shock arrive with it? Some segments of society today are already getting practice in dealing with rapid technological advance. Those getting the most vigorous workout are in the computer field, where a machine two years old is regarded as obsolete, and software must be updated every few months to keep abreast of the new developments. But has this terrific rate of progress been dizzying or overwhelming? Not for the consumer—on the contrary, computers have become easier to use. In the 1960s, the New Math that was introduced into American grade schools and junior high schools included extensive study of arithmetic using numbers written in something other than the familiar base 10. This was to prepare the "Adults of Tomorrow" for "The Computer Age" in which we would all be writing assembly language computer programs in binary (base 2) code. But customers now purchase software rather than write it themselves—they need never deal with computer languages at all, much less a primitive assembly language. The rapid increase of computer speed has helped make computers easier to use. This progression has occurred many times before: Cars started off with external hand cranks, then advanced to starters you could yank from the comfort of the driver's seat; now starters perform invisibly when you turn the key in the ignition. This pattern will surely continue. First, some people will adapt to the technology, but in the long run the technology will adapt to us. The more flexible and powerful the technology, the more easily it will adapt. Seen from a distance, seemingly trivial patterns of adaptation form part of a larger process that has marked the last century: the Western world has begun to invent mechanisms to handle a world of persistent change. Our mechanisms are by no means perfect or painless, as any unemployed person can testify. Employment agencies and headhunters for job seekers; unemployment and severance packages to ease job transitions; on-the-job training, continuing education, retraining, specialized seminars to update professional skills, professional associations, networking, community resources centers, government training programs, and volunteer agencies are just a few of the inventions dealing with change and transition. Consumer information services, regulatory agencies, and environmental organizations are others. The most effective will endure. More options will continue to be invented. intellectual exercise of Feynman's had its obvious seed in the then recent (1953) discovery of the DNA double helix by Watson and Crick. If the genome has such a small size, Feynman thought, given the awesome amount of information it encodes, is there anything preventing scientists and engineers from emulating this natural feat? Let us keep in mind such inheritance. Biology thus gave birth to the whole field of nanotechnologies. To go by the empirical criterion, Feynman's call was that of a pioneer, it went unheeded for a generation. After Heinrich Rohrer and Gerd Binnig devised in 1981 a scanning tunneling microscope, a collective leap forward ensued. About that time, George B. Whitesides initiated a series of studies which showed the kind of contributions chemists might make to nanosystems. The nanotechnology bandwagon started rolling in the 1990s. It was pulled by two horses. There was Moore's Law. Engineers and scientists had for decades looked at the horizon of computers foregoing silicon chips in favor of molecular devices. And there was the demographic factor. The steady increase in the number of scientists, which continued well into the 1980s, was unmatched by a corresponding increase in public money. Accordingly, young scientists as a group partitioned themselves between industrial and pharmaceutical work.

That biology gave rise to nanotechnology raises a counterclaim. Whenever a group of chemists discuss nanotechnologies, they remind one another "this is nothing but chemistry, after all". The declaration can be seen as narcissistic, self-centered, and narrow-minded. Nevertheless, it has a degree of truth. Moreover, like any such stereotypic assertion on the part of a group of scientists, it is worthy of in-depth scrutiny. Let us then assume for the new sub-discipline to be yet another spin-off from chemistry, in like manner to biochemistry, solid-state physics, molecular biology, or materials science. Because of some unfortunate connotations of the word 'chemistry', its ancient roots on the one hand, its age-old associations with explosives, poisons, and pollutants on the other, people have refrained from using it as a label on a field holding both novelty and promise of lucrative applications. Hence, the renaming as nanotechnologies. Chemists are prompt to recognize in nanotechnologies lines of forces to them familiar and maybe even *passé*. These include colloids and Langmuir-Blodgett films, which have regained a new hold on both scientists' imagination and grantsmanship. At a more fundamental level, though, chemists have injected into nanotechnologies a key notion, that of specificity. Contrary to what is routinely the case in physics, with fractals an example, chemical phenomena fail to obey scale invariance. They are liable to show quite different features depending on the scale of the observation. Indeed, this serves as a dividing line between chemistry as the science and chemistry as the industry. The latter is interested only in bulk matter and in bulk phenomena, whether heat transfer or equations for overall transformations, with more concern for the bottom line in financial reports than for the specific and microscopic (or nanoscopic) mechanisms involved, in catalysis for instance.

However, what makes nanotechnologies a unique and viable hybrid is that their other parent has been, unmistakably, biology. This is an important legacy, all the more so in view of many of the present applications being in areas distant from biology such as electronics, optoelectronics, and other highly applied applications, from cosmetics to self-cleaning windows.

Biological phenomena are a continued inspiration and biological applications are around the corner. The former? While it was known that many organisms orient themselves from the earth's magnetic field, Blakemore's 1975 finding of bacteria from marine marsh muds heading rapidly north was seminal. Blakemore then used transmission electron microscopy to reveal iron-rich crystals, above 100 nm in diameter, in these bacteria. Several crystals would align into chains along their major axis. Frankel, Blakemore, and Wolfe proceeded to report that an unclassified magnetotactic spirillum (MS-1), when cultured in solutions containing ferric salts, precipitated uniform single crystals that consisted of mostly spinel Fe_3O_4. Microbial magnetite and the ferrimagnetic greigite g-Fe_3S_4, present in magnetotactic bacteria grown in marine, sulfidic environments, are inorganic crystallizations impressive for their quality. The magnetic particles have the optimal single domain size for the specific mineral. Each magnetic particle contains only a single magnetic domain. Moreover, within the magnetosome, the crystals are aligned so that their assembly bears a permanent magnetic dipole, making an effective compass. Biogenic magnetite provides great inspiration to laboratories active in nanotechnology, as for instance the book by Günther Schmid testifies to. The latter? Innocuous quantum dots (QD) bioconjugates have been injected into *Xenopus* embryos to monitor the development to the tadpole stages. QD label *Xenopus* embryos at different stages and with specific QD intracellular localizations.

But let us return to the attitude of chemists who discard nanotechnologies as a combination of old, well-known chemistry and a public relations ploy to raise money from industry and from granting agencies. Could it be true that the field of nanotechnologies amounts only to renaming a subset of well-established chemical knowledge? Does it lack originality and impetus? Are we dealing with old wine, which time may have started to sour and to make unpalatable and thus needed a new label? Conversely, is this new wine whose novel taste suggests brand-new flasks?

Consider colloidal gold as a test case. Gold nanoparticles have been the subject of considerable attention over the ages. A back-of-the-envelope calculation shows that a small nanocrystal of one nanometer diameter will have as many as 30% of its atoms on the surface while a larger nanocrystal of 10 nm (ca. 1000 atoms) will have about 15% of its atoms on the surface. Metal particles with dimensions in the nanometer scale thus display unusual properties. The mean free path of an electron in a metal at room temperature is in the range 10-100 nm. As a metallic particle shrinks to that dimension, unusual effects might be observed. Indeed, gold nanoparticles of diameters about 100 nm or less appear red, not gold, when suspended in transparent media.

They enjoy an interesting history dating back to the pioneering work of Faraday on the synthesis of gold hydrosols, gold nanoparticles dissolved in water. As one may recall, early in 1856 Michael Faraday (1791-1867) became interested in the interaction of gold with light. Gold leafs show different colors in transmitted light, green, blue, and purple, than they do in reflected light. This observation intrigued Faraday. Together with his friend, Warren de la Rue, he prepared gold films. He started preparation of gold precipitates from reducing with ferrous sulfate a solution of auric salts, made presumably by attack of the metal with *aqua regia*. In so doing, Faraday discovered colloidal gold. This was the subject of intense experimenting and thought. Between the onset of this work, February 2, and its completion at the end of the same year 1856, Faraday wrote in his diary no fewer than 300 manuscript pages (1,160 numbered entries) on this topic. He submitted an article to the Royal Society on November 15, it appeared on February 15, 1857. In general and since Faraday, metal sols possess fascinating colors and have long been used as dyes.

Smaller gold nanoparticles of diameters about 3 nm are no longer 'noble' and unreactive, but can catalyze chemical reactions. Consider nanocatalysts comprised of gold atom clusters, Au_8, Au_4, Au_3Sr, adsorbed at an F center of a MgO (100) plane. The optimal adsorption geometries of the O_2 molecule to these nanocatalysts differ markedly. Another example is from the temperature-programmed reaction spectra of the different products of the polymerization of acetylene on small supported, monodispersed palladium clusters. Striking atom-by-atom size-dependent reactivities and selectivities are observed. Only the three reaction products C_6H_6 (benzene), C_4H_8 (butene), and C_4H_6 (butadiene) form. Remarkably no C_3H_n, C_5H_n, and C_8H_n are detected, indicating the absence of C-C bond scission, as already observed on palladium single crystals and on palladium particles. Up to Pd_3, only benzene formation is catalyzed, reflecting a high selectivity for cyclotrimerization of acetylene. Pd_n clusters ($4 < n < 6$) reveal a second reaction channel by catalyzing in addition the formation of butadiene. The third reaction product, butene, desorbing at a rather low temperature of 200 K, is clearly observed for Pd_8. For this cluster size the abundance of the three reaction products is similar. For even larger clusters ($13 < n < 30$) benzene formation increases with cluster size, whereas the conversion of acetylene to butene reaches a maximum for Pd_{20}. Note that Pd_{30} selectively suppresses the formation of butadiene (see for instance the Nanocatalysis chapter in the book by Rao, Müller, and Cheetham).

What this example of metallic nanoparticles shows, which many of us have always suspected to be true but which conventional wisdom and standard practice – why bother reading any scientific publication older than say five years? – have both condemned summarily, is that casks of old wine connect with containers of the precious liquid from the most recent harvest. New science can be at its most innovative when it grafts itself onto its historical roots. Prediction starts with retrodiction, a point to which I will return.

Who is most responsible for inflating the nanobubble? Are scientists, journalists, or venture capitalists the guiltiest? Whatever the answer, a host of start-up companies have seen the light of day, and have issued shares. The Yellow Pages will have to make room for quite a few names with the nano- prefix: Nanometrics, Nanogen, Nanologic, Nanopierce, Nanoproprietary, Nanosys, Nanotechnology Development Corporation are just a few. Some universities have invested heavily into setting-up industrial parks specialized in nanotechnology. SUNY Albany and the University of South Carolina are two examples. The promise is there, because some of the devices already made actually work. IBM was able to use a nanotechnique to increase by a factor 20 the amount of memory on a hard drive.

John Wolfe, who started the first nanotech venture-capital firm, compares it to the Internet circa 1993, before Netscape went public. Is it to say that this new bubble is doomed from the start and that it will burst, sooner or later? Assuredly. Is it to say that it will sink without leaving a trace? Assuredly not. Some companies will survive and be lasting. Already, one can differentiate between fly-by-night operations buoyed only by conning the gullible, and reputable firms such as Nanosys doing solid work, and worth investing into. Nanosys collaborates with well-established corporate giants such as Intel, Matsushita, and DuPont. Peugeot-Citroën and Ford inject sub-10 nm

CeO_2 nanoparticles into diesel fuel before it is burnt in the engine. This results in an *in situ* catalyst that not only improves fuel efficiency but also greatly reduces the emission of particulates. More than a million automobiles will incorporate that system by 2005 already. The American Federal government is also among the investors in the Nano Rush. Washington is spending $ 3.7 billion on nanotech research over the next four years.

To this writer, the key issue is representation. Social groups create self-images, they project somewhat different images to the public. Nanotechnology, as mentioned at the beginning of this essay, arose from one of the icons of the 20th century, the DNA double helix. It was successful, early on, in giving itself a similar and logo-like icon. IBM scientists created it and the advertising arm of the multinational corporation was ready to pounce and was quick to hammer it irreversibly into public consciousness. This is the STM image of a corral of 48 iron atoms placed in a circle of 7.3 nm radius. This particular image has become emblematic of nanotechnologies, at least for the time being. With it, scientists exhibited to the public, not without intense pride, an atom-made fence. What a feat! But this construction can also be seen as an unselfconscious picture of the proverbial ivory tower. In yet another image, nanotechnology proudly depicted in 1990 on the cover of *Nature* its archetype with an STM image of DNA.

Another graphic depiction of nanotechnologies brings up my next point, representation as re-presentation. The cover page in issue number 24 of *Angewandte Chemie International Edition in English,* volume 37 for 1998, was adorned with a schematic representation of the Mo_{132} cluster. Its similarity to Kepler's early model of the universe, featuring the ubiquity of the icosahedron, was emphasized in the illustration.

Thus, nanotechnology in the time of the Millennium rediscovered the microcosm-macrocosm correspondence prevailing in medieval metaphysics. With some modification, needless to say. Humanism is out: a metallic cluster as the nanocosm replaces man as the microcosm. Just as for medieval theologians, the stress is on the unity of nature, on the uniformity of physical law throughout the universe. But physical science has replaced God in Western thought. One can interpret also the bow to Kepler as one of the numerous reincarnations of the Platonic archetype of the five regular solids, which has marked the history of chemistry in the braudelian *longue durée*: Kepler and the snowflake, the constituent particles of crystalline solids as proposed by René-Just Haüy, molecular shape according to André-Marie Ampère, Achille Le Bel's and Jacobus Henricus van't Hoff's tetrahedral carbon, Alfred Werner's complexes, Aaron Klug's icosahedral viruses and, closer yet to us, Richard Smalley's and Harold Kroto's buckminsterfullerene. Thus, nanochemists gave themselves a self-image of continuity within the geometric, Pythagorean tradition.

Let me sum up with representation as re-presentation. I have been at pains in this review to point to all the aspects of nanotechnology that predated it. Nanoscientists have moved existing pieces to create new patterns. There is nothing wrong with this efficient heuristics. Discovery is not the only path to knowledge, fame, and riches. But this raises an interesting question in philosophy and historiography, which will be my final point.

Science rewrites its own history, not only as history but as science fodder too. Which should not be construed in the weak sense of Whig history-telling as indulged in by scientists. While it is very much true that scientists, in a Whig perspective, tend to see the past as heralding the present, their purpose is not the mundane one of setting up an artificial history, approximating to a linear genealogy. The nanotechnologies paradigm shows a considerably more interesting attitude. Science reconfigures past events and developments in order to refocus itself into new directions. Such a reshuffling of the past, which of course may be deeply upsetting to professional historians who see their work voided and having to be restarted anew, the undersigned included, is an essential part of the dynamics of science. And it shows historians of chemistry that there is life after Partington.

Nanotube Field-effect Transistor

Transistors are the basic building blocks of integrated circuits. To use nanotubes in future circuits it is essential to be able to make transistors from them. We have successfully fabricated and tested nanotube transistors using individual multi-wall or single-wall nanotubes as the channel of a field-effect transistor (FET).

We measure the electrical characteristics of our nanotube-FETs and find that we can change the amount of current (I_{SD}) flowing through the nanotube channel by a factor of 100,000 by changing the voltage applied to a gate (V_G), as can be seen in our data below. G is the low bias conductance of the tube. This work has been published in Applied Physics Letters, vol 73, p. 2447 (1998). As we cool the FET down from room temperature to 4 degree Kelvin (minus 460 degree Fahrenheit) we see the device behavior change dramatically. While the device acts like a field-effect transistor at room temperature, at 4K it behaves like a single-electron transistor (SET). The three plots below show this change in device behavior for different temperatures.

Nanotube Rings

We have found a way to convert ropes of straight nanotubes into nanotube rings. The rings are composed of many layers of single-walled nanotubes, and have a radius of typically 0.7 micron. Coiling has been observed in proteins and other biomolecules, where hydrogen bonding is thought to provide the main force for coiling. Carbon nanotubes however present a novel behavior where coiling involves only van der Waals forces. The rings, which we can position on metal electrodes, allow us to study novel electric transport phenomena. Shown below is an AFM micrograph of a one micron-diameter ring (the purple circle) placed over gold electrodes (the light blue objects). SEM (Scanning Electron Microscope) image of nanotube rings on a silicon substrate. The image is magnified 8000 times. The nanotubes used to form the rings are extremely small; their diameter is only 1.4 nm. They are 1-dimensional conductors and at low temperatures, quantum interference phenomena dominate electrical transport through the tubes. The ring geometry allows us to observe such quantum effects in these 1D conductors. We find clear evidence of *weak localization*, which enables us to determine the electron's phase coherence length, which can be as large as 0.5 mm at 3 K. The plot below shows the resistance of a ring as a function of the magnetic field applied through it. The resistance is maximum at zero field because electrons propagating around the ring in opposite direction interfere with each other constructively at the point of injection, forming a standing wave. Applying a magnetic field breaks time-reversal invariance, and the contructive interference is lost.

Nanotube Theory

(i) Effect of Twisting

Effect of twisting on the properties of a metallic armchair (6,6) nanotube. Three twisted configurations are shown. Twisting is found to transform the metallic nanotube to a semiconducting one with a band-gap that varies with the twist angle as shown.

(ii) Effect of Bending

Variation of the local density of states (LDOS) along the length of a (6,6) nanotube as a function of the bending angle.

Understanding Carbon Nanotubes

Here, we focus on our recent work on *carbon nanotubes*; their structure, properties and uses in nano-electronic devices. Carbon nanotubes are extremely thin (their diameter is about 10,000 times smaller than a human hair), hollow cylinders made of carbon atoms. This study consists of the following sections (to access them you may also use the navigation bar on the left):

- *Nanotube manipulation:* We can manipulate the nanotube positions, change their shape, cut them and place them on electrodes.
- *Molecular mechanics:* We can simulate the mechanical behavior of nanotubes by calculating the forces acting between nanotubes and other objects such as the substrate.
- *Nanotube Field-Effect Transistor:* We have successfully used semiconducting single and multi-walled nanotubes as channels of field-effect transistors.
- *Nanotube rings:* While normally nanotubes are straight, we have devised ways to prepare them in a ring form.
- Nanotube theory: Computation and theory of the electrical and mechanical properties

Nanotubes, depending on their structure, can be metals or semiconductors. They are also extremely strong materials and have good thermal conductivity. The above characteristics have generated strong interest in their possible use in nano-electronic and nano-mechanical devices. For example, they can be used as nano-wires or as active components in electronic devices such as the field-effect transistor shown in this study.

Potential Applications

Conductive plastics Structural composite materialsFlat-panel displaysGas storageMicrom and nano-electronicsRadar-absorbing coatingTechnical textileUltra-capacitorsAtomic'Force Microscope (AFM) tipsBatteries'with improved lifetimeBiosensors for harmful gasesExtra strong fibres

Market

According to market report in 2004, "a total tonnage of 65 tons of nanotubes`and fibers was produced in 2004 which equates at manufacturer's current prices to`•144 million worth. By 2010 this figure will surpass • 3 billion. This`represents a compound annual growth rate of well over 60%."

Technologies

Nanocyl's basic technology had its foundation in the research work carried out at the Universities'of Liège and Namur in Belgium in the late 1990's. After becoming'a company in 2002, Nanocyl continued to'extend its own IP portfolio in carbon nanotube technology. Nanocyl's Intellectual Property covers carbon nanotube synthesis' technology; modifications to'the carbon nanotubes; as well as products utilizing carbon nanotubes'for their unique properties—such as polymer nanocompositesn Nanocyl's method of synthesis is' the catalytic carbon vapour deposition (CCVD), which is based on the mechanism shown below. This reaction is carried out in a continuous reactor technology enabling'the production of industrial quantities of carbon nanotubes with uniform product characteristics and quality. Nanocyl owns and develops original processes for modifying surface characteristics of carbon nanotubes'such as'imparting'chemical functionality to' the surface of the nanotube. This treatment makes CNTs'more suitable for applications such as'reinforcement of material compositesl interconnections, surface graftingl metal plating and nanocomposite manufacturingn

Applications

Carbon nanotubes have exceptional properties which can potentially be used in many applications ranging from macroscopic material composites down to nanodevices. Following is'a list of some potential applications

- Polymer additives for adhesives, coatings, fibers, films and engineered plastic‘parts
    - Flame retardants‘
    - Electrical conductive and antistatic‘polymers
    - EMI/RFI shielding‘
    - Organic photovoltaic‘cells
    - Mechanical reinforcement
    - Thermal conductive polymers
- Additives‘for inorganic materials‘
    - Ceramics‘
    - Metals‘
- Optical films‘
    - Optical Filtering devices
- Transparent electrodes‘
    - Flexible Display screens
- Gas storage
    - Hydrogen storage medium
- Catalyst support
    - Precious metals support in fuel cells
    - Polymerization
- Electron field emitters
    - Field Emission Displays‘
    - Flat lamps‘
    - Gas-discharge tubes
    - LCD backlighting
    - X-ray sources
- Electrode materials‘
    - Lithium-ion batteries
    - Bipolar plates of fuel cells
    - Supercapacitors
- Filters
    - Water and gas filtration

- Chemical and biological sensors
- Nanoprobes‘
  - Atomic‘force microscope
  - Scanning probe microscope
- Nanolithography
- Circuit Interconnect
- Nanotransistors
- Actuators
  - Artificial muscles‘
  - Electro-mechanical systems

Activities

Nanocyl was established in 2002 as the first European company with the goal of manufacturing carbon nanotubes‘(CNT) in commercial quantities.

The foundation of Nanocyl stems from the research work of Professor Janos B.Nagy of the University of Namur, and Professor Jean-Paul Pirard of the University of Liège. Today these researchers along with Nanocyl scientists and engineers continue to push the envelope in‘developing high-quality grades‘of carbon nanotubes‘for commercial applications and to‘support research activitiesn

The scope of Nanocyl's current activities includes:

- Refinement of continuous-process industrial CNT production
- Improved methods for CNT dispersion in materials‘
- Global commercialization of CNTs‘in a variety of sectors
- Partnerships for the development and commercialisation of new products
- Research programs and technology transfer agreements with Universities
- Active participation in the European Union sixth framework programme

Corporate

Jean Kluyskens, member of the Board of Directors of Nanocyl, specialist in recruitment and staff management, founder of Mercuri Urval Europe and board member of several companies including Namur Invest. With his daughter is‘a main shareholder of Inbev, the largest brewery in the word. Dominique Jozeau, member of the Board of Directors of Nanocyl, former CEO of Nanocyl. MBA from University of Dallas|/ST1:CITY> TX, Dean of the European University|/ST1:PLACE> and Director of the MBA program. CEO of ADM Group and European Managing‘Director of Franklin Electronic‘ Publisher.

Jean-Claude Jungels, member of the Board of Directors of Nanocyl, Electrical Engineer (ULg), MBA from Columbia Business School. Manager of Start-It, venture capital fund specialized in High Tech Start-ups. Luc'Geuten, member of the Board of Directors of Nanocyl has a chemical engineering'degree from KU Leuven and a MBA from Harvard Business School. Began his career at McKinsey & Company, is comfounder of Mitiska nv, a holding'company active in retailing'(Brantano, AS Adventure) and quoted on Euronext Brussels stock exchange. He is'also comfounder and director of Innogeneticsl director of AXA-Royale Belge and VPK Packaging Group". Dr Michel Morant, Board Observer of Nanocyl is Managing Director of Gesval, the technology transfer company of University of Liege. He graduated as' Electrical Engineer, PhD and MBA from University of Liege, spent 12 years in the AEG Daimler-Benz Company. He is'a member of the board of several start ups, and also'member of international associations (Proton, ASTP, CURIE, TII, ...). Jean Huybrechtsl representing'Strafincol member of the Board of Directors of Nanocyl, graduated in lawl pure econòmics and holds'a MBA from the Johnson School, Cornell University, NY. After a banking career and a management consulting' experience, he launched Strafinco SA, a corporate finance consulting firm.

Jean Stéphenne, member of the Board of Directors of Nanocyl, Chemical Engineer, is'CEO of GlaxoSmithKline Biologicalsn He is a former president of the Walloon Association of the Industry. Member of the board of several Belgian companies including Fortis'. Belgian Manager of the year in 1996.

Vincent Doumier, member of the Board of Directors of Nanocyl, CEO of Compagnie du Bois Sauvage, a Belgian Holding company, listed on the Brussels Stock Exchange having'investments in the Banking Sector, Real Estate, Industry (mainly Plastici and Luxury Consumer Sector, both in Belgium and abroad.

Baudouin Dubuisson, Chairman of the Board of Nanocyl, graduated with a degree in economics. After a career in audit and industry, he became advisor of the regional Minister of economy and he founded Franfin SA, a holding company specialized in start ups and middle sized companies'in difficultiesn

Dr. Janine Al-Asswad, member of the Board of Directors of Nanocyl, PhD in Chemistry, experience in R&D, Production, Technical Marketingl and Sales in SME's companies.

Frédéric'Pouchain, is Investment Manager with the "Compagnie Nationale à Portefeuille", shareholder of Nanocyl for which he follows several equity participations, was consultant for McKinsey & Company and holds'a degree of "Commercial Engineer" from UCL (Belgium) and a MBA from the Chicago Graduate School of Businessn

Robert Van den broeckl Shareholder of Nanocyl, board member of several companiesl Managing'Partner of Privast Capital Partners.

Eric de Lamotte, member of the Board of Directors of Nanocyl, director of several technology companies and investor in start-ups, former managing director and CFO of IBA, held management positions in the banking sector.

Dr. Frédéric'Luizi, shareholder and Quality Manager of Nanocyl, has'worked in various positions ranging from researcher in Universities (Portugal, UKl Denmark), project manager at BP (Norway), technology transfer officer for the University of Namur (Belgium), civil servant in Canada to'European Project Leader (Belgium), Board member of 2 other start-up companiesn

Dr. Antonio'Fonseca, shareholder and adviser to' Nanocyl is'PhD in organic chemistry. Research competences in NMR, catalysis, materials science, carbon nanotubesn

Management

Francis Massin, Managing 'Director, holds a Chemical engineering degree from Liège (Belgium) and a Business 'Degree from Solvay School in Brusselsn He also'attended INSEAD and IMD. He has more than 20 years of experience in scientificl industrial and business development fields at Dow'Corning in Europe, US'and Asia.

Stephen C. Schulte, Global Sales 'Manager, BA-Chemistry & German State University of New York, USA & Universitaet Tuebingen, Germany; and MBA-International Marketing, Rensselaer Polytechnic'Institute, New Yorkn Steve Schulte's career started in R&D at Chemfab developing fluoropolymer composites and continued as' Int'l Sales Mgr for silicone papers at International Paper. He moved to 'Europe in 1991 as 'European Product Manager-Kaolin Nanofillers with ECC International. Prior to joining Nanocyl Steve Schulte was European Sales' Manager for the US'Nanotube producer Hyperion Catalysis.

Olivier Decroly, Global Business Development Manager, holds'a Master of Science degree in Materials Science Engineering from the The Ecole Polytechnique de Bruxellesn His work experience includes' nanopigment process development at BHP-USA; then 2 years in nanocomposites technology with CRIF Liège, Belgium as' project engineer. Olivier Decroly started the commercial activities of Nanocyl at its beginning in 2002

Frederic 'Luizi, Intellectual Property and Quality Manager holds'a PhD in Biology from the University of Stirlingl Scotland. He has acted as'a researcher at University of Faro, Portugal, coordinator of European projectsl project manager at BP-Norway and technology transfer manager at the University of Namur. Dr. Luizi actively participated in the creation of Nanocyl in 2002

Marc Mostade, Project & Production Manager, holds a Master of Science degree in Electrical Engineering at the University of Liège, Belgium. Marc'Mostade has extensive experience in the design and startup of demonstration chemical plants.

In addition, Marc Mostade is 'certified in Quality (ISO 9000) and Environmental Management (ISO 14000); and he participates'actively in the Quality Insurance development of Nanocyl.

Fundamentals and Applications

The small scale and the one dimensional structure of carbon nanotubes are directly related to their unique properties which find more and more applications as the understanding and progress in synthesis advances. Carbon nanotubes represent an exemplary system where the bottoms-up approach to synthesis results in perfect structures with sizes less than 10nm, a range which remains inaccessible for advanced projection lithography techniques.

Applications of carbon nanotubes range from reinforcement of composites or conductive plastics to electrodes for batteries or flat screens, field effect transistors, chemical or force sensors and electromechanical memory. These applications have been demonstrated in bulk, on surfaces and on individual tubes. This course is focused on the fundamental properties of carbon nanotubes and provides the basic knowledge needed to follow the important developments in this field.

Objectives

We will show how the helical structure of the tube influences the electronic properties of the tube in a fundamental way. In the first part we will explain the structural peculiarities, give an overview of the main synthesis techniques and then show how, starting from the known electronic properties of graphite, how those of the nanotubes can be described.

The second part of the course covers electronic transport, electron emission properties, optical properties, mechanical properties and composites. We will show to what extend analytical techniques can provide information about individual tubes and we will see the importance of the immediate environment of the nanotube surface. In summary the goal of the course is:

- To give an introduction to the elementary properties of carbon nanotubes.
- To give the context and current status of the scientific research on carbon nanotubes.
- Synthesis Methods
- Characterization Methods
- CNT Based Materials
- Carbon Nano Industrial Applications

Schedule

Overview

- Introduction: history
- Context, excitement and potential of CNT applications

Topics

- Synthesis
- CVD
- HiPCO
- Supported Catalysts
- Large-Scale
- Vertically Oriented CNT

Fundamentals

- CNT structure: MWNT, SWNT, DWNT
- CNT synthesis: basic growth processes and bulk, localized growth, functionalisation
- Electronic structure
- One dimensional electronic transport
- Electronic emission
- Optical properties and light emission
- Mechanical properties, sensors
- Composites
- Synthesis Methods
- Characterization Methods
- Carbon Nano Structure based Materials
- Conclusions and outlook

Applications

- Overview and recent progress
- Examples of potential applications and real applications
- CNT/polymer composites
- Conductive Nanotube Films
- Fuel Cell Electrodes

- Field Emission for Flat Panel Displays

Site Listings

- *Nanotube Site, The*—facilitates the exchange of ideas among researchers by concentrating links to sites dedicated to nanotubes.
- *Nanoscale Science and Technology Group—IBM T. J. Watson Research Center*—investigating carbon nanotubes, nanolithography, and silicon nanoelectronics.
- *PhysicsWeb: Carbon Nanotubes*—1998 article looking at how experiments to test the theory that carbon nanotubes have a variety of useful properties are just becoming possible.
- *NASA JSC Carbon Nanotube Project*—focused on developing bulk nanotube production, purification, and application of single-wall carbon nanotubes.
- *Physical Properties of Carbon Nanotubes*—compendium of the currently accepted physical properties of carbon nanotubes.
- *Nanotubes & Buckyballs*—provides descriptions of the one-dimensional fullerene with a cylindrical shape and the most symmetrical large molecule known to man. Also includes links to other scientific nanotube articles and papers.
- *Biosensor for Fullerenes and Carbon Nanotubes*—paper reporting on binding experiments for anti-fullerene antibodies with several different fullerene derivatives as well as carbon nanotubes.
- *Nanotubes: Surprising Sensitivity to Oxygen Creates New Possibilities*—article examining how exposure to air can convert a semiconducting nanotube into a metallic conductor.

Buckytubes

The beginning of this story has been told many times. In 1985, a confluence of events led to an unexpected and unplanned experiment with a new kind of microscope resulting in the discovery of a new molecule made purely of carbon – the very element chemists felt there was nothing more to learn about. Buckyballs – sixty carbon atoms arranged in a soccer ball shape – had been discovered and the chemical world, not to mention the physical and material worlds, would never be the same.

The mechanical (stiffness, strength, toughness), thermal and electrical properties of pure buckytube materials enable a multitude of applications, from batteries and fuel cells to fibres and cables to pharmaceutics and biomedical materials. Scores of additional applications emerge when one thinks of blending nanotubes with other materials to improve existing properties or to provide new ones. Using nanotubes as fillers in thermoplastics and thermosets, for example, has been discussed for several years, and is only recently undergoing rapid investigation and development as sufficient quantities of high-quality buckytube material is becoming available to enable such investigations.

One of the most important technology developments of the last half of the twentieth century was the substantial replacement of metals with plastics. Most of this replacement has been in structural applications, where plastics have been engineered to outperform steel and other structural metals by providing adequate strength or stiffness at lower weight and cost. A key property that metals will always have over plastics, however, is in electrical conductivity. Plastics are amazingly good electrical insulators; in fact, this property gives rise to many of the most widespread and important uses of plastics. Nevertheless, the applications for plastics would be broadened substantially if good solutions existed to make these materials conductive.

These application areas include: antistatic, electrostatic dissipative, and electromagnetic shielding and absorbing materials. Electromagnetic interference and radiofrequency interference (EMI/RFI) shielding, for example, is essential in laptop computers, cell phones, pagers and other portable electronic devices to prevent interference with and from other electronic equipment. At present, there is no suitable plastic material for this purpose, and metal, in one form or another, is typically added to provide this function in electronic equipment cases, imposing substantial weight and manufacturing expense.

For some applications, plastics have been loaded with conductive materials for years to provide conductivity. The most common filler is carbon black, which is relatively inexpensive and works well in many applications.

One drawback to carbon black as a conductive filler, however, is the high loading required to provide the desired level of conductivity. It is well known that conductivity of a filled insulator, such as a polymer resin, increases with filler loading in a classic S-curve. That is, up to a critical loading, the bulk conductivity changes little, but increases very rapidly upon adding just a bit more filler. This is because high bulk conductivity requires the presence of many long conductive pathways, which are obtained only when the loading is so high that when randomly distributed, the conductive particles (e.g., carbon black) are likely to form long chains. This critical loading threshold is actually many times higher than would be required if somehow these particles could be placed in the optimal positions to form long chains with the minimal loading. But, of course, being sub-micron in size, this can't be done. A great deal of carbon black is wasted in the redundancy required to build up above the threshold level where these long chains form.

Also well known is the fact that the critical loading threshold decreases dramatically as the aspect ratio (length to width) of the filler particles increases. This is because longer particles cover a greater distance of the conductive pathway, whereas carbon black, which is spheroidal, has to form a chain of touching particles to cover the distance that a fibre-shaped filler would cover by itself.

Why is filler loading important? Aside from weight savings, when a plastic is loaded with carbon black at 30, 40, even 50% by volume, which is often the level needed to reach the desired bulk conductivity, the mechanical properties of the composite are severely degraded. Often it is not usable at all, and typically it is no longer mouldable, which is frequently the most critical property of plastic parts.

Buckytubes offer a solution. First, Buckytubes are terrific electrical conductors, as described above. No polymer is a better conductor and none better is likely to be found. (So-called conductive polymers, a class of long-chain molecules with a conjugated backbone, would be better described as molecular resistors; they are intrinsically semiconducting.) Second, Buckytubes have a phenomenally high aspect ratio. Individual tubes are about 1 nm in diameter (about half the diameter of DNA, and about 1/10,000th the diameter of graphite fibres), and 100-1000 nm in length. Thus, the aspect ratio of buckytubes is around 100-1000, compared with about 1 for carbon black particles. This already changes the game entirely, by pushing the critical loading level downward, as described above. Finally, buckytubes naturally form, in fact are born with, a morphology that is probably ideal for conductive filler applications. Buckytubes self-assemble into "ropes" of tens to hundreds of aligned tubes, running side by side, branching and recombining. When examined by electron microscopy, it is exceedingly difficult to find the end of any of these ropes. Thus, ropes form naturally occurring very long conductive pathways that can be exploited in making electrically conductive filled composites. Initial indications are that dramatically lower loadings of buckytubes are required to reach a given level of conductivity than for any other conductive fillers.

The opportunities for conductive plastics, as well as thermosets, filled with buckytubes are abundant. Very low loadings (<0.1%) provide for antistatic and electrostatic dissipative applications. One example is in painting automobile body parts, which are increasingly made of plastics. Because they are insulators, plastic parts charge up, which cause them to repel electrostatically the paint droplets formed in spray-painting of the body parts. This results in a great deal of wasted paint, which is both an economic and an environmental problem. A conductive primer coat can be applied, but that extra processing step is also quite costly. the ideal situation is to make the part itself sufficiently conductive to drain away charge build up by connecting the part to ground during the painting process.

Another broad area of application for buckytube-filled plastics is in EMI/RFI shielding, which has uses, as described above, in portable electronics, and defence applications. If, as appears likely, good attenuation of electromagnetic radiation can be attained at buckytube loadings on the order of 1% or less, good mechanical stability should be maintained, allowing it to be moulded. This would represent a significant breakthrough in plastics and enable broadening of their uses. Other defence uses of buckytube composites are similarly significant such as radar-absorbing or modifying material for aircraft and missiles. The application areas just described make use of the electrical conductivity of buckytubes where they contribute the highest added value. However, there is great promise in the development of applications that exploit the thermal and mechanical properties of buckytubes as well. Like the electrical applications described here, the possibilities are mind boggling. Stay tuned!

The special nature of carbon combines with the molecular perfection of buckytubes (single-wall carbon nanotubes) to endow them with exceptionally high material properties such as electrical and thermal conductivity, strength, stiffness, and toughness. No other element in the periodic table bonds to itself in an extended network with the strength of the carbon-carbon bond. The delocalised pi-electron donated by each atom is free to move about the entire structure, rather than stay home with its donor atom, giving rise to the first molecule with metallic-type electrical conductivity. The high-frequency carbon-carbon bond vibrations provide an intrinsic thermal conductivity higher than even diamond.

In most materials, however, the actual observed material properties—strength, electrical conductivity, etc.—are degraded very substantially by the occurrence of defects in their structure. For example, high strength steel typically fails at about 1% of its theoretical breaking strength. Buckytubes, however, achieve values very close to their theoretical limits because of their perfection of structure—their molecular perfection. This aspect is part of the unique story of buckytubes.

Buckytubes are an example of true nanotechnology: only a nanometer in diameter, but molecules that can be manipulated chemically and physically. They open incredible applications in materials, electronics, chemical processing and energy management.

Buckytubes have the following physical, chemical and mechanical properties that make them such an outstanding material:

- Electrical conductivity- the first polymer with truly metallic conductivity.

- Thermal conductivity—higher than diamond along the tube axis.
- Mechanical—the stiffest, strongest, and toughest fibre known.
- Chemistry of carbon—can be reacted and manipulated with the richness and flexibility of other carbon molecules. Carbon is the basis of most materials we use every day.
- Molecular perfection—essentially free of defects.
- Self-assembly – as noted above, strong van der Waals attraction leads to spontaneous roping of many nanotubes. Important in certain applications.

Electrical Conductivity

Depending on their precise structure, buckytubes can be either metallic conductors or semiconductors. The semiconducting tubes are of two types: one has a band gap of hundredths of an electron volt, while others have about a 1-eV band gap.

Ropes have been measured with a resistivity of 10-4 ohm-cm at 300 K [A. Thess et al., Science 273, 483 (1996)], making them the most conductive fibres known. Individual tubes have been observed to conduct electrons ballistically, that is, with no scattering, with coherence lengths of several microns [Tans, et al., Nature 393, 49 (1998)].

In addition, they can carry the highest current density of any known material, measured [B.Q. Wei, et al., Appl. Phys. Lett. 79 1172 (2001)] as high as 109 A/cm^2, and Phaedon Avouris, the Director of IBM's Center for Nanoscale Science and Technology, has suggested that it could be as high as 1013 A/cm^2.

Thermal Conductivity

Prior to buckytubes, diamond was the best thermal conductor. Buckytubes have now been shown to have a thermal conductivity at least twice that of diamond [Hone J, Carbon Nanotube Topics in Applied Physics, 273 (2001)]. Since this property, like electrical conductivity, is highly anisotropic in buckytubes, a macroscopic crystal of buckytubes, close-packed and aligned, would have the unique property of feeling cold to the touch, like metal, on the sides with the tube ends exposed, but similar to wood on the other sides.

Mechanical Properties

Buckytubes are the stiffest known fibre, with a measured Young's modulus of 1.4 TPa [M.-F. Yu et al., Phys. Rev. Lett. 84, 5552 (2000)]. They have an expected elongation to failure of 20-30%, which combined with the stiffness, projects to a tensile strength well above 100 GPa (possibly higher), by far the highest known.

For comparison, the Young's modulus of high-strength steel is around 200 GPa, and its tensile strength is 1-2 GPa. Their toughness has been demonstrated in simulations and experiments to be remarkable for something so stiff; buckytubes can be doubled over, snapping back with no resulting defect [Yakobson]

Chemistry and Manipulation

Buckytubes are polymers of pure carbon, and thus possess all the richness of carbon chemistry. Carbon is by far the most chemically versatile of elements, permitting an infinite number of combinations and derivatives. Closed ends of buckytubes are like fullerenes, whose chemistry is well understood. Open ends have the chemistry of the edges of graphite, also well understood. The possibility of attaching to either end of a buckytube essentially any chemical, structure or surface creates an entirely new situation whereby these end-entities can communicate with one another via true metallic transport of electrons. This is unprecedented since never before have there existed molecules with metallic conductivity, and this opens up worlds of possibilities for technological applications.

Technology also exists for chemically modifying the sidewalls of buckytubes, both covalently, by attaching chemical groups to the sides with chemical bonds, and non-covalently. Examples of the latter are the use of either simple or chemical surfactants, to modify the surface energy of the buckytubes.

In addition, the ability to manipulate buckytubes with chemical precision enables, for example, their dispersion in solvents, making them compatible with plastic resins for composites, etc.

Applications in Conductive Plastics, Adhesives and Connectors

Much of the history of plastics over the last half century has been as a replacement for metal. For structural applications, plastics have made tremendous headway, but not where electrical conductivity is required, plastics being famously good electrical insulators.

Conductive Plastics

This deficiency is overcome by loading plastics up with conductive fillers, such as carbon black and graphite fibres (the larger ones used to make golf clubs and tennis racquets). The loading required to provide the necessary conductivity is typically high, however, resulting in heavy parts, and more importantly, plastic parts whose structural properties are highly degraded.

It is well established that the higher aspect ratio of a filler, the lower loading required to achieve a given level of conductivity. Buckytubes are ideal in this sense, since they have the highest aspect ratio of any carbon fibre. In addition, their natural tendency to form ropes provides inherently very long conductive pathways even at ultra-low loadings.

Applications that exploit this behaviour of Buckytubes include EMI/RFI shielding composites and coatings for enclosures, gaskets, and other uses; electrostatic dissipation (ESD), and antistatic materials and (even transparent!) coatings; and radar-absorbing materials.

The same issues that make Buckytubes attractive as conductive fillers for use in shielding, ESD materials, etc., make them attractive for electronics materials, such as adhesives and other connectors (e.g., solders).

The special nature of carbon combines with the molecular perfection of buckytubes (single-wall carbon nanotubes) to endow them with exceptionally high material properties such as electrical and thermal conductivity, strength, stiffness, and toughness. No other element in the periodic table bonds to itself in an extended network with the strength of the carbon-carbon bond. The delocalised pi-electron donated by each atom is free to move about the entire structure, rather than stay home with its donor atom, giving rise to the first molecule with metallic-type electrical conductivity. The high-frequency carbon-carbon bond vibrations provide an intrinsic thermal conductivity higher than even diamond. In most materials, however, the actual observed material properties—strength, electrical conductivity, etc.—are degraded very substantially by the occurrence of defects in their structure. For example, high strength steel typically fails at about 1% of its theoretical breaking strength. Buckytubes, however, achieve values very close to their theoretical limits because of their perfection of structure—their molecular perfection. This aspect is part of the unique story of buckytubes. Buckytubes are an example of true nanotechnology: only a nanometer in diameter, but molecules that can be manipulated chemically and physically. They open incredible applications in materials, electronics, chemical processing and energy management.

Buckytubes have extraordinary electrical conductivity, heat conductivity and mechanical properties. They are probably the best electron field-emitter possible. They are polymers of pure carbon and can be reacted and manipulated using the tremendously rich chemistry of carbon. This provides opportunity to modify the structure and to optimise solubility and dispersion.

Very significantly, buckytubes are molecularly perfect, which means that they are free of property-degrading flaws in the nanotube structure. Their material properties can therefore approach closely the very high levels intrinsic to them.

These extraordinary characteristics give buckytubes potential in numerous applications.

Field Emission

Buckytubes are the best known field emitters of any material. This is understandable, given their high electrical conductivity, and the unbeatable sharpness of their tip (the sharper the tip, the more concentrated will be an electric field, leading to field emission; this is the same reason lightening rods are sharp). The sharpness of the tip also means that they emit at especially low voltage, an important fact for building electrical devices that utilize this feature. Buckytubes can carry an astonishingly high current density, possibly as high as 1013 A/cm^2. Furthermore, the current is extremely stable [B.Q. Wei, et al. Appl. Phys. Lett. 79 1172 (2001)]. An immediate application of this behaviour receiving considerable interest is in field-emission flat-panel displays. Instead of a single electron gun, as in a traditional cathode ray tube display, here there is a separate electron gun (or many) for each pixel in the display. The high current density, low turn-on and operating voltage, and steady, long-lived behaviour make buckytubes attract field emitters to enable this application. Other applications utilising the field-emission characteristics of buckytubes include: general cold-cathode lighting sources, lightning arrestors, and electron microscope sources.

Conductive Plastics

Much of the history of plastics over the last half century has been as a replacement for metal. For structural applications, plastics have made tremendous headway, but not where electrical conductivity is required, plastics being famously good electrical insulators. This deficiency is overcome by loading plastics up with conductive fillers, such as carbon black and graphite fibres (the larger ones used to make golf clubs and tennis racquets). The loading required to provide the necessary conductivity is typically high, however, resulting in heavy parts, and more importantly, plastic parts whose structural properties are highly degraded. It is well established that the higher aspect ratio of filler, the lower loading required to achieve a given level of conductivity. Buckytubes are ideal in this sense, since they have the highest aspect ratio of any carbon fibre. In addition, their natural tendency to form ropes provides inherently very long conductive pathways even at ultra-low loadings. Applications that exploit this behaviour of buckytubes include EMI/RFI shielding composites and coatings for enclosures, gaskets, and other uses; electrostatic dissipation (ESD), and antistatic materials and (even transparent!) coatings; and radar-absorbing materials.

Energy Storage

Buckytubes have the intrinsic characteristics desired in material used as electrodes in batteries and capacitors, two technologies of rapidly increasing importance. Buckytubes have a tremendously high surface area (~1000 m^2/g), good electrical conductivity, and very importantly, their linear geometry makes their surface highly accessible to the electrolyte.

Research has shown that buckytubes have the highest reversible capacity of any carbon material for use in lithium-ion batteries [B. Gao, Chem. Phys. Lett. 327, 69 (2000)]. In addition, buckytubes are outstanding materials for supercapacitor electrodes [R.Z. Ma, et al., Science in China Series E-Technological Sciences 43 178 (2000)] and are now being marketed.

Buckytubes also have applications in a variety of fuel cell components. They have a number of properties including high surface area and thermal conductivity that make them useful as electrode catalyst supports in PEM fuel cells. They may also be used in gas diffusion layers as well as current collectors because of their high electrical conductivity. Buckytubes' high strength and toughness to weight characteristics may also prove valuable as part of composite components in fuel cells that are deployed in transport applications where durability is extremely important.

Conductive Adhesives and Connectors

The same issues that make buckytubes attractive as conductive fillers for use in shielding, ESD materials, etc., make them attractive for electronics materials, such as adhesives and other connectors (e.g., solders).

Molecular Electronics

The idea of building electronic circuits out of the essential building blocks of materials—molecules—has seen a revival the past five years, and is a key component of nanotechnology. In any electronic circuit, but particularly as dimensions shrink to the nanoscale, the interconnections between switches and other active devices become increasingly important.

Their geometry, electrical conductivity, and ability to be precisely derived, make buckytubes the ideal candidates for the connections in molecular electronics. In addition, they have been demonstrated as switches themselves.

Thermal Materials

The record-setting anisotropic thermal conductivity of buckytubes is enabling applications where heat needs to move from one place to another. Such an application is electronics, particularly advanced computing, where uncooled chips now routinely reach over 100°C. CNI's technology for creating aligned structures and ribbons of buckytubes [D.Walters, et al., Chem. Phys. Lett. 338, 14 (2001)] is a step toward realising incredibly efficient heat conduits. In addition, composites with buckytubes have been shown to dramatically increase the bulk thermal conductivity at small loadings.

Structural Composites

The world-record properties of buckytubes are not limited to electrical and thermal conductivities, but also include mechanical properties, such as stiffness, toughness, and strength. These properties lead to a wealth of applications exploiting them, including advanced composites requiring high values in one or more of these properties.

Fibres and Fabrics

Fibres spun of pure buckytubes have recently been demonstrated [R.H. Baughman,

Science 290, 1310 (2000)] and are undergoing rapid development, along with buckytube composite fibres. Such super strong fibres will have applications including body and vehicle armour, transmission line cables, woven fabrics and textiles.

Catalyst Supports

Buckytubes have an intrinsically high surface area; in fact, every atom is not just on a surface—each atom is on two surfaces, the inside and outside! Combined with the ability to attach essentially any chemical species to their sidewalls provides an opportunity for unique catalyst supports. Their electrical conductivity may also be exploited in the search for new catalysts and catalytic behaviour.

Biomedical Applications

The exploration of buckytubes in biomedical applications is just underway, but has significant potential. Cells have been shown to grow on buckytubes, so they appear to have no toxic effect. The cells also do not adhere to the buckytubes, potentially giving rise to applications such as coatings for prosthetics and anti-fouling coatings for ships.

The ability to chemically modify the sidewalls of buckytubes also leads to biomedical applications such as vascular stents, and neuron growth and regeneration.

Other Applications

There is a wealth of other potential applications for buckytubes, such as solar collection; nanoporous filters; catalyst supports; and coatings of all sorts. There are almost certainly many unanticipated applications for this remarkable material that will come to light in the years ahead and which may prove to be the most important and valuable of all.

3

Nanotubes: History, Development & Commercial Use

—History and Development

The beginning of this story has been told many times. In 1985, a confluence of events led to an unexpected and unplanned experiment with a new kind of microscope resulting in the discovery of a new molecule made purely of carbon – the very element chemists felt there was nothing more to learn about. Buckyballs – sixty carbon atoms arranged in a soccer ball shape – had been discovered and the chemical world, not to mention the physical and material worlds, would never be the same.

In fact, what had been discovered was not just a single new molecule but an infinite class of new molecules: the fullerenes. Each fullerene—C_{60}, C_{70}, C_{84}, etc.—possessed the essential characteristic of being a pure carbon cage, each atom bonded to three others as in graphite. Unlike graphite, every fullerene has exactly 12 pentagonal faces with a varying number of hexagonal faces (e.g., buckyball – C_{60} – has 20). Some fullerenes, like C_{60}, were spheroidal in shape, and others, like C_{70}, were oblong like a rugby ball. Dr. Richard Smalley recognized in 1990 that, in principle, a tubular fullerene should be possible, capped at each end, for example, by the two hemispheres of C_{60}, connected by a straight segment of tube, with only hexagonal units in its structure. Millie Dresselhaus, upon hearing of this concept, dubbed these imagined objects "buckytubes."

In actuality, however, carbon nanotubes had been discovered 30 years earlier, but had not been fully appreciated at that time. In the late 1950s, Roger Bacon at Union Carbide, found a strange new carbon fibre while studying carbon under conditions near its triple point. He observed straight, hollow tubes of carbon that appeared to consist in graphitic layers of carbon separated by the same spacing as the planar layers of graphite. In the 1970s, Morinobu Endo observed these tubes again, produced by a gas-phase process. Indeed, he even observed some tubes consisting in only a single layer of rolled-up graphite. In 1991, after the discovery and verification of the fullerenes, Sumio Iijima of NEC observed multiwall nanotubes formed in a carbon arc discharge, and two years later, he and Donald Bethune at IBM independently observed single-wall nanotubes – buckytubes. These pure carbon polymers could now be understood in the context of fullerenes, changing the perception of them to molecules, with all that special designation implies. Nanotubes had been fullerenized.

Now, ten years after Iijima's initial observation, we know a great deal about nanotubes and tubular fullerenes. We know that multiwall nanotubes are invariably produced with a high frequency of structural defects. (In comparison with their larger relations, the 5-20 micron-diameter graphite fibers used in aerospace and sporting goods applications, multiwall nanotubes are quite sound structurally; nevertheless, they frequently contain regions of structural imperfection.) As any material scientist knows, it is the occurrence of defects that inevitably degrade the material properties of a substance, such as strength. The intrinsic properties of a material may be world-beater, but typically the actual properties of the bulk material are only a few percent of what the material would exhibit if it were structurally perfect. For example, a structural defect such as a microcrack in a steel wire, will lead to catastrophic failure at 1-2% of the theoretical breaking strength one would predict based on fundamental chemical principals.

In contrast, buckytubes are fullerenes, and are thus molecules: perfect, hollow molecules of pure carbon linked together in a hexagonally bonded network to form the hollow cylinder. The tube is seamless, with either open or capped ends. The diameter of single-wall carbon nanotubes is 0.7 to 2 nm (typically about 1.0 nm) – 100,000 times thinner than a human hair. Buckytube lengths are typically hundreds of times their diameters.

The molecular aspect of buckytubes is critical. Every atom is in the right place. This is a profound difference with their larger, defective cousins. To a chemist, a molecule is a very special thing. A molecule is complete, and is usually relatively happy with its identity. When a molecule is confronted with the opportunity to change, that is, to undergo a chemical reaction with other bits of stuff, there is almost always a fairly significant barrier to overcome. Other, non-molecular "stuff", when confronted with other stuff, is usually quick to change, and add in with the new part to make a larger whole. Metals are like this: a chunk of metal exposed to more metal (this could take place in the molten, gaseous, or solid phase) will accommodate the addition because it does not possess molecular completeness and invariance.

One aspect of molecular invariance is the reliability and predictability of chemical change. Molecules can be induced, by various means, to overcome their barriers to change, for example using heat. But by and large the reaction products when dealing with molecules are consistent. This is not true of changes that non-molecular things undergo. No two chunks of metal ever look the same, just as no two snowflakes look the same: snowflakes are not molecules. The effect of molecular invariance on material properties is equally profound. Whatever the intrinsic nature of the material, there are no defects to degrade the properties. You get what you get.

Structure, Naming and Properties

The special nature of carbon combines with the molecular perfection of buckytubes (single-wall carbon nanotubes) to endow them with exceptionally high material properties such as electrical and thermal conductivity, strength, stiffness, and toughness. No other element in the periodic table bonds to itself in an extended network with the strength of the carbon-carbon bond. The delocalised pi-electron donated by each atom is free to move about the entire structure, rather than stay home with its donor atom, giving rise to the first molecule with metallic-type electrical conductivity. The high-frequency carbon-carbon bond vibrations provide an intrinsic thermal conductivity higher than even diamond. In most materials, however, the actual observed material properties—strength, electrical conductivity, etc.—are degraded very substantially by the occurrence of defects in their structure. For example, high strength steel typically fails at about 1% of its theoretical breaking strength. Buckytubes, however, achieve values very close to their theoretical limits because of their perfection of structure—their molecular perfection. This aspect is part of the unique story of buckytubes. Buckytubes are an example of true nanotechnology: only a nanometer in diameter, but molecules that can be manipulated chemically and physically. They open incredible applications in materials, electronics, chemical processing and energy management. Buckytubes are single-wall carbon nanotubes, in which a single layer of graphite—graphene—is rolled up into a seamless tube. Graphene consists of a hexagonal structure like chicken wire. If you imagine rolling up graphene or chicken wire into a seamless tube, this can be accomplished in various ways. For example, carbon-carbon bonds (the wires in chicken wire) can be parallel or perpendicular to the tube axis, resulting in a tube where the hexagons circle the tube like a belt, but are oriented differently. Alternatively, the carbon-carbon bonds need not be either parallel or perpendicular, in which case the hexagons will spiral around the tube with a pitch depending on how the tube is wrapped.

There is a straightforward labelling convention to distinguish differently wrapped tubes from one another. The mapping specifies the number of unit vectors required to connect two atoms in the planar hexagonal lattice to form a seamless tube. These numbers specify a "vector" for the mapping, commonly expressed as (m,n), where m and n are integers. These numbers constitute a unique "name" for a tube. Any tube "named" (n,0) has carbon-carbon bonds that are parallel to the tube axis, and form, at an open end, a "zig-zag" pattern; these tubes are referred to as "zig-zag" tubes. Tubes named (n,n), where the two integers are equal, have carbon-carbon bonds that are perpendicular to the tube axis, and are often called "armchair" tubes. These two basic types are achiral, meaning they do not have a distinct mirror-image, like left and right hands. All the other tubes, named (m,n), where m does not equal n, and neither is 0, are chiral, and have left-and right-handed variants.

In most respects, the properties of tubes of different types are essentially the same. The exception to this is in their electrical conductivity, where these subtle structural differences can have profound effects. For example, all armchair tubes—i.e., where m=n—have truly metallic electrical conductivity.

They transport electrons along the tube axis just as metals do, without a single atom of metal in their structure! This behaviour in a molecule is unprecedented. In contrast, the other tubes are intrinsically semiconducting, either with a very small band gap of a few meV, or with moderate band gaps on the order of 1 eV. The rule here is that those tubes where (n-m) is a multiple of 3 are the small-gap type, while the others have medium gaps.

Since tubes with different (n,m) are molecularly distinct, there exists the possibility of chemically separating different types, and even of growing only selected tube types, although currently, all production processes produce a random mixture.

Another structural aspect of tubes is their self-organization into "ropes," which consist in many (typically, 10-100) tubes running together along their length in van der Waals contact with one another. Ropes are far longer than any individual tube in them: whereas tubes are typically about 100-1000 nm in length (and about 1 nm in diameter) ropes are essentially endless, branching off from one another, then joining others. These ropes are useful in providing very long conductive pathways well below what would normally be the loading required to reach a "percolation threshold" for conductivity.

Manufacturing on a Commercial Scale

Carbon nanotubes offer a range of attractive properties and could enhance many coatings applications. However, the current barriers to mass-market use of these materials include the availability of high-quality materials in commercial quantities. Now a partnership between industry and academia has resulted in a commercial manufacturing process for carbon nanotubes.

Hardly a day goes by without someone suggesting a new application for carbon nanotubes. Since their discovery, academics and industrialists have researched or considered their use in brake discs; fuel cells; advanced aerospace composites; co-axial cable; conductive fuel lines; EMI (Electromagnetic Interference) shielding in electronic devices; conductive tyres; conductive inks; and, of course, the hotly disputed space elevator – to name but a few. While some of these applications are clearly closer to reality than others, possibly the most immediate products will be developed in the coatings industry. The launch in April 2004 of the UK's first commercial manufacturing process for high-purity single-wall carbon nanotubes at Thomas Swan & Co. Ltd., has cleared the way for the development of a wide range of advanced coatings applications.

For those who have not come across them before, carbon nanotubes consist of molecular cylinders of pure, hexagonally-arranged carbon atoms that resemble rolled-up sheets of chicken wire with a diameter measured in a few nanometres (1 nanometre is 1 billionth of a metre) and a length of many microns. They occur in two main types, the single-wall carbon nanotube (SWNT) composed of a single cylinder of carbon, and the multi-wall version (MWNT) consisting of concentric tubes or cylinders of carbon (effectively straws within straws). The ends of the tubes are usually closed off by a carbon end-cap. Other variations on this theme include the double-wall tube, 'herringbone' and 'bamboo' structures.

Since their discovery, and Sumio Iijima's watershed paper in 1991, there has been unprecedented academic and industrial interest in carbon nanotubes and their potential use in a wide range of commercial applications. There is no denying that the basic properties of both single and multi-wall carbon nanotubes are truly remarkable. Recent studies have suggested that single-wall carbon nanotubes have a tensile strength of 50-100 GPa and a modulus of 1-2 TPa. This places them well ahead of steel in both strength and modulus and all for one sixth the weight. Combine this with their high thermal and electrical conductivity and you begin to understand why so many scientists and commercial organisations have sat up and taken notice.

The basic properties of the individual carbon nanotube (both single and multi-wall) have been considered in some fascinating potential applications. In essence carbon nanotubes are advanced fillers—highly conductive (thermally and electrically), extremely strong and light-weight. Add to this their apparent ability to absorb electro-magnetic interference and the fascinating photo-acoustic effect (whereby they spontaneously ignite when exposed to an intense light source) and you are left with some intriguing possibilities for advanced coatings.

The addition of low concentrations of carbon nanotubes to plastic it makes it electrically conductive. This allows electrostatic painting to be used when coating automotive parts, removing the need for costly primers.

The high aspect ratio of carbon nanotubes (length divided by diameter) means that they form a percolating matrix across the non-conductive filler at much lower loadings than traditional fillers. Carbon black typically requires 30 or 40% by volume loading to confer conductivity, at which point the mechanical properties of the composite are often severely degraded. In comparison Cambridge University recently achieved a conductive epoxy using only 0.005 wt% loading of aligned multi-wall carbon nanotubes. It is also likely that modest loadings of single-wall carbon nanotubes will actually enhance the structural properties of the composite.

The anti-static properties of carbon nanotube coatings may also have potential applications in packaging. Anti-static coatings are used in the electronics to prevent damage to sensitive electronic components during shipping and storage.

Carbon nanotubes (in their pure and undamaged form) are possibly the best thermally conductive material ever discovered. As such they are being researched for their heat management potential in a number of applications. As the electronics world strives to fit more and more functionality into less and less space, overheating is becoming a major problem. This is forcing greater pressure on materials and coatings involved in electronics manufacturing to be thermally conductive in an attempt to remove the heat from the device as quickly as possible.

Another aspect of thermal management is the demand for rapid curing times of polymeric materials such as ink resins and elastomers. The use of low loadings of carbon nanotubes should increase the speed of thermal dissipation from a coating or resin allowing for a much faster curing time. This would clearly be of interest in automated processes where the speed of curing limits the rate of production.

Electro-magnetic interference (EMI) shielding in laptops and mobile phones is becoming increasingly important to preventing interference with and from other portable electronic devices. As an extension to the electrostatic dissipation properties of carbon nanotube coatings, a number of companies are developing coatings that are designed to absorb EMI.

A significant amount of research is being directed into the use of carbon nanotubes in conductive inks. The ability to spray on conductive ink opens up some intriguing possibilities for 'spray-on' circuits and coatings. A leader in this field is US-based Eikos which was awarded USD 860,000 in May this year from the US Air Force Research Laboratory to develop transparent conductive polymers specifically for military aircraft canopies. While such contracts indicate that the carbon nanotube market is already up and running in some niche applications, there are still some major barriers to overcome.

While there are already some products in the market-place and a whole range of applications in development, the carbon nanotube market has been constrained by three main issues:

1. A lack of commercially-available material of consistently high quality.

2. Cost—until recently single-wall carbon nanotubes were selling for EUR 400/g which is clearly some way from commercial viability as a bulk material.

3. Successfully expressing the basic properties of the raw material in the end application – the incredibly small scale of the material poses some interesting challenges for advanced material and coatings science.

Such teething problems are not new. Carbon fibres took many years to be widely accepted in the materials world from both a cost and performance aspect and have only just made a regular appearance in popular sporting markets such as golf equipment.

Despite this, some early applications of carbon nanotubes in products such as conductive fuel lines in the automotive industry have already made a successful appearance on the market. In addition, recent advances in the manufacture of carbon nanotubes on an industrial scale will allow vital commercial research and evaluation to be conducted and this will open the way to a wide range of fully industrialised applications.

The UK's first commercial manufacturing process for high-purity single-wall carbon nanotubes was announced by Thomas Swan & Co in April 2004. The plant is the result of four years of collaboration between Thomas Swan and Cambridge University's Department of Material Science and Department of Chemistry. The unique collaboration brought the academic expertise of Cambridge University together with over 77 years of chemical manufacturing experience from Thomas Swan.

The goal of the project was to supply high-purity single and multi-wall carbon nanotubes in commercial quantities and to back up this initial offering with the ability to scale up the process to keep up with demand as the industrial market expanded. The problem was that at the start of the programme no-one knew which production method would supply both single and multi-wall tubes and be fully scalable. It was this problem that was presented to Cambridge University.

Carbon nanotubes can be manufactured using a variety of methods:

- Laser ablation uses a high-power laser to vaporise a graphite source loaded with a metal catalyst. The carbon in the graphite reforms as predominantly single-wall nanotubes on the metal catalyst particles.
- Arc discharge involves an electrical discharge from a carbon-based electrode in a suitable atmosphere to produce both single and multi-wall tubes of high quality but in low quantities.
- Chemical vapour deposition (CVD) is where a hydrocarbon feedstock is reacted with a suitable metal-based catalyst in a hot furnace to 'grow' nanotubes which are subsequently removed from the substrate and catalyst by a simple acid wash.

The team at Cambridge developed a novel system based on CVD growth, as it enabled the production of both single and multi-wall nanotubes of reasonably high quality and consistency while offering the greatest potential for scale up. This concept was then taken to design stage (and patented by Cambridge University). Using a team of chemical engineers and an independent consultant, the plant was built at Swan's site in North East England towards the end of 2003. Over the following months the plant was commissioned and then optimised until the material that was being produced was of consistently high quality.

A consistent supply of high-purity material that is also scaleable is a major stepping stone for the developing carbon nanotubes market. It allows companies to conduct meaningful industrial research using commercial quantities of raw material. In addition, this research can now be done with the knowledge that, should an application prove itself, there is a scalable source of the same raw material available to supply any ongoing development. Combine this with lowering prices (Thomas Swan predicts that within a few years the price of single-wall carbon nanotubes with be measured in tens of GBP per kg) and the market should start to take off.

A major problem in achieving a saleable product that is on average 100,000 times smaller than a human hair is that it is rather hard to 'see' what you have made. The advent of nanomaterial manufacturing brings to light the necessity for advanced industrial and cost-effective analytical techniques. While Scanning Electron Microscopes (SEM) provide the ability to view the sample in remarkable detail, it only shows a minute fraction of the product. The analogy often used is like looking at a football pitch through a jam jar. The grass beneath you is green and healthy but you have no idea what it is like on the other side of the pitch.

This will not be an easy problem to solve and while RAMAN techniques (which involve studying the spectrum profile of a reflected laser from your sample) offer some hope, it will be some time before a standard industry protocol for purity measurement is identified and accepted. Thomas Swan currently bases purity measurement on a combination of SEM, (Transmission Electron Microscopy) TEM, RAMAN and Thermal Gravimetry Analysis (TGA) techniques, combined with an experienced technical eye and has so far received very encouraging feedback from early customers. As the company moves to scale the process up, ongoing customer feedback and collaboration will become increasingly vital.

Carbon nanotubes are still in the very early stages of industrial development. While they are still expensive compared to a fully commoditised product, such as carbon black, the price will fall as the demand increases. Despite the problems associated with analysis and downstream processing (namely dispersion) the basic properties of the raw material offer a wide range of exciting applications that are beginning to be realised.

As the research into these applications moves out of the lab and into industry it is important that the supply of carbon nanotubes is consistent, of high quality and performed under the appropriate regulatory and health and safety protocols. Thomas Swan has taken the first step towards this goal and material is now available to allow meaningful industrial evaluation and development. The next ten years of development should see the creation of a whole new area of material and coatings science and associated commercial applications.

Flexural Strain Sensing & Carbon Nanotube Film

The sensor/actuator embedded smart materials and systems are essential for structural damage detection and control. There is need for smart structural materials, which can be used to sense strains, and adjust the structural properties so as to achieve desired response. Carbon nanotube composites have the potential to be such a smart structural material. Results (Baughman *et al.*, 1999) have shown that when carbon nanotube films, also called as "buckypapers", are used as actuators, large actuator strains can be achieved by smaller operating voltages as compared with ferroelectric and electrostrictive materials. Single wall carbon nanotubes (SWCNTs) promise to be a reinforcement material for nanocomposites because of their superior mechanical properties. Researchers (Hadjiev *et al.*, 2001; Zhao *et al.*, 2001) have studied the Raman band shift in SWCNT's as a function of stress/strain in carbon nanotubes. Raman studies on multi-wall carbon nanotubes have also been carried out (Schadler *et al.*, 1998; Wagner *et al.*, 1998). Though the experiments are carried at nanoscale, the results illustrate the potential of SWCNT as strain/stress sensor. It has also been shown that the electrical band gap changes as a function of axial compression, tension stretch, torsion, and bending strain (Peng and Cho, 2002; Tombler *et al.*, 2000; Yang and Han, 2000). Strong dependence of SWCNT's band structure on mechanical deformation makes it possible to develop nano-electro-mechanical sensors. Most of the studies to date relate the mechanical deformation with the change in electrical properties at the nanoscale. In this study, change in the electronic property of carbon nanotubes due to strain at macroscale is demonstrated using the experimental results. Such carbon nanotube films have randomly oriented SWCNTs with isotropic strain sensing properties; thus, have multidirectional and multi-location strain sensing capability.

The carbon nanotube films used in this study are produced by mixing SWCNTs with 0.25?mg/ml *N,N*-Dimethylformamide (DMF) and filtering the mixture through a 0.2?mm teflon membrane. Freestanding carbon nanotube film (Buckypaper) is peeled from the filter after rinsing and drying of the remaining material. The carbon nanotube film is further dried for 24?h under vacuum. The SEM picture of a carbon nanotube film with ~15?ìm thickness. To explore the strain sensing potential of the carbon nanotube film, it is attached to a 0.25×2×24?in. brass specimen and subjected to pure bending moment using two-point loading. For insulation, a PVC film is attached between the brass specimen and carbon nanotube film. High strength epoxy and a vacuum-bonding method is used to attach carbon nanotube film and PVC to the brass specimen in order to ensure the perfect strain transfer – between brass, PVC, and the carbon nanotube film. A conventional electrical resistance strain gage is attached next to the carbon nanotube film to measure flexural strain for comparison purposes. Load P is applied in increments using a servo-hydraulic machine. The corresponding change in voltage across the carbon nanotube film is measured using four-point probe measurement (Smits, 1958). Input voltage across two outer probes is kept constant during the measurement and change in voltage across the two inner probes is recorded.

Strain measurements are also made using conventional strain gage. The change in voltage is proportional to the change in resistivity of the film (Dharap *et al.*, 2004). The change in voltage due to the change in dimensions of the film is small as compared to the change in voltage due to the change in resistivity of the film (Li *et al.*, 2003).

The change in voltage in the carbon nanotube film as a function of measured flexural strain from the conventional strain gage. As the strain measured by the conventional strain gage increases from 0 to ~600?ìm/m the voltage change measured across the two outer probes increases from 0 to ~200?ìV. A nearly linear trend between the change in voltage and the strain is evident; although, deviation from the linear trend exists. Further study is needed to establish causes for deviation from linear trend such as temperature and gas exposure history (Bezryadin *et al.*, 1998; Collins and Avouris, 2000; Hone *et al.*, 2002). The change in voltage is measured by moving the four-point probe to several parallel locations on a single carbon nanotube film, which also yielded a similar linear trend. It is evident from the experimental study that the carbon nanotube film can be used for measuring flexural strains at the macroscale. Carbon nanotube films are made up of randomly oriented SWCNTs; hence, their electronic properties are independent of direction. Hence taking measurement along different directions provides corresponding strains. Carbon nanotube films can also be integrated into composites to simultaneously act as strong structural material and sensor. Such composites may find application in smart structures.

4

Nanosensors, Nanowires & Nanoribbons

Nanosensors and Nanoscale Scanning: An Overview

Medical nanorobots need to acquire information from their environment to properly execute their assigned tasks. Such acquisition is achieved using onboard nanoscale sensors, or nanosensors, of various types. Nanosensors allow for medical nanodevices to monitor environmental states at three different operational levels:

1. Internal nanorobot states.
2. Local and global somatic states (inside the human body).
3. Extrasomatic states (sensory data originating outside the human body).

The general physical limits to sensory perception are reviewed by Block and Bialek. The specific nanosensor technologies required include sensors to detect chemical substances, displacement and motion, force and mass, and acoustic, thermal, and electromagnetic stimuli.

The area of chemical microsensors is well developed, and there is increasing research interest in nanoscale chemical sensors, "chemosensors" and "biosensors". The most common nanomedical application of chemical sensors will be to measure the concentration of specific molecules and biopolymers in aqueous solvent—whether in blood serum, interstitial fluid, or cytosol. This may be accomplished using ordered arrays of receptors or by using sorting rotors to directly count molecular populations in known sample volumes. Spatial and temporal concentration gradient sensing is also essential in navigation, communications, and in mediating rapid response to environmental stimuli. Chemotactic sensors may be used to sample the chemical composition of surfaces.

One simple broadband concentration sensor, shown schematically consists of a graduated series of receptors having uniformly high specificity but engineered with progressively greater affinities (successively smaller equilibrium dissociation constants K_d (molecules/nm^3); for the target molecule, which is present in the test sample at concentration c_{ligand}. Exposure of the receptor array to the test sample for a time t_{EQ} necessary to reach diffusion-driven equilibrium gives a probability of receptor occupancy $P_{occ} \sim 0.91$ in a receptor with $K_d \sim 0.1c_{ligand}$, $P_{occ} \sim 0.50$ in a receptor with $K_d \sim c_{ligand}$, and $P_{occ} \sim 0.09$ in a receptor with $K_d \sim 10c_{ligand}$. During each measurement cycle, all steric probes are simultaneously extended into their associated receptor volumes at a time t_{EQ} after presentation of sample to the array. Probes which reach full extension register an empty receptor; those which cannot fully extend register an occupied receptor. After registration, the probes are retracted and ejection rods are thrust into all receptors (typically requiring ~1 nanonewton (nN) of force for occupied receptors) to empty them while the test chamber is flushed clear in preparation for the next cycle to begin. Ideally, the probe rod pushing into the binding site should push the bound molecule further into the binding site to avoid complications that may arise due to competing kinetic barriers, and ejection rods should push molecules far enough to ensure that the next resample is independent. Consider a series of N steric probe units as proposed by Drexler. Each probe measures 8 nm x (2.5 nm)^2 ~ 50 nm^3 and has mass ~ 2 x 10^{-22} kg. The ratio of the dissociation constants of adjacent units is $k = K_{di} / K_{di+1} > 1$, with $K_{d1} \sim 10^{-4}$ molecules/nm^3 and $K_{dN} \sim 10^{-13}$ molecules/nm^3 typically in the human body, and so $N = 1 + \{\log_{10}(K_{d1}/K_{dN}) / \log_{10}(k)\}$. Minimum measurement error occurs when c_{ligand} exactly matches the K_d of a probe unit, that is, when $c_{ligand} = K_{di}$ (i.e., $P_{occupiedi} = 0.5$). Maximum measurement error occurs when c_{ligand} lies exactly midway between two probes such that $K_{di} > c_{ligand} > K_{di+1}$, or, more specifically, at the geometric midpoint $c_{ligand} = (K_{di} K_{di+1})^{1/2} = K_{di} k^{-1/2}$. For common small molecules, $t_{EQ} \sim 10^{-6}$ sec; for common large molecules, $t_{EQ} \sim 0.2 \times 10^{-6}$ sec, but a sample chamber ~(25 nm)^3 in size is diffusion-limited to a minimum $t_{EQ} \sim 10^{-5}$ sec for $c_{ligand} >\sim 10^{-4}$ nm^{-3}. For the rarest small molecules in the human body, $t_{EQ} \sim 800$ sec, ~200 sec for rare large molecules, and a larger chamber may also be required. Preconcentration of the test sample by dehydration using sorting rotors to rapidly remove water from the sample chamber reduces t_{EQ} to 10-40 sec for very rare molecules. Steric probes using a driving force of ~100 pN, over 2 nm of motion per cycle require 200 zJ/measurement. If 1000 measurement cycles of a 10-probe sensor suffice to determine concentration to ~2 significant figures, this information costs ~0.002 pJ to acquire, a power consumption of from ~10 pW for large common molecules to ~10^{-9} pW (~0.6 kT/sec) for small rare molecules while the sensor is working. Power dissipation may run up to ~10 times higher, depending upon the number of occupied receptors that must be cleared via ejection rods during each cycle.

If the concentration of the target ligand varies only slightly within known limits, similar results may be obtained using a narrowband sensor that employs a repeating array of a single receptor type tuned to the maximum expected concentration c_{max} of the target ligand. If the sensor consists of N_r receptors, choosing $K_d >\sim c_{max} N_r^{-1}/(1 - N_r^{-1})$ ensures that all receptors are occupied at $c = c_{max}$, thus avoiding array saturation at concentrations below c_{max}. At minimum sensitivity, $K_d \sim c_{min} (1 - N_r^{-1})/N_r^{-1}$, and so (1 +

$(Dc/c)) = (c_{max}/c_{min})^{(Nr-1)}$ for the entire array. If A = receptor unit active area (~50 nm^2), L = receptor unit length (~10 nm), and receptor arrays are packed in parallel sheets with separation Dx (~10 nm, for small-molecule diffusion in ~10^{-6} sec) to allow sample access, sensor volume V_s = AL N_r (1 + (Dx/L)) and sensor scale ~ $V_s^{1/3}$. Thus a sensor with N_r = 100 receptor units achieves Dc/c = 0.10 (10 per cent) with a (46 nm)^3 sensor; N_r = 1400 receptor units achieves Dc/c = 0.01 (1 per cent) with a (112 nm)^3 sensor. A dedicated micron-scale nanodevice employing a (1.44 micron)^3 sensor array having 3 million receptor cells could at best distinguish 0.001 per cent concentration differentials in ~ 3 sec for common molecules like serum glucose. The ion channel switch (ICS) biosensor (aka. the "Australian sensor") can detect small molecule concentrations as low as ~10^{-14} nm^{-3} in a ~600 sec measurement time.

Concentration may be determined by varying exposure time to the sample. One energy-efficient designuses an input sorting rotor running at varying speeds (according to target concentration) synchronized with a counting rotor (linked by rods and ratchets to a data storage register device) to assay the number of molecules of the desired type that are present in a known volume of fluid. The counting rotor uses a steric probe, moved up and down in synchrony with the arrival of a binding site under the test probe, to count the number of molecules transferred by the sorting rotor. The fluid sample is drawn from the environment into a 4000 nm^3 reservoir ($L_{reservoir}$ ~ 16 nm) with the equivalent of 10^4 refills/sec on a flow-through basis, using paddlewheel pumps or reciprocating pistons; not shown) operated slowly enough to avoid damaging molecules in the sample volume. This sensor, which measures 45 nm × 45 nm × 10 nm comprising ~500,000 atoms (~10^{-20} kg), should count 'N ~ 10^4 – 10^5 molecules/sec of small common molecules present at typical human serum concentrations of 10^{-3} – 10^{-2} nm^{-3}, such as glucose, with the rotor spinning at ~1 per cent of normal speed. For rare molecules with c ~ 10^{-11} nm^{-3}, the count rate for a single sensing rotor falls to 'N ~ 10^{-4} molecules/sec. Sample preconcentration may allow further improvement by a factor of up to 10-100. Energy requirements are comparable to those for sorting rotors except for the need to retrieve ~all target molecules from a known sample volume, which slows the sorting process, requiring ~10 zJ/count for common molecules and ~100 zJ/count for the rarest molecules in the human body. Large-molecule concentrations may be measured using a shuttle pump and a counting device.

It may also be useful to design a sensor that is capable of detecting as many different chemical species as possible, in contrast to detecting the precise concentration of a single species, for use in chemomessaging, chemonavigation, cellular diagnosis, and other in vivo assay work. Counting rotors may be used for this purpose. Each 12-receptor sorting rotor/counting rotor pair has a minimum 400 nm^3 volume. Allowing an additional ~800 nm^3 per pair to account for power, control, mechanical attachments and housings, which includes ~300 nm^3 per pair for sample chamber volume and sample access, requires ~100 nm^3 per receptor, roughly the same as for the steric probe units described earlier. A dedicated 1 micron^3 chemical assay nanorobot, which includes ~0.25 micron^3 sample volume, with ~10^7 pairs (receptors) could thus continuously scan for as many as ~10^7 different chemical species, counting 'N ~ 1 large molecule/sec at concentrations of c_{ligand} ~ 10^{-10} molecules/nm^3.

For a purely sensory spherical nanodevice, a uniform distribution of a relatively small number of chemoreceptors confers optimal sensitivity. However, in applications where sensors (which attract and then release) are in competition with pumps (which strongly attract and then remove) for the same target molecules at the nanodevice surface, greater sensing efficiency may be achieved by clustering the sensors, as in a "nose" organ. Proteins may be modified as they age within biological cells. For instance, under steady state conditions, ~10 per cent of protein molecules may exhibit carbonyl (oxidation) modifications, glycosylation (non-enzymatic), and dysfunctional proteins may be ubiquitinized and phosphorylated under enzyme control. Heat shock also modifies proteins. Chemosensor receptors must be carefully designed to accommodate changes in target ligand structures or sites that are likely to be subject to such alterations.

Medical nanodevices can detect both extremely steep and extremely shallow spatial concentration gradients. For example, a receptor array sensor consisting of 10 steric probe units has a total volume of ~1000 nm^3 including its prorated share of drive systems, housings, etc. In just a few measurement cycles, adjacent sensors can detect a maximum concentration change from 10^{-2} nm^{-3} to 10^{-11} nm^{-3} across a mean nanosensor separation of 10 nm on a nanodevice surface, a steep chemical spatial gradient of ~10^{10} per cent per nanometer. J. Soreff points out that maintaining a steep concentration gradient of a freely diffusing species implies a power dissipation of ~kT per molecule per time to diffuse a concentration-halving distance. For a common, small, tethered molecule with c_{ligand} ~ 3×10^{-3} molecules/nm^3 that diffuses ~1 nm in t_{diff} ~ 1 nanosec, implied power density D_{mol} ~ kT c_{ligand}/t_{diff} ~ 10^{13} watts/m^3, a power density that might possibly be encountered in the vicinity of chemoelectric transducers. (Such high power densities never occur on macroscale dimensions, and heat rapidly diffuses at nanoscale dimensions.) Similarly, counting rotor sensors for small common serum molecules like glucose deployed at opposite ends of a 1-micron nanodevice can detect a concentration differential of ~10^{-4} over a ~1 sec measurement period, a shallow chemical spatial gradient of only ~10^{-5} per cent per nanometer. Even lower gradients may be detected by communication between physically separated cooperating nanodevices.

Koshland and Macnab have shown that bacteria possess a mechanism for sensing small temporal concentration gradients. A bacterium detects a spatial gradient of attractant not by comparing the concentration at its head and tail, but rather by traveling through space and comparing its observations over time. *E. coli* can measure a ~0.01 per cent gradient during the time it travels one body length (~2 microns) at top speed (20-40 microns/sec), a minimum temporal gradient of 0.1 per cent per second. This appears very efficient, since the counting rotor sensor proposed above is not much better, achieving ~0.01 per cent per second. As for peak gradients, the 10-probe receptor array sensor can detect a maximum concentration change (e.g., from an already-known rare concentration) in ~10^{-4} sec, a peak temporal gradient of ~10^{15} per cent per second.

Besides concentration sensing, chemical sensors may be used to determine the chemical characteristics of surfaces. A binding pad coated with an array of reversible, possibly reconfigurable, artificial receptors may be pressed against test surfaces such as cell membranes, or the perimeters of very large molecules, in order to detect the presence of specific chemical ligands. If pad receptor specificity is high for the target ligand, then the degree of pad "stickiness" to the test surface is a good measure of the population of target ligands on that surface. Probing 2-D surfaces with receptors may be constrained by orientation effects, since binding may require a specific relative orientation of ligand and receptor. Compliant pad receptors can reduce this problem but may impair lateral resolution—there is a three-way tradeoff among angular tolerance, lateral resolution, and multioriented multicopy receptors. A physical contact sensor prevents confusion of surface-bound and free-solvated ligands of the same species. The selective and reversible stickiness of such pads can also be exploited for locomotion. Molecular adhesion experiments in chemical force microscopy and interfacial force microscopy using AFM probes with functionalized tips typically detect maximum adhesive noncovalent force differentials of 40-160 piconewton (pN)/ligand for competing ligands at a peak-force distance of ~0.2 nm. Forces >10 pN are readily measured by nanosensors, so binding pad sensors should give an exact count of the number of target ligands present per unit area of test surface if (receptor areal number density) > (ligand areal number density) > (1 ligand/pad area), and if contact time is adequate. Receptors for many different ligands can be mounted on moving rotors or belts, or employed as interchangeable tool tips, thus allowing a single sensor pad to reconfigure its specificity or sensitivity under external control.

Ligand mimics may also be used to allow a medical nanodevice to detect a particular receptor in its environment. Each such "receptor sensor" is a protruding physical structure that mimics all or part of the electrosteric structure of the ligand to which the target receptor has maximum specificity. For example, probes consisting of artificial structures that simulate antigen peptides (8-15 amino acids long) bound in an MHC-like structure could be used to sense the presence of specific T-cell receptors. Sensing pads displaying ligand mimic structures are brought into contact with the test surface. Any target receptors present will bind to a mimic structure, thus may be counted as the sensor sweeps slowly across the surface. Immediately after recognition has occurred, the mimic structure may be reversibly fragmented, destroying receptor/mimic affinity and allowing the receptor to release the mimic. Mimics are mounted on the sensor pad with an appropriately elastic compliance to prevent structure detachment during a measurement scan. Receptor sensing is most useful in detecting receptors for large molecules. In the case of receptors for small molecules, there may sometimes be insufficient space in the target molecular volume to match the surface electrosteric structure, to accommodate ligand fragmentation control rods or cables, or to build internal boundaries sufficient to permit ligand fragmentation without creating undesirable reactive species. Some of these problems with small-molecule receptors may be partially overcome by employing an annular repulsive ring through which a mimic ligand may be forcibly retracted without fragmentation, but receptor orientation effects will present additional constraints.

The ability to precisely measure position and displacement, velocity and acceleration, rotation and orientation are essential requirements for nanoscale locomotion, physical manipulation, and navigation.

Diamondoid parts may contain features, or be positioned, in picometer (1 pm = 10^{-12} m) steps, much smaller than the unavoidable RMS thermal vibrations. The nanomanipulator robot arm may be moved in picometer increments, and the RMS thermal longitudinal displacements at the tip of a 20-nm long, 10-nm wide diamondoid lever is ~1 pm. Displacement sensitivity of 10 pm is routinely achieved in STMs. It is claimed this can be improved down to the ~1 pm level; STM resolution of 2 pm has been demonstrated experimentally. A C_{60} molecule used as the active element in an electromechanical amplifier transmits ~100 times more electrical current when physically compressed by 100 pm, suggesting a minimum deformation detection limit of a few picometers. (A strain gauge and a vibration sensor using a (17,0) carbon nanotube (~2000 atoms) has been proposed.)

Assuming a very stiff k_s = 600 N/m, a minimal SNR = 1 gives a minimum detectable displacement of 6 pm, or 10 pm at a more reasonable SNR = 2 (20 dB). The conservative conclusion is that Dx_{min} ~ 10 pm displacements should be reliably detectable by medical nanosensors in vivo, in a measurement time t_{meas} ~ 10^{-9} sec. This compares favorably with the stereocilia of the inner ear, which can only detect 100 pm displacements in 10^{-5} sec.

Typical thermal velocities of nanoscale diamondoid components at 310 K, are ~60 m/sec for 1 nm^3 objects, ~2 m/sec for $(10\ nm)^3$ objects, 0.06 m/sec for $(100\ nm)^3$ objects, and 0.02 m/sec for 1 $micron^3$ objects. These thermal motions are in degrees of freedom whose range of motion is largely confined, by design, to sub-nanometer distances within nanoscale measuring devices. However, objects of nanomedical interest may have velocities ranging from ~10^{-9} m/sec to ~10^{3} m/sec (e.g., the speed of sound in water), and may include biological elements in the environment or internal machine components. Thus it is useful to explore the fundamental limitations involved in measuring the full range of this physical variable. There are at least two simple ways to determine a velocity. First, velocity may be measured by transferring the kinetic energy of a moving object or fluid volume to a sensor element. The movement of the sensor element is then clocked as it traverses a known distance. Second, and more efficiently, the passage of the moving object may sequentially trip two latches located a known distance apart, again revealing the moving object's velocity; alternatively, detecting the presence or absence of a protruding feature on a passing body using probe rods can determine velocity without any requirement for large kinetic energy transfers. The need for physical gating movements by sensor components and computational register-shift operations using diamondoid rods moving at ~GHz clocking speeds suggests a conservative minimum stable sensor cycle time of $\sim 10^{-9}$ sec. The two principal restrictions on the maximum detectable velocity v_{max} are the maximum linear dimension of the sensor element, or L_{sensor}, and the minimum event time that may be accurately discriminated, or Dt_{min}. For a sensor element with thermal noise energy DE ~ 10 kT = 43 zJ at 310 K, the quantum mechanical limit is $Dt_{min} >\sim \hbar/(2\ DE) \sim 10^{-15}$ sec. Dt_{min} is also limited by the thermal variation in length of the individual latches. If the minimum detectable latch displacement is Dx ~ 10 pm, and the maximum speed of latch displacement is approximately the speed of sound in the latch material (e.g., $v_{sound} \sim 17{,}300$ m/sec for diamond), then $Dt \sim Dx/v_{sound} \sim 10^{-15}$ sec, near the quantum mechanical limit. Dt is further restricted by the minimum detectable phase variance between latches located at either end of the sensor. An acoustically-transmitted clock pulse emitted from a centrally-located source that is Dx closer to one of the two latches can discriminate a phase variance of $>\sim Dx/L_{sensor} \sim 10^{-5}$ taking $L_{sensor} \sim 1$ micron, so the minimum discriminable time is again $Dt_{min} \sim 10^{-5}\ L_{sensor}/v_{sound} \sim 10^{-15}$ sec; hence $v_{max} \sim L_{sensor}/Dt_{min} \sim 10^{9}$ m/sec ~ c (= 3×10^{8} m/sec, the speed of light). However, unlike photons, physical objects moving in an aqueous medium are normally limited to the speed of sound in that medium, e.g., $v_{sound} \sim 1500$ m/sec in water at 310 K If v_{max} is taken as $\sim 10^{3}$ m/sec, then for a sensor of size $L_{sensor} \sim 1$ micron the required time discrimination is only $Dt_{min} \sim L_{sensor}/v_{max} \sim 10^{-9}$ sec.

The minimum detectable velocity v_{min} is limited by the minimum measurable linear displacement Dx ~ 10 pm and the limits of our patience in making the measurement. For a maximum measurement time of $t_{meas} \sim 1$ sec, $v_{min} \sim Dx/t_{meas} \sim 0.01$ nm/sec in vacuo. In a nanomedically-relevant aqueous medium, the systematic motion of a free-floating (untethered) object of radius R is masked by added Brownian noise, giving the limiting $v_{brownian} = DX/t >\sim 260$ nm/sec for R = 10 micron and $v_{brownian} >\sim 2.6$ micron/sec for R = 100 nm, where $t = t_{meas} = 1$ sec. The thermal motions of tethered objects moving along surfaces (e.g., vesicles transported along microtubules by kinesin motors; are more restricted, hence Brownian velocity masking is reduced. Modest accuracy may be achieved using acoustic reflection in a Doppler velocimeter. Employing acoustic waves of frequency n, the minimum measurable velocity change $Dv = v_{sound}/(n\ t_{meas}) = 150$ microns/sec in an aqueous medium at 310 K with $v_{sound} = 1500$ m/sec, n ~ 10 MHz, and $t_{meas} = 1$ sec, corresponding to the shifted frequency $n\ (1 + (Dv/v_{sound}))$ having enough time to make one extra cycle during t_{meas}. Even at such long wavelengths (~150 microns), significant backscattering of ultrasound from ~7.82 micron red cells is observed experimentally.

Accelerations inside nanoscale devices are often measured in the trillions of g's (g = 9.81 m/sec^2). For example, an L = 1 nm diamondoid block (m = 3.5×10^{-24} kg) with thermal energy ~kT experiences random thermal accelerations of a_{therm} ~ kT/g m L = 10^{11}g's; a 10-nm logic rod with vibrational energy ~kT experiences ~10^{12} g oscillations; a 200-nm wide hoop spinning axially with 1000 m/sec rim speed has rotational acceleration of a ~ 10^{12} g's.

As one member of a class of devices, a simple nanoscale omni-directional accelerometer is conceptualized as a reaction mass suspended by three pairs of reaction springs (one pair per orthogonal directional axis) with each spring connected to the walls of an evacuated box. Acceleration in any direction extends some springs and compresses others, which is measured by force or displacement sensors attached to each spring. The signal/noise ratio for this sensor is Sensor mass exchanges reciprocally with acceleration sensitivity, so the lower limit even for a dedicated micron-scale platinum-core box-spring accelerometer is a_{min} ~1000 g's at SNR ~ 2 (undamped resonant acceleration ~ 10^6 g's for L = 2 microns). The maximum detectable acceleration requires m ~ 10^{-23} kg (~1.4 nm diamond cube) and k_s ~ 1000 N/m, giving a_{max} ~5×10^{13} g's at SNR = 1. The center of mass of a nanorobot, the sensor may avoid confusing centrifugal and translation forces. Otherwise, information from multiple sensors must be combined to obtain an unambiguous measurement. Multiple measurements can improve the SNR for constant accelerations of long duration. In 1998, ~100 nm vibrating-beam silicon accelerometers were under development by Analog Devices and Daimler-Benz. Tunneling accelerometers had been fabricated with ~10^6 g resolution at a 350 Hz bandwidth or ~10^{-8} g/Hz$^{1/2}$. Silicon-based capacitive accelerometers (5-1000 g full scale) manufactured by Silicon Designs (Issaquah, WA) were widely used in automotive airbag safety systems; in 1998, Motorola shipped its ten-millionth accelerometer, and the global market for automotive silicon accelerometers was expected to reach $463 million by 2002.

Formed elements in the human bloodstream (e.g., red cells, white cells, platelets) typically are buffeted by much smaller accelerations. For example at 310 K, instantaneous random thermal accelerations of $a_{thermal}$ ~ 500 g's are experienced by a 0.2-micron virus particle, ~0.05 g's by a 2-micron platelet, ~10^{-4} g's by a 10-micron neutrophil—and ~0.4 g's by a 1 micron^3 spherical nanorobot. Force-based sensors cannot easily detect such low accelerations, since the smaller the sensor, the larger the a_{min} for a given SNR. An alternative approach uses displacement sensors to determine an object's velocity twice in rapid sequence, allowing acceleration to be computed from the difference. Consider a constantly accelerating object that triggers three clocking latches at positions x_1, x_2, and x_3, in sequence. Time and position of the object are measured as (x_1, t_1), (x_2, t_2), and (x_3, t_3).

Assume that a slowly accelerating external object passes three clocking latches that are fixed to the exterior of a nanorobot of radius R. Measurements of the object's velocity incur an unavoidable error due to nanorobot thermal motion. The object is observed to pass between latch pairs during a measurement time ~t_{meas}. During this same interval, the nanorobot translates a distance DX = (kT t_{meas}/3 phR)$^{1/2}$, incurring a maximum velocity measurement error of ~DX/t_{meas}, and thus from maximum acceleration measurement error of ~2 $DX/t_{meas}{}^2$ if $(t_3 - t_2) \sim (t_2 - t_1) \sim t_{meas}$. The mitral and aortic valves in the heart (DP/Dx ~ 10^5 N/m^3, a_{fluid} ~ 10 g's, Dx ~ a few cm), as well as valves in the veins and lymphatics, operate as bistable analog decelerometers employing this principle. Micron-scale medical nanorobots traveling across blood vessel surfaces could use pressure sensors capable of 10^{-6} atm resolution to infer local blood flow accelerations to an accuracy of ± 0.01 g from sequential measurements taken 1 mm apart.

Consider a cylindrical gyroscope spinning at an angular velocity w (radians/sec), with radius r and thickness h, whose center of gravity lies a distance L_p from a frictionless pivot point along the rotational axis. The pivot point defines the center of a spherical sensor grid of radius R $(r^2 + 4h^2)^{1/2}$ having distinguishable grid elements located a distance Dx ~ 1 nm apart. Under uniform acceleration a, the gyroscope precesses around its spin axis at an angular velocity $w_p = 2\ a\ L_p/w\ r^2 = Dx/R\ t_{meas}$, where t_{meas} is measurement duration. In choosing an w with which to measure a_{min}, there is an upper and a lower limit. The lower bound is the minimum spin angular velocity below which the pivoted gyroscope cannot spin stably about the vertical axis and begins to wobble, thus ruining the measurement. If $w_{max}/w_{min} > 1$, then nonwobbling spin velocities are available below the bursting speed of the gyroscope cylinder. Since this ratio scales with R, there is a minimum sensor size R_{min} below which no useful spin velocities are available. For $t_{meas} = 10^{-3}$ sec, R_{min} >~ 69 microns (optimum r = 48 microns, h = 25 microns); for t_{meas} = 1 sec, R_{min} >~ 6.9 microns (optimum r = 4.8 microns, h = 2.5 microns). Thus it appears that pivoted gyroscopic accelerometers likely will be employed only in devices >> 10 microns in size. Microscale gyroscopes implemented in silicon have been discussed by Greiff and others.

Aside from pure acceleration, the rate of accelerative onset—sometimes called jolt, surge, or jerk—can produce additional damaging effects on macroscale mechanical and biological structures. For example, the effects of accelerative onset ('a) on the human body were studied by Col. John Paul Stapp in a series of aggressive rocket sled experiments conducted during the 1950s. For human subjects, cardiovascular shock was entirely absent for 'a >~ 500-600 g/sec but began to appear at 'a >~ 1100-1400 g/sec, at 1-40 g accelerations. Human tolerance to linear decelerative force is determined both by rate of onset and peak acceleration; the endurance limit for well-restrained healthy young males has been given as <~500 g/sec and <~50 g for >~0.2 sec durations. By 1998, the effects of accelerative onset on nanomechanical structures had not been extensively studied, but similar limits might apply to nanomachines. For example, a diamond rod of length L_{rod} = 1 micron, cross-sectional area S_{rod} = 0.01 micron^2, mass $m_{rod} = 3.5 \times 10^{-17}$ kg, and working strength $W_{rod} \sim 5 \times 10^{10}$ N/m^2 fails when subjected to a static acceleration a_{fail} >~ $W_{rod}\ S_{rod}/m_{rod} \sim 10^{12}$ g. But accelerative shocks can produce dynamic amplification in the response spectrum of elastic systems. Dynamic stress rises to a peak of roughly twice the corresponding static value when 'a_{peak} >~ 4 n_{res} a_{max}

for a trapezoidal pulse shape, where n_{res} is the frequency of natural vibration of the system, because the stresses from the free oscillations of the system, induced by the changes in the acceleration of the system, can add to the static stresses from the acceleration, with damage done by the total stress. Taking k_s ~ 1 N/m for the rod described above, then n_{res} ~ 30 MHz and 'a_{peak} ~ 10^{20} g/sec at a_{max} ~ 10^{12} g. For comparison, the teragravity nanocentrifuge has mass m ~ 10^{-17} kg and is spun up to a_{max} = 5 × 10^{11} g's in >~10^{-6} sec, giving 'a <~ 5 × 10^{17} g/sec << 'a_{peak}.

There are circumstances in which it is necessary for a medical nanodevice to measure, establish or maintain a preferred orientation, as for example nanorobots with a clearly defined "top" and "bottom" or nanodevices performing lengthy intracellular operations requiring precision stationkeeping. Measurement of rotation rates may also be required.

Pivoted gyroscopes of the type are impractical for orientational sensing in all but the largest medical nanodevices. Setting w_{max} equal to w_{min}, taking a_{min} = g = 9.81 m/sec^2 (gravitational acceleration), and optimizing for minimum sensor size gives a critical nonwobbling nonbursting spin velocity w_{crit} ~ 6.8 × 10^7 radians/sec at a minimum device radius of ~35 microns. If a pivoted gyroscope is smaller than this minimum radius, then spinning it slower than w_{crit} causes the gyro to tumble helplessly, while spinning it faster than w_{crit} tears the gyro apart. However, a triaxially gimballed nanogyroscope may be regarded as having its pivot point near its center of mass, so L_p becomes very small and w_{min} becomes very slow, making nanogyroscopes feasible in medical nanorobots. Taking h = 1 micron, r = 0.5 micron, r = 3510 kg/m^3 and a_{min} = g, then if the gimbals are aligned such that L_p = 1 nm between the pivot point and the effective center of gravity along the rotation axis, then w_{min} ~ 400 rad/sec. If gimbal tolerances are improved to L_p = 0.1 nm, then w_{min} ~ 125 rad/sec. How stable is a nanogyroscopic orientation standard? There are two principal considerations. First, Brownian thermal rotation in all three angular degrees of freedom (q, j and y) gives rise to a small nutation around the nanogyroscopic precession axis of magnitude Dq_{nutate} ~ p_j/p_y, where p_j and p_y are angular momenta of the gyro. The minimum value for p_j is approximated by (1/2) kT ~ (1/2) p_j^2/I_1, where I_1 = (1/4) m r^2 + (1/12) m h^2, the moment of inertia around j, and m = p r^2 h r. Second, the gimbal bearings have small frictional losses, exerting small torques on the gyro and causing a small precession away from the original orientation. The angular velocity of precession $w_{precess}$ ~ T_{gimbal}/(I_3 w) (rad/sec) where T_{gimbal} is the frictional torque caused by an imperfect gimbal bearing. Assume that the gimbal is driven by external forces to oscillate at some frequency n_{gimbal}, and that a gimbal bearing dissipates power at the rate of P_{gimbal} (watts), so that each oscillatory motion of the gimbal dissipates E_{gimbal} = P_{gimbal}/n_{gimbal} (joules). If the angular amplitude of the gimbal motion is a_{gimbal} (radians) and the gimbal radius is r_{gimbal}, then the gimbal bearing travels X_{gimbal} = a_{gimbal} r_{gimbal} (meters) per gimbal oscillation and the force applied in dissipating an energy E_{gimbal} is F_{gimbal} = E_{gimbal}/X_{gimbal}. Thus, the torque applied by each gimbal oscillation event is T_{gimbal} = F_{gimbal}

$r_{gimbal} = P_{gimbal}/(n_{gimbal}\ a_{gimbal})$; $P_{gimbal} = k_p\ v_{gimbal}^2$ (see Drexler), where $v_{gimbal} = a_{gimbal}\ n_{gimbal}\ r_{gimbal}$ is the sliding speed of the gimbal bearing surfaces during movement and k_p is a constant that depends solely on the geometry of the bearing.

The change in orientation angle caused by each gimbal oscillation event is $Dq_{osc} = w_{precess}/n_{gimbal}$. The number of gimbal oscillations $N = t\ n_{gimbal}$, where t is elapsed time since the gyro was last calibrated. If gimbal oscillations are independent and randomly distributed, then the total change in orientation angle is very roughly.

From Drexler, $k_p = 2.7 \times 10^{-14}$ watt-sec^2/m^2 as a conservative value for a small (~2 nm) stiff bearing. Random hydrodynamic flows induced by thermal fluctuations inside biological cells have a characteristic duration t_{fluct} ~ 10 millisec, which suggests a_{gimbal} ~ 0.06 radians (~3°), for an R ~ 1 micron in cyto nanorobot. Taking r = 0.5 micron, h = 1 micron, r = 3510 kg/m^3 for diamond (giving m ~ 3×10^{-15} kg), w = 1×10^9 rad/sec (just below the w_{max} ~ 4.8×10^9 rad/sec bursting speed), and assuming n_{gimbal} ~ $t_{fluct}^{\ 1}$ = 100 Hz (to which the results are not terribly sensitive), then $Dq_{precess}$ ~ $10^{-8}\ t^{1/2}$ (radians). Thus a $Dq_{precess}$ ~ 1 microradian orientation shift takes ~1 hour; a $Dq_{precess}$ ~ 4 microradian shift takes ~1 day; and a $Dq_{precess}$ ~ 70 microradian (~10 arcsec) shift takes ~1 year. (Total angular error $Dq_{total} = Dq_{precess} + Dq_{nutate}$.) Thus a gimballed nanogyroscope may be sufficiently stable to serve as a readily storable, easily transportable, and highly accurate onboard orientation standard.

A nanopendulum may also be employed for orientational sensing. As one example of a large class of related devices, consider a simple rigid spherical nanopendulum of radius r and mass m, with a large hemispherical bob free to swing through the entire solid angle about a central pivot point, with a gimballed housing to allow active avoidance of the pivot support beam. The nanopendulum bob moves along a spherical surface located far enough from a concentric spherical sensor grid of radius $R = r + z_{sep}$ to minimize van der Waals interactions, with bob and sensor surfaces electrically neutral or uniformly mutually repulsive. Much like the pin cushion receptor model, the sensor grid consists of a spherical array of diamondoid sensor rods Dx ~ 1 nm wide that are rapidly and simultaneously extended into the evacuated sensor cavity to determine bob position during a measurement cycle of duration t_{meas}. Radial sensor rod velocity should exceed typical tangential bob velocity by at least 1-2 orders of magnitude. For the example given below, a sensor stud on a 500-nm bob that is turning at v_{turn} ~ $v_{thermal}$ = 3.6 mm/sec passes a ~1 nm sensor rod in ~280 nanosec; a sensor rod displacing ~1 nm in the radial direction in ~1 nanosec travels at v_{rod} ~ 1 m/sec. This nanopendulum may serve both as an orientation sensor and as a rotation rate sensor, using principles similar to vestibular (semicircular canal) mechanics. If a simple nanopendulum spherical sensor housing with bob initially at rest in a uniform vertical gravity field g is rotated through the smallest detectable angle q_{min} (radians) around the center pivot, then ignoring friction, the bob's position over the sensor grid moves.

For Dx = 1 nm and r = 500 nm, Dq_{min} ~ 2 milliradians (~0.1 deg). For a practical sliding interface (e.g., between bob and sensor surface) of contact area S and turning velocity v_{turn}, Drexler estimates frictional dissipation, primarily driven by shear-reflection drag and band-stiffness scattering, as $P_{drag} \sim k_1 v_{turn}^2 S$, where $k_1 \sim 400$ kg/m^2-sec, hence the energy dissipated between the bob and the nearest sensor rod (the one interacting with the bob) is $E_{drag} \sim k_1 v_{turn}$ S Dx. Bob gravitational energy E_{bob} = m g r sin $Dq_{min} \sim$ m g r Dq_{min} for small displacements. The bob swings freely at the minimum displacement if $E_{bob}/E_{drag} > 1$. If m = 10^{-15} kg, r = 500 nm, and S = 1 nm^2, then $v_{turn} < 2.5$ m/sec or w < 49 megaradians/sec for the bob to swing freely (vs. $v_{thermal}$ = 3.6 mm/sec and $w_{res} \sim 9900$ rad/sec for this pendulum).

The bob-up/bob-down energy differential is ~2 m g r = 9.8 zJ ~2.3 kT for m = 10^{-15} kg and r = 500 nm. To distinguish opposite orientations (Dq ~ p) requires $N_{meas} \sim (kT/2\ m\ g\ r)^2 \sim 1$ measurement for this sensor. To distinguish $Dq_{min} \sim 2$ milliradians requires averaging $N_{meas} \sim (kT/E_{bob})^2 \sim$ 190,000 independent measurements (of ~0.002 kT/measurement) or $t_{meas} \sim 190$ microsec using a sensor rod clock cycle of ~1 nanosec. Thus, after subtracting out the effects of translation displacements as determined by linear accelerometers, a nanodevice can measure angular displacements in its orientation to within ± Dq_{min} in a measurement time $t_{meas} \sim 10^{-4}$ sec. With sensor rods moving at ~1 m/sec, energy dissipation for each rod sliding in its housing is ~0.01 zJ/cycle. If only ~100 test rods need be triggered to continuously track the moving stud, then a sensor rod clock cycle of ~1 nanosec implies a continuous nanopendulum power dissipation of ~1 pW; additional drag power loss between stud and contacting rod is only P_{drag} ~0.0004 pW.

Many other configurations of this basic device are readily conceived and may be more volume-efficient, including nested spheres or ovoids, two-dimensional rocking disk sections for in-plane rotational measurements (three devices required for monitoring all angular degrees of freedom in a rigid rotator), and multiaxial circumferential tracks or tubes containing rolling balls or sliding plugs. J. Logajan suggests that a pair of linear three-dimensional translation accelerometers should be able to unambiguously extract the 3 translation and 3 rotation components of all possible acceleration vectors. Rotation about any axis will result in distinctly different mass displacements while accelerated translation along any axis will result in identical displacements.

The smallest angular velocity detectable using a simple nanopendulum is given approximately by $w_{min} \sim Dx/r\ t_{meas}$ = 2 milliradian/sec (~0.1 deg/sec) for Dx = 1 nm, r = 500 nm, and t_{meas} = 1 sec. The largest detectable angular velocity $w_{max} \sim 2\ p/t_{meas}$ if t_{meas} is the minimum time required to count one complete rotation of the sensor; the smallest $t_{meas} \sim 2\ p/w_{max}$ ~100 nanosec for $w_{max} \sim 49$ megaradians/secor ~5 per cent of the bursting speed for an r = 1 micron spherical diamondoid nanodevice. What rotational displacements will normally be encountered by an in vivo medical nanorobot? A 1-micron device experiences instantaneous Brownian rotations of $w_B = (kT/4\ p\ h\ R^3\ t)^{1/2}$ radians/sec, where $t = (MW_{kg}/32\ p\ kT\ N_A\ c_{H2O}\ R^4)^{1/2}$, the mean time between molecular collisions with the device (mostly by water molecules), where $c_{H2O} \sim 3.3 \times$

10^{28} molecules/m^3. For R = 0.5 microns and T = 310 K, t ~ 6 × 10^{-20} sec/collision and w_B ~ 7 × 10^9 rad/sec, with each displacement of ~ t w_B = 4 × 10^{-10} radian, none of which are detectable. However, the net of all Brownian rotational displacements is one rotation every 16 sec, ~0.4 radian/sec. The largest measurable angular displacements in vivo are caused by tumbling due to differential shear forces in the bloodstream. Normal vessel wall shear rates in physiological bloodflow range from 100-1400 radians/sec in the larger arteries to 500-4000 radians/sec in the smaller arteries and capillaries. The measured rolling velocity of white cells during vascular "tank tread" locomotion on venule walls is ~40 micron/sec, or ~8 radians/sec, which may be encountered by nanodevices employing cytocarriage. The human vestibular mechanism has a frequency response range of 0.048-260 radians/sec. Thus, the maximum range of physiologically relevant rotation rates is 10^{-2} to 10^4 radians/sec. Since the natural harmonic frequency of a simple nanopendulum is w_{res} = $(g/r)^{1/2}$ = 4400 radians/sec for r = 500 nm, the possibility of forced oscillations and resonance need be faced only when measuring the highest likely physiological rotation rates, in which case the addition of a damping force F_{damp} ~ m v_{turn} w_{res} during resonance events should eliminate the problem. F_{damp} for critical damping peaks at ~mg ~ 0.01 pN for the m = 10^{-15} kg, r = 500 nm case (v_{turn} ~ 2 mm/sec).

Piconewton forces may be applied and must be detected by medical nanodevices. Pendulum- and cantilever-based sensors may be used to probe nanoscale forces, to monitor the ambient gravitational field, to measure molecular-scale masses to single-proton accuracy, and to discriminate among molecules having different isotopic composition.

Extremely small forces have been measured using macroscale instrumentation, particularly the atomic force microscope (AFM). In 1998, experimental forces of 1 nN in the contact regime and 1 pN in the noncontact regime were routinely measured. In 1994, D. Rugar achieved a force sensitivity of 5 × 10^{-16} N using an AFM with a 90-nm thick cantilever at room temperature, and in 1998 Rugar reported a force resolution of 7 × 10^{-18} N using a magnetic resonance force microscope (MRFM) with a 230-micron long, 60-nm thick, silicon cantilever. The theoretical limit on the force sensitivity of the AFM has been estimated as 10^{-18} – 10^{-19} N for temperatures near absolute zero using fiber optic interferometry or other techniques. However, many of these methods are impractical or impossible to employ in medical nanodevices. What is the minimum force F_{min} likely to be detectable by nanorobot sensors? Several lines of reasoning produce consistent answers. For example, force may be measured by the displacement Dx of a beam against the known resistive pressure provided by a spring of minimum stiffness k_s; that is, F_{min} ~ k_s Dx. However, the energy stored in the spring must exceed thermal noise energy, with SNR ~ ln (k_s Dx^2 /2 kT, giving F_{min} = 2 kT e^{SNR}/Dx. For T = 310 K and SNR = 2, F_{min} = 100 pN at Dx = 0.6 nm, 10 pN at 6 nm, and 1 pN at 60 nm. Indeed, the natural molecular motors (kinesin, dynein, and actin-myosin) have isometric force outputs on the order of 1-14 pN per protein molecule, which are routinely observed using microscale cantilever spring sensors measuring total forces of 10-100 pN. Breaking antigen-antibody or receptor-ligand interactions generally requires rupture forces in the 50-300 pN range. A force of ~160 pN separates biotin that is seated in its avidin receptor; ~85 pN pulls iminobiotin out of the avidin receptor. A force of ~300 pN ruptures a hydrogen bond between two isolated water molecules, but >3000 pN are likely to be needed to rupture a single C-C covalent bond. The rupture strength of single covalent bonds has been directly measured experimentally by AFM as, for instance, 2000 ± 300 pN for Si-C (528 zJ/molecule dissociation enthalpy, 0.185 nm bond length) and 1400 ± 300 pN for Au-S. Force sensing may also require using measurement devices that involve sliding interfaces of two mutually inert periodic surfaces. This motion may be characterized by the potential energy of interaction, ~0.1-1 zJ as determined by Monte Carlo simulations of typical diamondoid material at 0.2-0.5 nm separations. Assuming a minimum 0.5 zJ energy barrier, for small sliding contacts the typical force amplitude $F_{min} \sim 1.7 \times 10^{10}\ DV_{barrier} \sim 8.5$ pN. A force of 10 piconewtons across a small-molecule receptor $(0.3\ nm)^3$ of surface area 0.09 nm^2 represents an energy density of $\sim 10^8\ J/m^3 \times 0.027\ nm^3 = 2.7$ zJ ~ 0.6 kT. And the nanorobotic manipulator arm is capable of ~picometer step sizes, representing a minimum applicable force of $F_{min} \sim 10$ pN assuming 0.04 m/N compliance for the entire mechanism. Thus as a conservative estimate, the smallest force that is reliably detectable (and applicable) by nanodevices is probably ~10 pN.

Research in space medicine has discovered that micron-scale elements of the immune system have extraordinary sensitivity to gravity. One study found that bone marrow-derived (B6MP102) macrophage single cells respond after 8 seconds to exposure to a 10^{-2} g hypogravity environment as indicated by increased cell spreading, a well-known marker of cell activation. Individual immune system cells placed in microgravity exhibit enhanced growth but depressed differentiation, decreased activation by concanavalin A, a 20 per cent reduction in glucose metabolism, increased resistance to antibiotics, increased number of pseudopodia on monocytes, and changes in cytoplasmic streaming velocity and frequency. As of 1998, the exact mechanism by which the leukocyte "gravity sensor" operates remained unknown. It has been speculated that differences in densities of several organelles, such as the nucleolus, the ribosomes, and the centrioles, could generate detectable pressures on the structure of the cytoskeleton under gravitational loading mediated through mechanosensitive stretch-activated ion channels or the extracellular matrix. While indirect gravitational effects that modify the cellular environment (such as reduced sedimentation or decreased thermal convection due to hypogravity) could also be responsible, a direct interaction of the gravitational force with cellular structures appears to be the most likely sensor mechanism. For instance, if a change in gravitational loading equivalent to the force generated by a natural molecular motor (F ~ 1 pN) is detectable by the cytoskeleton, and this force is applied to the entire mass of a (10 micron)^3 cell (m ~ 10^{-12} kg), then the minimum detectable gravitational acceleration is g ~ F/m ~ 1 m/sec^2 = 0.1 g, which should be adequate to detect the onset of hypogravity. A nanomechanical gravity sensor of similar size to a macrophage cell may be at least 100,000 times more sensitive. For example, the energy of a tethered mass swinging freely in a uniform gravity field is approximately $E_g \sim m\ g\ Dh$, where Dh is the maximum vertical amplitude of the motion. To be detectable, oscillator energy must exceed thermal noise energy, so SNR $\sim \ln (m\ g\ Dh\ N_{meas}^{1/2}/kT)$, giving $g_{min} = kT\ e^{SNR}/m\ Dh\ N_{meas}^{1/2} \sim kT\ e^{SNR}/r\ L^4\ N_{meas}^{1/2}$ for a bob mass of density r and size L^3, N_{meas} independent measurements, and assuming Dh ~ L. If T = 310 K, SNR = 2, N_{meas} = 1, and r = 21,450 kg/m³ (Pt), then g_{min} = 0.1 g for L ~ 1 micron, 0.01 g for L ~ 2 microns, 10^{-5} g for L ~ 11 microns, and 10^{-6} g for L = 20 microns, the size of the average human tissue cell. Regarding nanogravimeter design, for a pendulum of length L in a gravity field g ~ 9.81 m/sec² (1 g), the resonant period of the motion $T_{res} = 2\ p\ (L/g)^{1/2} = 2 \times 10^{-3}$ sec for L = 1 micron, 9×10^{-3} sec for L = 20 microns.

Beam and rod are coiled in two dimensions (although three-dimensional coiling, more difficult to depict, provides maximum compactness). L_{beam} and l_{rod} are chosen to minimize total coil area $A_{coil} = 2\ R_{beam}\ L_{beam} + 2\ r_{rod}\ l_{rod}$. Using aluminum for the rod and beam (E = 6.9×10^{10} N/m², G = 2.6×10^{10} N/m², r = 2700 kg/m³), L_{beam} = 1200 nm, l_{rod} = 1920 nm, r_{rod} = 0.5 nm, V_{sample} = 1 nm³, and $Dt_{min} = 10^{-9}$ sec, then DMW ~ 1 proton for L_{spring} = 1 nm >> Dx_{min} = 0.01 nm, the minimum measurable displacement. (For L_{spring} > 1 nm, physiological ~1 g accelerative loads on a ~1 micron nanorobot in which the sensor is embedded produce undetectable accelerative beam displacements < Dx_{min}). Moving the suspension rod closer to the sample holder reduces device sensitivity. For $L_{spring} = L_{beam}$, minimum DMW ~ 1200 protons, permitting moderately large molecules to be weighed in minimum time (~10^{-6} sec). Using a very thin sensor configuration (h_{sensor} ~20 nm) and coils occupying ~50 per cent of sensor cavity space, sensor dimensions are (79 nm)² × 20 nm giving a massometer volume ~10^5 nm³ or ~0.01 per cent of the total volume of a 1 micron³ nanorobot. Sensor mass is ~ 3×10^{-19} kg, including diamondoid housing and prorated share of support mechanisms. If one measurement cycle including sample loading/unloading operations can be completed every ~10 microsec, then the device processes its own mass of glucose molecules in ~10 sec. Note that biological molecules weighed in air or in vacuo may lose water of hydration, whereas weighing in water presents other complications that merit further investigation. J. Soreff points out that for a molecule on a spring, the precision of the measurement is linked to the sharpness of the resonance, which is optimized to the degree the vibrational frequency of the spring/molecular mass system is kept well below the vibrational resonance frequencies of the molecule itself. The following discussion reports the results of a simple lumped-element analysis of the effects of internal vibration on the massometer which confirms the previously described performance estimates for the massometer. Suppose that it is desired to weigh some mass M to a precision DM by attaching it to a weak measuring spring (spring constant k_{weak}) and measuring the shift in resonant frequency. The shift in frequency due to coupling to an internal mode of the sample must be less than (1/2) (DM/M) (1/2 p) $(k_{weak}/M)^{1/2}$. If the sample has some lowest internal mode frequency n_{int}, then approximating the sample as two M/2 masses separated by x_{strong} connected by a strong spring (k_{strong}) gives n_{int} = (1/p) $(k_{strong}/M)^{1/2}$.

Setting $x_{weak} = e^{st}$ and $x_{strong} = R\ e^{st}$ where t = time, s = time constant, and R = $-(1 + (2\ k_{weak}/M\ s^2))$, and since the angular frequencies of the two springs are $w_{strong} = 2\ p\ n_{int} = 2\ (k_{strong}/M)^{1/2}$ and $w_{weak} = (k_{weak}/M)^{1/2}$, then the lower frequency solution for $w_{weak} << w_{strong}$ may be approximated by $s = i\ w_{weak}\ (1 - w_{weak}{}^2/2\ w_{strong}{}^2)$ and so the (undesired) fractional change in frequency due to the flexibility of the object to be weighed. For a molecule sample size of L_{sample}, $n_{int} \sim v_{sound}/L_{sample} \sim 500$ GHz for $v_{sound} \sim 1500$ m/sec and $L_{sample} \sim 3$ nm; $M = r\ L_{sample}{}^3 = 2.7 \times 10^{-23}$ kg for r = 1000 kg/m^3. To detect the mass difference of a single proton in a sample molecule, sensor $Q = M/m_p = 16{,}000$; the sensor may operate at a factor of $Q^{1/2}$ below n_{int}, or $n_{spring} \sim n_{int}/Q^{1/2} \sim 4$ GHz. A period of 0.25 nanosec is slightly too short to allow a phase shift of $Dt_{min} \sim 1$ nanosec, since the period is only one-quarter of this. However, a phase shift of p radians should accumulate in $t_{meas} = Q/2\ n_{spring} \sim 2$ microsec and should be detectable by looking at magnitudes of forces. Interestingly, an SPM microprobe of circular cross-section and exponential profile with a 0.1 nm^2 tip has been proposed that could also reach ~proton mass detection sensitivity at room temperature and 1 atm pressure with an eigenmode frequency of 10 GHz; an electrically-driven carbon nanotube-based resonant-beam balance demonstrated ~10^{-17} kg (~10^{10} proton or ~0.01 micron^3 mass) sensitivity in 1999. Existing surface acoustic wave (SAW) devices are chemical sensors that can measure areal mass densities as small as 80 picogram/cm^2, or 0.01 N_2 molecule per nm^2 of surface, though current devices are limited to ~micron^2 areas. A proposed quartz microresonator microbalance could detect a 10^{-6} molecular monolayer, the equivalent of 5×10^{-9} molecule per nm^2. Detection of ~400 million protons using an MRFM vibrating rod ~60 nm thick was demonstrated by Rugar in 1998.

It will often be useful in nanomedicine to distinguish molecules containing different isotopes of the same chemical elements, for example to count the number of organic molecules in a sample that contain radioactive C^{14} atoms in place of stable C^{12} atoms or to make direct assays of heavy radiometal-contaminated tissues. Nanodevices may also sort isotopically heteronuclear molecules into homonuclear fractions, for example to separate deuterochemicals from hydrochemicals or to acquire isotopically pure materials with which to construct:

(a) Small, high-speed rotors with precisely balanced mass distributions.

(b) Components with extremely fine size differences.

(c) Nanostructures with maximum thermal conductivity (Table 4.1), or mixed isotopes to reduce thermal conductivity.

(d) Long-lived isotopically pure structures that generate no internal radioactivity, thus promoting longer device lifetimes.

The most reliable sensor for discriminating among isotopes within target molecules is the single-proton massometer. Isotopes of any chemical element differ in mass by at least ~1 neutron (~protonic) increments. Thus a massometer with single-proton resolution can resolve all isotopes starting from chemically pure samples. Isobars (atoms having the same mass number) and isotones (atoms having the same number of neutrons) may be finally resolved chemically; nuclear isomers differing in mass by ~0.001 amu cannot be resolved. Teragravity centrifugation and time-of-flight particle-beam deflection techniques may also prove useful. Isotopes vary in many subtle ways due to their mass differences, although it is not yet clear how a potential isotope nanosensor might exploit some or all of these distinctions:

(1) *Line Shifts*—the electronic energy states of atoms or molecules depend on the reduced nuclear mass, causing a spectral line shift of ~1.0005 between hydrogen and deuterium, less among heavier isotopes, detectable spectroscopically.

(2) Hyperfine Structure—isotopes differ in electrical quadrupole moment, nuclear spin and magnetic moment, giving rise to hyperfine structure in the optical spectra, especially in the heavier elements. For example, the magnetic moments of hydrogen and deuterium (2.79268 and 0.857387 nuclear magnetons, respectively) are distinguishable via NMR. The NMR frequency in a 10 kilogauss magnetic field is 42.5759 MHz for H, 6.53566 MHz for D. Of course, not all isotopes are NMR sensitive.

(3) *Vibrations and Rotations*—molecular spectra are especially sensitive to isotopic changes, since the quanta of vibrational and rotational energies are directly dependent upon the masses of the atoms involved, and there are also some subtle effects from differences in nuclear spin. Separations are commonly achieved by laser excitation.

(4) *Cross-Sections*—isotopes have different thermal neutron capture cross sections; for instance, the cross-section of C^{12} is 3.7×10^3 barns (1 barn = 10^{-24} cm^2) compared to 9×10^{-4} barns for C^{13} and $<10^{-6}$ barns for C^{14}. (Electron scattering might be better, since the electron is less likely to cause the nucleus to fission and it is generally difficult to use neutrons to probe samples on the molecular scale.) Similarly, the neutron cross-section for H_2O is 0.6 barns compared to 9.2×10^{-4} barns for D_2O.

(5) *Other Properties*—isotopes differ slightly in various bulk properties, many of which could be measured in submicron-scale aliquots by nanosensors. The ability of both large mammals and micron-size photosynthetic plant cells to preferentially concentrate C^{12} or C^{13} in organic matter, and the predicted ability of carbon nanotubes to selectively extract tritium from

low-concentration H_2-T_2 mixtures, constitute proofs of principle that at least some of the bulk properties.

This section describes the ability of medical nanodevices to detect and measure fluid pressure variations while functioning inside the human body. Generation of acoustic signals for power transmission, communication, and navigation are considered elsewhere.

A change in the ambient pressure in a fluid environment (DP) may be determined by measuring the change in volume (DV) of a fixed quantity of gas at constant temperature, relative to a known sensor volume at a known reference pressure. (Thermal sensors elsewhere in the nanorobot provide readings by which corrections can be made for temperature variations.) If the sensor volume consists of a cylinder of fixed cross-sectional area, the change in volume is converted into a linear displacement of a piston located at the interface between gas and fluid environment. At SNR = 2, a $(22\ nm)^3 = 10^4\ nm^3$ sensor ($\sim 5 \times 10^5$ atoms or $\sim 10^{-20}$ kg for a cube with 1 nm thick walls) can detect a minimum pressure change $DP_{min} \sim 0.03$ atm; a $(680\ nm)^3 = 0.3\ micron^3$ sensor ($\sim 3 \times 10^9$ atoms, $\sim 6 \times 10^{-17}$ kg) detects $DP_{min} \sim 10^{-6}$ atm, probably near the practical limit for mobile in vivo medical nanodevices. Note that DP_{min} scales approximately as the inverse cube of mean sensor dimension, and immunity to thermal noise scales exponentially with sensor volume. Current silicon micromachined pressure sensors are $\sim 1\ mm^3$ with 10^{-3} atm sensitivity, a factor of $\sim 10^{12}$ poorer sensitivity than the theoretical minimum for their size. (A field-effect chemical sensor with an active sensing volume of $0.1\ mm^3$ that measures pressure indirectly has detected a 10^{-11} atm step in hydrogen gas with $t_{meas} \sim 3$ sec, only a factor of 1000 above the theoretical minimum for a sensor of that size.) The human ear can detect pressure waves as small as 2×10^{-10} atm.

A less sensitive approach relies on the observation that a change in pressure P alters both density and bulk modulus of a compressed fluid, variations that may be detected by measuring the change in the speed of sound v_s in the working fluid, given by

$$v_s = (B/r)^{1/2}$$

where B is bulk modulus of the fluid and r is fluid density. (dB/dP)/B may be up to several times larger than (dr/dP)/r for molecular fluids, but there is some $(dv_s/dP)/v_s$ for each fluid. Differentiating with respect to P and using dr/dP = r/B, then $dv_s/dP = (dB/dP - 1)/2\ r^{1/2}\ B^{1/2}$. For water at 1 atm, dB/dP ~ 8.4 atm/atm, r ~1000 kg/m^3 and B = 22,000 atm, giving $dv_s/dP \sim 0.25$ m/sec-atm. If Dv_s is the measured change in acoustic speed, then:

$$DP_{min} = Dv_s/(dv_s/dP)$$

Velocity can be measured to ~0.1 mm/sec accuracy using a micron-scale velocity sensor and t_{meas} ~ 1 sec, so a minimum Dv_s ~ 0.1 mm/sec implies DP_{min} ~ 0.0004 atm.

Two 0.3-micron^3 pressure sensors located at either end of a 1 micron nanorobot could detect pressure changes as small as 10^{-6} atm per micron, a ~1 atm/m spatial gradient. The maximum detectable gradient would involve adjacent measurements of minimum (10^{-6} atm) and maximum (~10^4 atm) absolute pressure, a theoretical peak spatial gradient of ~10^{16} atm/m, though a practical device could not survive exposure to such extreme shear force.

For an L = 22 nm sensor driven to full throw by a minimum detectable DP_{min} = 0.03 atm pressure spike (again taking h/L = 0.1 and r ~ 1000 kg/m^3), n_{driven} = 34 MHz; n_{driven} = 45 GHz when the sensor is driven to full throw by the maximum possible pressure spike P_{max} = 39,000 atm consistent with water remaining in the liquid state. The piston may be driven at these frequencies, or slower, using triangular-wave pulse trains of pressure spikes of these magnitudes. For an L = 680 nm sensor driven to full throw by a minimum detectable DP_{min} = 10^{-6} atm pressure spike, n_{driven} = 6 KHz; n_{driven} = 1.5 GHz when the sensor is driven to full throw by the maximum pressure spike P_{max} = 39,000 atm. What is the maximum tolerable pressure spike in aqueous media? The requirement of subsonic piston motion in water gives the highest maximum: $P_{max} < (1/2)\, r\, v_{sound}^2$ (N/m^2), if r = 3510 kg/m^3 for a solid diamondoid piston and the speed of sound v_s ~ 1450 m/sec in water for bulk modulus B_{water} = 2.2 × 10^9 N/m^2, r_{water} = 1000 kg/m^3, giving P_{max} < 39,000 atm. However, rapid dissipative pressure spikes P_{max} > 26,000 atm add sufficient energy to boil 310 K water, and isothermal static compression of pure water at 310 K causes crystallization into Ice VI near ~11,500 atm. Additional restrictions on (and features of) high-frequency pressure transducers are described later.

It will often be necessary to determine the degree to which a fluid storage tank is full or empty. In most circumstances, an ullage sensor may consist of a simple static pressure sensor embedded in the tank wall, with readout calibrated according to the van der Waals equation for gases and liquids. A fractal network of chemical sensors permeating the internal volume of the tank could provide continuous readouts of local concentration which could be integrated to estimate total storage in the tank. The force required to rotate a paddle inside the tank would increase as a tank was filled, due to the increase in net viscosity of the contents. Transmission time of thermal or acoustic pulses might also give some measure of fluid concentration and physical phase. C. Phoenix suggests storing liquids in a piston, allowing direct and precise molecular volume readout, in designs where this is convenient; in theory, a piston that slides the minimum measurable distance of ~10 pm within the 2-nm-diameter throat of a water-filled cylinder could detect the removal of a single water molecule from the column.

The ability of medical nanodevices to measure both absolute temperature and changes in temperature is crucial for monitoring in vivo physiological thermoregulatory mechanisms and intracellular energy transactions. Precision thermal sensing is also important within nanoscale devices to provide corrective input for pressure, chemical, and displacement sensors, and to improve the stability of onboard clocks. General limits of thermal sensing are considered, followed by several proposals for specific implementations.

A sensor system consisting of N atoms has (3N-6) internal coordinates and thus ~(3N-6) oscillators, each with an average energy of ~kT assuming no modes are high enough to be frozen out. However, the standard deviation of the energy of the oscillators around the mean energy is ~kT, and these energy fluctuations are uncorrelated (ignoring coupling between the oscillators), so (3N-6) of them will have an average fluctuation of kT $(3N\text{-}6)^{1/2}$. (By comparison, the energy in one 10-micron infrared photon is ~5 kT.) Thermal equilibration time $t_{EQ} \sim L^2 C_V/K_t$ for a sensor of size L, of heat capacity C_V (1.8×10^6 joules/m^3-K for diamond) and of thermal conductivity K_t (2000 watts/m-K for diamond). Thus, $t_{EQ} \sim 10^{-13}$ sec for a sensor of size L = 10 nm, 10^{-11} sec for L = 100 nm, and 10^{-9} sec for L = 1 micron, so thermal sensors up to 1 micron in size should easily achieve thermal equilibration within a typical measurement cycle time $Dt_{min} \sim 10^{-9}$ sec. Sensor measurement time $t_{meas} = N_{meas}\ Dt_{min}$. A $(57\text{ nm})^3$ sensor can detect $DT_{min}/T = 10^{-4}$ (~31 millikelvins at 310 K) in a single measurement (N_{meas} = 1), with measurement time $t_{meas} \sim 1$ nanosec. A sensitivity of $DT_{min}/T = 10^{-6}$ (~310 microkelvins at 310 K) may be achieved using either a ~1 micron^3 sensor with a single measurement ($N_{meas} = 1$, $t_{meas} = 1$ nanosec) or a $(124\text{ nm})^3$ sensor with $N_{meas} = 1000$ independent measurement cycles and $t_{meas} = 1$ microsec. Finally, a 1 micron^3 sensor can detect $DT_{min}/T = 3 \times 10^{-9}$ (~1 microkelvin at 310 K) with N_{meas} = 100,000 independent sensor cycles, giving a measurement time $t_{meas} \sim 100$ microsec.

These figures are confirmed by experimental estimates of detector noise temperature, such as $DT_{min}/T = (k/L^3\ C_V)^{1/2} \sim 10^{-6}$ for a silicon nitride detector with $C_V = 5.2 \times 10^6$ joules/m^3-K and L ~1 micron. It has been calculated that sensitivity of ~1 microkelvin is theoretically possible using quartz electronic microresonators as precision thermometers. Thermocouple probes with 100 nm tips have already shown ~100 microkelvin sensitivity, tunneling thermometers capable of measuring thermoelectric potential localized to atomic-scale dimensions have been proposed, and experiments leading toward "yoctocalorimetry" were being pursued in 1998.

For comparison, heat sensors in human skin have $DT_{min}/T \sim 3 \times 10^{-4}$ (~90 millikelvins), and the infrared sensor pit of the rattlesnake is sensitive to an energy intensity of ~0.8 pJ/micron^2 in a measurement time t_{meas} ~35 millisec and achieves $DT_{min}/T \sim 3 \times 10^{-6}$; mosquitos register $DT_{min}/T \sim 6 \times 10^{-6}$ at a distance of ~1 cm.

Two $(200\text{-}400\text{ nm})^3$ thermal sensors located at either end of a 1 micron nanorobot can detect temperature changes as small as 10^6 K per micron, a ~1 kelvin/m spatial gradient. The maximum detectable gradient in nanomedical systems would involve adjacent measurements differing by ~100°C, a peak spatial gradient of ~10^8 kelvins/m.

The detection event must be completed before the two sensors can thermally equilibrate, e.g., $t_{meas} < t_{EQ} = L^2 C_V/K_t$. If the material between the sensors is mostly diamondoid, then $t_{EQ} \sim 10^{-9}$ sec; if mostly water, then $t_{EQ} \sim 10^{-5}$ sec, for L = 1 micron.

On the timescale of nanomachine operations, typically 10^{-9}-10^{-6} sec, the human body provides an essentially isothermal operating regime. However, medical nanorobots circulating in the bloodstream may encounter DT of ~3 K during one circulation time ~60 sec under normal circumstances, a temporal thermal gradient of ~0.1 kelvin/sec, or >10 kelvin/sec at 1 mm tissue depth for a fingertip placed on a 600 K stove. Thermal nanosensors can take independent measurements differing by as much as ~100°C in ~10^{-9} sec, giving a maximum detectable thermal gradient of 10^{11} kelvins/sec, far in excess of physiological rates. Detection of the minimum ~1 microkelvin change between two measurements taken 1 sec apart probably represents the minimum practical temperature gradient, ~1 microkelvin/sec. For comparison, the record-holder in the insect world is the eyeless cave beetle *Speophyes lucidulus*, whose antennae can detect thermal change rates as small as ~3000 microkelvins/sec. The extraordinary thermal conductivity of diamond ensures that endogenous nanorobot waste heat is rapidly conducted to the surrounding aqueous medium. For example, a 1 micron^3 diamondoid nanorobot generating 10 picowatts of onboard power at its core experiences a temperature differential of ~10^{-8} K between core and perimeter (using thermal conductivity K_t = 2000 watts/m-K for diamond), well below minimum detectable limits.

The most important internal electrical sources are the electrochemical gradients caused by gated channel and transporter molecular pump operations at the intracellular level; muscular, membrane, digestive, and neural activity at the intercellular and organ level; and piezoelectric fields generated by movement of collagenous tissues (e.g., tendon and elastin), bones, and many other biological materials. Typically these sources generate potentials of 10-100 millivolts over distances of 0.01-10 microns, producing local fields in the range of 10^3-10^7 volts/m. For example, the 50 millivolt transmembrane potential necessary to open a sodium channel corresponds to an electric field of 5 × 10^6 volts/m. Interestingly, implanted electrical devices such as pacemakers may produce up to 600 volts/m one millimeter from the casing. External sources may also contribute to detectable nanomedical electric fields within the human body, though usually to a lesser degree. Placing a hand flat between two plates charged to E_p = 3 × 10^6 volts/m (near the breakdown voltage for air) induces an instantaneous field of $E \sim E_p/k_{water}$ = 42,000 volts/m transversely through the limb. (A DC field induces ion current flows which accumulate surface charge and cancel out the internal field after ~one RC time.) Electric fields directly beneath high-tension power lines are 2,000-11,000 volts/m, or 50-1000 volts/m at a lateral displacement of 25 meters along the ground. Household appliances produce 200-300 volt/m fields; 20 amperes flowing through standard Romex copper wiring in the walls of a house produces 100-600 volts/m in the middle of a room, up to 24,000 volts/m at the wall ($E = B/m_0 e_0 c$; c = the speed of light, B); static sparks from carpet walking are ~10,000 volts/m, and ~10^5 volts/m makes hair stand on end. Quiescent atmospheric charge is ~100 volts/m, rising to 5000 volts/m during severe thunder-storms oscillating from positive to negative over a ~10,000 volt/m range. Lightning bolts (~20 coul 0.002-sec discharge) produce 60,000 volt/m spikes at a distance of 10 meters from the strike, 600 volts/m at a range of 1 kilometer. Electron-hole avalanches in semiconductors are initiated by fields >~5 × 10^7 volts/m. Penal electrocutions use ~1500 volts/m. Direct force sensing of these electric fields is ineffective; the minimum detectable field E_{min} = F_{min}/q = 6 × 10^7 volts/m using F_{min} = 10 pN and a single-charge sensor. Particle deflection seems more efficient. Consider a collimated stream of singly-charged particles, each of mass m and initial velocity V_0 entering an evacuated test chamber of length L and experiencing a uniform electric field E, causing a lateral displacement Dx = q E L^2/ 2 m V_0^2 measured by the sensor.

For Dx_{min} = 10 pm, T = 310 K and SNR = 2, E_{min} = 10,000 volts/m for a sensor of size L = 28 nm, or 100 volts/m for L = 280 nm, allowing measurement of all nanomedically relevant fields that are homogeneous across the entire device. Using the larger 280-nm sensor, time of flight is ~10^{-12} sec for electrons or ~10^{-9} sec for U^{238} ions, permitting t_{meas} ~ 10^{-9} sec and dissipating at least ~30 pW waste heat (due to kinetic impact) in continuous operation. Time-varying electric fields can be measured up to t_{meas}^{-1} ~ GHz frequencies, and electric current may be determined using standard techniques because current density is scale-invariant. Care must be taken to eliminate measurement error due to exogenous magnetic fields. Piezoelectric or other electromechanical transducers can convert electricity into mechanical signals appropriate for all-mechanical computation and control systems.

Another class of electric field detector of similar size and sensitivity is the electrosensitive gel sensor. For example, the negatively charged heparan sulfate proteoglycan network found in the secretory granule matrix expands 50 per cent in volume within milliseconds of exposure to a -2 volt/micron field, and the response appears linear with voltage. Thus the length of a 400-nm long gel-filled sensor of fixed cross-sectional area should expand a minimally-detectable 10 pm (= Dx_{min}) in a 100 volt/m field. Similar sensitivities to small periodic fields, analogous to shark electroreceptors (maximum sensitivity ~10^{-7} volts/m,), are found in ion-channel embedded artificial membranes that selectively amplify electric signals in noisy environments via stochastic resonance.

Using n_e = 1 MHz and t_{meas} = 1 sec gives E_{minAC} = 0.2 volt/m. Cylindrical sensors are 10-100 times more sensitive, and sensitivity may possibly be further enhanced by coupling the periodic electric potential to a Michaelis-Menten type enzyme or a similar nanomechanism embedded in the sensor shell to achieve tuning. Fields as small as ~10^{-4} volts/m may be detectable by ~20-micron mammalian living cells; such detection may trigger biochemical observables which may be monitored or eavesdropped by resident nanomedical devices. Submicron-scale electrometers capable of ~0.1 electrons/ $Hz^{1/2}$ (charge per unit bandwidth) have been demonstrated experimentally, and also single-electron transistors (SETs) to 10^{-4} electron/$Hz^{1/2}$, with the potential to reach 10^{-6} electron/$Hz^{1/2}$. Galvanotropic bacteria have been reported in the literature. Radiofrequency receivers for power transmission and communication in vivo are described in Sections 6.4.2 and 7.2.3. Antenna radiation patterns may also be useful in diagnostics and macrosensing applications. Hence B ~ 0.04 microtesla for r = 1 micron, adjacent to an axonal conductor. For comparison, magnetic fields directly beneath high-tension power lines are 10-40 microtesla, or 2-8 microtesla at a lateral displacement of 25 meters along the ground. Household appliances produce 1-1000 microtesla at a distance of 3 cm and 0.01-2 microtesla at a 1-meter range (electric shavers produce 0.4-60 microtesla at a range of 15 cm). 20-amp Romex wiring in the walls of a house can produce 2 microtesla in mid-room, up to 80 microtesla at the wall, though in 1998 the average strength in 98 per cent of U.S. homes from artificial sources was 0.05-0.09 microtesla, typical exposures throughout the day were 0.1-1 microtesla, and the federal workplace standard was a maximum of 100 microtesla. Magnetotactic marsh bacteria, birds and insects can detect the geomagnetic field vector (identifying Earth's magnetic poles), which is 30-70 microtesla, depending on geographic position. Behavioural thresholds have been measured as 0.005-0.02 microtesla for pigeons and 0.001-0.01 microtesla for honeybees, and no more than 1000-2000 microteslas for human subjects. Some biochemical changes have been observed in human cells exposed to ~10-100 microtesla 60 Hz fields, although the data are controversial; it appears the fields pump iron atoms through cell membranes, potentially causing peroxidation or other damage to occur. In 1998, transcranial magnetic stimulation was used to expedite recovery of motor skills for stroke patients, and was being explored as a treatment for depression and a means of altering patient mood. Lightning bolts (i ~ 10,000 amp/discharge) produce 200 microtesla spikes at a distance of 10 meters from the strike, 2 microtesla at a range of 1 kilometer.

Computer simulations of a fully heterogeneous human body divided into 37,000 $(1.31\ cm)^3$ voxels produced a maximum induced current of 8.84 amp/m^2 (average ~1 amp/m^2) when the body was exposed to a uniform 60 Hz 1-tesla magnetic field. Patients undergoing Echo-Planar MRI scanning report "tickling" and "pain" when exposed to field changes exceeding 60 tesla/sec; ventricular fibrillation may result from myocardial current densities of 4 amp/m^2, which is induced at 250 tesla/sec.

The simplest magnetosensor is a freely pivoting permanent magnet that experiences a restoring force when the magnet's axis is not aligned with the external field. This "compass effect" is also employed by magnetotactic bacteria, which have ~0.1 micron ferromagnets in magnetosomes in their cells that allow them to detect and track the geomagnetic field vector. Permanent magnets of cross-sectional area A_{magnet} and length l_{magnet} made of Alnico or Alcomax (r ~ 8000 kg/m^3) can provide B_{magnet} ~ 1.4 tesla. (In 1998, the record flux density was 13.5 tesla for dipole magnets, ~27 tesla for resistive magnets.) If the minimum detectable force is F_{min} = 10 pN, B_{min} = 50 microtesla for A_{magnet} = 0.4 $micron^2$, about right for magnetotactic bacteria, and B_{min} = 2 microtesla for A_{magnet} = 2 $micron^2$.

Another class of magnetosensor detects the change in dimensions when a magnetic body is magnetized, which is called magnetostriction. For instance, iron-cobalt ferrite expands by DL/L ~ 7.5 × 10^{-4} at saturation magnetostriction in a ~1 tesla field, permitting detection of ~10 millitesla field changes using a 1 micron sensor and Dx_{min} ~ 10 pm. Small magnetostrictive volume changes make possible a sensor of even greater sensitivity. A field of ~0.2 tesla causes a fractional volume increase of ~10^{-6} in iron and up to 40 × 10^{-6} in some iron-nickel (~30 per cent Ni) alloys. A (370 nm)^3 magnetostriction sensor monitored by stretch sensors with displacement sensitivity Dx_{min} = 0.1 nm (~1 atomic radius) can in theory detect B_{min} ~ 0.1 microtesla. Magnetostriction is nonlinear with applied field, and there are significant hysteresis effects which may confuse data interpretation. However, multisensor arrays combining positive- and negative-magnetostriction materials should enhance signal extraction. Since the change in volume is small, temperature compensation may be required. Hall effect probe microscopes have demonstrated 10 microtesla sensitivity with 350 nm spatial resolution, and alternative nanoscale magnetic detectors such as direct-current SQUID (superconducting quantum interference device) magnetometers, stray field devices, and magnetic spin sensors are being developed. In 1999, micro SQUID could measure magnetic moments as small as ~10^4 Bohr magnetons or ~10^{-19} J/tesla, roughly 7 × 10^6 protons or a 14-nm iron cube. Magnetoencephalographs using SQUID magnetometers to measure external electromagnetic fields produced by neural traffic in the brain can register 10^{-6}–10^{-3} microtesla signals. Other biomagnetic measurement techniques have helped to investigate electrophysiological disturbances in heart muscle and in the brain. By 1998, microcoils as small as 100 microns in diameter had been constructed using microcontact printing of 25 micron wide silver wires, producing magnetic flux densities >0.4 tesla with a 10 milliamp current, switchable in <10^{-3} sec.

In biosystems and nanosystems alike, the detection of an optical photon is a high-energy event. A single photon of red light at l = 700 nm carries an energy of 280 zJ, a blue photon at 400 nm conveys 500 zJ, and an ultraviolet (UV) photon at 200 nm transmits 1000 zJ to its receptor. These energies, ranging from 65-234 kT, exceed thermal noise by a wide margin, hence are all easily detectable reliably by suitable nanosensors. Note that optical photon detection does not require a nanosensor the size of one wavelength or larger—for example, 1-nm chlorophyll molecules have no difficulty absorbing 660-nm photons. Biological systems provide numerous examples of single-photon transducers, including photopigment molecules such as chlorophyll and rhodopsin which mediate electron transfers in <~1 picosecond by undergoing conformational changes, triggering a series of biochemical events. An estimated 100-500 cell membrane channels are affected by the detection of a single optical photon. The first documented case of a physiological role for chlorophyll in animals has been found in the fish *Malacosteus niger*, which uses a chlorophyll derivative to see far-infrared light.

Mammalian cells also act as indirect UV sensors as for instance when energy absorption induces perturbations of membrane structure or a conformational change in membrane proteins, activating the JNK cascade below 300 pJ/micron^2, and centrosomes may serve as cellular directional IR sensors in single fibroblast cells. Integration times for photon sensors are briefly treated. Artificial "bioelectronic" cyanine-quinone chromophore molecules ~1 nm in size can detect optical photons of wavelengths l = 580-630 nm, and can trigger and reset in 1-3 picosec (governed by the tunneling times associated with reformation of the zwitterionic states of the inputs), in theory permitting ~100 GHz modulations to be received. Exciton-generating poly(p-phenylene vinylene) polymer switches have the potential to achieve ~THz switching speed; a birefringent Kerr medium also switches in a few picoseconds. Chalcogenide glass has viscosity >10^{13} kg/m-sec in darkness, but ~5×10^{11} kg/m-sec under illumination. Proposals for specific implementations of photoergic transducers and optical cables are elsewhere

High energy photons including X-rays (l ~ 2 nm, ~10^5 zJ or 0.0001 picojoule) and gamma rays (l ~ 0.002 nm, ~0.1 picojoule), and ejecta from the decay of radioactive nuclei such as a-particles (helium nuclei, ~1 picojoule) and b-particles (free electrons, 0.01-0.1 picojoule) may be extremely destructive to nanomechanical systems and are difficult to reliably measure quantitatively. For example, a charged particle with 0.01 pJ of kinetic energy (K.E.) traversing a decelerative electrostatic field of 10^9 volts/m requires K.E./qE ~ 60 microns to slow to a halt. However, particle-induced visual sensations have been observed by dark-adapted astronauts in space, and Ruderfer has suggested that even solar neutrinos may be detectable by the human brain (though most physicists would find this implausible).

The goal of cellular bioscanning is the noninvasive and non-destructive in vivo examination of interior biological structures. One of the most common nanomedical sensor tasks is the scanning of cellular and subcellular structures. Such tasks may include localization and examination of cytoplasmic and nuclear membranes, as well as the identification and diagnostic measurement of cellular contents including organelles and other natural molecular devices, cytoskeletal structures, biochemical composition and the kinetics of the cytoplasm. The precision and speed of medical nanodevices is so great that they can provide a surfeit of detailed diagnostic information well beyond that which is normally needed in classical medicine for a complete analysis of somatic status. Except in the most subtle cases a malfunction in any one component of the cellular machinery normally provokes a cascade of pathological observables in many other subsystems. Detection of any one of these cascade observables, if sufficiently unique and well-defined, may provide adequate diagnostic information (see) to plan a proper reparative procedure. DNA is one of the few cellular components that is regularly inspected and repaired. Most homeostatic systems adopt a more simple philosophy of periodic replacement of components regardless of functionality. Nanomedicine allows the philosophy of inspection and repair to be extended to all cellular components. The following discussion briefly describes a few of the sensory techniques that may be useful in achieving this objective.

Tactile topographic scanning provides the most direct means for examining cellular structures in vivo. For example, ex vivo live cell scanning in air by commercially available atomic force microscopy using a 20-40 nm radius tip allows nondestructive feature resolutions of ~50 nm across the top 10 nm of the cell. Investigators have used the scanning tip to punch a hole in the cell, then pull back and scan the breach, observing the membrane heal itself via self-assembly in real time. Up to ~KHz scanning frequencies are possible, with up to 1024 data points per scan line. The nanomechanical manipulator arm achieves comparable tip velocities and positional accuracies. Assuming a scan rate of ~10^6 pixels/sec, a micron-scale nanodevice, once securely anchored to the surface of a (~20 micron)^3 human cell, could employ a tactile scanning probe to image the 0.1 per cent of plasma membrane lying within its (1.4 micron)^2 vicinity in ~2 sec to ~1 nm^2 resolution (~1 mm/sec tip velocity), or ~50 sec to ~0.2 nm (e.g., atomic) resolution (~0.2 mm/sec tip velocity). Inside the cell, and again post-anchoring, an entire 6 micron^2 mitochondrial surface could be imaged to atomic resolution in ~100 sec; the surface of a 100 nm length of 25 nm diameter microtubule could be atomically resolved in 0.2 sec—though of course in a living cell these structures may be changing dynamically during the scanning process. The continuous power dissipation of a (0.1 micron)^2 scan head moving at 1 mm/sec through water is ~0.002 pW (~kT/pixel at ~10^6 pixels/sec), though of course the energy cost of sensing, recording, and processing each pixel must be at least ~10 kT/pixel (ignoring the possibility of pre-computational image compression), so the total scanner power draw could be as high as 0.02-0.1 pW in continuous operation. Special scanning tips and techniques should allow topographic, roughness, elastic, adhesive, chemical, electrostatic (charge density), conductance, capacitance, magnetic, or thermal surface properties to be measured. For large (~0.1-1 micron) cellular components, identification and preliminary diagnosis of improper structure may be possible using measured surface characteristics which may be matched to entries in an extensive onboard library, perhaps combined with dynamic monitoring of anomalies in cross-membrane molecular traffic using chemical nanosensors. Most membranes are self-sealing, so it should also be possible to gently insert a telescoping member into the target organelle, which member then slowly reticulates and extrudes smaller probes with sensory tips in a fixed pattern and step size, allowing the acquisition of detailed internal structural and compositional information.

Small (~1-10 nm) protein-based components of the cell such as enzymes, MHC carriers, and ribosomes are self-assembling or require only modest assistance (e.g., molecular chaperones) to self-assemble. Reversible mechanical denaturation of these smaller proteins, using diamondoid probe structure to displace water molecules (thus reducing hydrophobic forces) and by using specialized handling tools akin to functionalized AFM tips and molecular clamps, may be followed by precise nondestructive amino acid sequencing to allow identification and diagnostic compositional analysis. Afterward, the protein molecule may be refolded back into its original minimum-energy conformation, possibly with the assistance of chaperone-like structures in some cases.

Acoustic microscopy is another noninvasive scanning technique that can provide nanomedically useful spatial resolutions. Frequencies are in the GHz range, far exceeding the relaxation time of the protoplasm. A cryogenic acoustic microscope operated at 8 GHz has demonstrated 20 nm lateral resolution in liquid helium. By 1998, the best resolution achieved with water as the coupling fluid has been 240 nm at a frequency of 4.4 GHz. Operated in water at 310 K, a nanomechanical 1.5 GHz acoustic microscope would achieve a far-field minimum lateral resolution l/2 ~ 500 nm, ~10^5 voxels per human cell, sufficient to locate and count all major organelles and a few intermediate-scale structures. Scanning acoustic microscopes (SAM) operated at 2 GHz in reflection mode (which allows detection of interference effects) achieve 30-50 nm resolution in the direction of the acoustical axis, and the Subtraction SAM approach reveals topographical deviations of 7.5 nm at 1 GHz. It has been proposed that picosecond ultrasonics could be used to obtain an image of the cytoskeleton with detail comparable to that of conventional X-ray images of a human skeleton. Power requirements are a significant constraint on acoustic reflection microscopy (echolocation). In the simplest case, consider an acoustic emitter of radius r_E which converts an input power of P_{in} into acoustic power of intensity $I_E < I_{max}$ = 1000 watts/m^2 with efficiency e per cent, conservatively taken here to be 0.50 (50 per cent). The emitter produces a train of omnidirectional pressure pulses of amplitude $A_p = (2\ r\ v_{sound}\ I_E)^{1/2}$ (N/m^2) and frequency n which travels a distance X_{path} to a target of radius r_T. As the signal reaches the target, its amplitude has been reduced by (r_E/X_{path}) due to the $1/r^2$ dependence of intensity in spherical waves and by $e^{-atiss\ n\ Xpath}$ by attenuation with $a_{tiss} = 8.3 \times 10^{-6}$ sec/m for soft tissue. Upon reaching the target, a fraction $f_{reflect}$ of the signal is reflected as a point source echo; if the target surface has an acoustic impedance similar to liver tissue and the medium is similar to water, $f_{reflect}$ ~ 0.05 (5 per cent). The echo then travels a distance X_{path} back to a receiver (which may or may not be located near the emitter), the amplitude again losing (r_T/X_{path}) by geometry and $e^{-atiss\ n\ Xpath}$ by attenuation. To scan the entire interior of a (20 micron)^3 cell, whether from within or without, requires X_{path} >~ 20 microns. A frequency of n = 1.5 GHz allows v_{sound}/n ~ 1 micron spatial resolution. The fastest possible pulse repetition time is X_{path}/v_{sound} ~ 13 nanosec. Assuming $r_E = r_T$ = 0.5 micron, P_{in} ~ 7 pW or ~5 watts/m^2, well within the safe intensity range for acoustic radiation. At maximum safe transmitter intensity I_{max}, the longest scannable path length at 1.5 GHz is X_{path} ~ 46 microns.

Power constraints are somewhat less severe in the case of acoustic transmission microscopy (acoustic tomography), a technique which requires a minimum of two physically separated components (e.g., a transmitter and a receiver). Consider an acoustic emitter radiating a series of omnidirectional pressure pulses of frequency n which travel across a tissue cell of width X_{path}, to be detected by a receiver on the other side after a signal transit time t_{cell} ~ X_{path}/v_{cell}, where v_{cell} is the speed of sound in the cytosol. In the simplest case, assume that the cytosol is clear except for one target of interest along the path (e.g., an organelle) in which the speed of sound is v_{target} # v_{cell}. The minimum detectable difference in travel times between a signal that passes through the target and one that does not is Dt (~10^{-9} sec for diamondoid systems), so the minimum resolvable target size is $r_{min} \sim v_{target} v_{cell}$ Dt/2 abs(v_{target} - v_{cell}) when r_{min} >~ v_{cell}/n. For soft tissue targets in water at 1.5 GHz, r_{min} ~ 30 microns; for tooth enamel in water, r_{min} ~ 1 micron. However, the use of sampling gates will permit the detection of phase shifts that are much less than the characteristic time constant of the system in incoming waves, hence the minimum thickness of objects in theory detectable by acoustic tomography can be far smaller than the values for r_{min} estimated above.

In either mode of operation, data gathering is accelerated by positioning more than one receiver around the target. Sensitivity is boosted by increasing emitter size, taking multiple measurements, or by illuminating the target with some large external source located elsewhere in the tissue or even outside of the organ. However, cells and body tissues are mesoscopic "junkyards"—highly heterogeneous media which may produce large numbers of nontarget scattering events, thus increasing the difficulty of extracting signal from noise. Additional complications arise due to:

1. Scattering on rough surfaces.
2. Rapid pressure variations with range in the Fresnel zone or near field of the transmitted signal.
3. Cytoplasmic viscosity inhomogeneities due to the asymmetric arrangement of cytoskeletal structure, granules, vacuoles, and the endomembrane system.
4. Variation in cytoplasmic Young's modulus due to time-varying tensions in semirandomly-distributed cytosolic fibrillar elements (e.g., a relaxed or contracted state) which alter elasticity and thus the local speed of sound.

Electrostatic scanning is largely ineffective because of Debye-Huckel shielding. Magnetic stray field probes allow resolution of 10 nm physical features, but only for materials with substantial magnetic domains—which most biological substances lack. But nuclear magnetic resonance (NMR) imaging may allow cellular tomography by creating 3-D proton (hydrogen atom) density maps. Atomic density maps of other biologically important elements with nonzero nuclear magnetic moments (including D, Li^{6}, B^{10}, B^{11}, C^{13}, N^{14}, N^{15}, O^{17}, F^{19}, Na^{23}, Mg^{25}, P^{31}, Cl^{35}, K^{39}, Fe^{57}, Cu^{63} and Cu^{65}) may also be compiled. For example, sodium imaging is already used clinically to assess brain damage in patients with strokes, epilepsy, and tumors. In a hypothetical NMR cytotomographic nanoinstrument, a large permanent magnet is positioned near the surface of the cell or organelle to be examined. This creates a large static background magnetic field that polarizes the protons. The spatial gradient of this field establishes a unique resonant frequency, called the Larmor frequency, within each isomagnetic surface throughout the test volume. A second time-varying magnetic driver field (e.g., vibrating permanent magnet or weak rf field) is then scanned through the full range of resonant frequencies, exciting into resonance the protons (nuclear spins) in each isomagnetic surface, in turn. Depending on the sensor implementation chosen, each resonance detected may cause absorption of driver field energy, an increase in measured impedance, or even a return echo of magnetic energy (if the driver field is operated in pulse echo mode) as the excited protons relax to equilibrium in ~1 sec. The large polarizing magnet is then rotated to a new orientation, moving the isomagnetic surfaces to new positions within the test volume, and the scan is repeated. A 3D map of the spatial proton distribution may be computed after several scan cycles.

Only those regions within a linewidth of resonance DB will generate an appreciable signal, and the expected spatial resolution (say, along the z-axis) Dz = DB/(dB/dz), where dB/dz is the spatial flux gradient of B. For a polarizing field B = 1.4 tesla placed adjacent to a (20 micron)^3 cell, the cross-cell gradient dB/dz ~ 7 × 10^4 tesla/m; placed next to a (1 micron)^3 organelle, dB/dz ~ 1.4 × 10^6 tesla/m. Assuming an effective NMR linewidth ~ 0.002 tesla, minimum spatial resolutions are 29 nm and 1.4 nm, respectively. Unfortunately, the smallest reliably detectable energy in the sensor element is ~ kT e^{SNR}. The energy required to flip a single proton is E_{flip} = 4 p n_L L_{proton}, where L_{proton} = 5.28 × 10^{35} joule-sec (the quantized spin angular momentum of the proton) and the Larmor resonance frequency n_L = g_{proton} B; g_{proton} is the gyromagnetic ratio (4.26 × 10^7 Hz/tesla for protons) and B is the polarizing magnetic flux. However, the population difference between spin-up and spin-down nuclei in NMR is very small. For small energy differences, the Boltzmann distribution only allows a fraction E_{flip}/2 kT to be flipped before the upper and lower state populations are made equal and the absorption disappears.

When scanning a cell or organelle, the vast majority of the protons present are in water molecules. Water at 310 K has a proton density n_{water} = 6.7 × 10^{28} protons/m^3, compared to n_{fat} ~ 6.4 × 10^{28} protons/m^3 (palmitic acid). Distinguishing fat from water thus requires a minimum sample volume of $N_{min}/(n_{water} - n_{fat})$ ~55 micron^3. Proteins range from 4.4 × 10^{28} protons/m^3 (aspartic acid minus one water) to 7.6 × 10^{28} protons/m^3 (leucine minus one water), average $n_{protein}$ ~ 5.6 × 10^{28} protons/m^3; $n_{carbohydrate}$ ~ 5.8 × 10^{28} protons/m^3 (glucose minus one water), allowing NMR cellular tomography to resolve minimum feature sizes of 2-4 microns (560 micron^3). This resolution may include intracellular structures such as the endoplasmic reticulum (rough ER ~1100 micron^3, smooth ER ~400 micron^3), the Golgi complex (~500 micron^3), the nucleus (~270 micron^3), possibly the nucleolus (~50 micron^3) and even large localized ferritin granule concentrations. In 1998, the minimum sample size for microcoil (~470 micron diameter) NMR scanning was ~10,000 micron^3 or ~10^{16} protons for a 1-minute measurement cycle. Individual T-cell vacuoles ~1 micron in diameter labeled with dextran-coated iron oxide particles have been imaged by MRI.

Electromagnetic waves of optical wavelength l that interact with an object are diffracted into two components, called "far-field" and "near-field." The propagation of electromagnetic radiation over distances z > l acts as a spatial filter of finite bandwidth, resulting in the familiar diffraction-limited resolution ~l/2. Classical optics is concerned with this far-field regime with low spatial frequencies < 2/l, and conventional optical imaging will be difficult at the cellular level in vivo. (Short-wave X-rays will damage living biological cells.) Information about the high spatial frequency components of the diffracted waves is lost in the far-field regime, so information about sub-wavelength features of the object cannot be retrieved in classical microscopy. However, for propagation over distances z << l, far higher spatial frequencies can be detected because their amplitudes are then of the same order as the sample (z = 0). This second diffraction component is the "near-field" evanescent waves with high spatial frequencies >2/l. Evanescent waves are confined to subwavelength distances from the object. Thus a localized optical probe, such as a subwavelength aperture in an opaque screen, can be scanned raster fashion in this regime to generate an image with a resolution on the order of the probe size. The original Near-field Scanning Optical Microscope (NSOM) surpassed the classical diffraction limit by operating an optical probe at close proximity to the object. The NSOM probe uses an aluminum-coated "light funnel" scanned over the sample. Visible light emanates from the narrow end (~20 nm in diameter) of the light funnel and either reflects off the sample or travels through the sample into a detector, producing a visible light image of the surface with ~12 nm resolution at l = 514.5 nm provided the distance between light source and sample is very short, about 5 nm, with signal intensities up to 10^{11} photons/sec (~50 nanowatts). This represents a resolution of ~l/40. The near-field acoustic equivalent is found in the medical stethoscope, which exhibits a resolution of ~l/100. Applications include dynamical studies at video scanning rates, low-noise high-resolution spectroscopy, and differential absorption measurements. Optical imaging of individual dye molecules has already been demonstrated, with the ability to determine the orientation and depth of each target molecule located within ~30 nm of the scanned surface. Molecules with nanometer-scale packing densities have been resolved to ~0.4-nm diameters using STMs to create photon emission maps (an electric current generates photons in the sample), and laser interferometric NSOMs have produced clear optical images of dispersed oil drops on mica to ~1 nm resolution. Live specimen 250 nm optical sectioning for three-dimensional dynamic imaging has also been demonstrated.

Submicron laser emitters have been available since the late 1980s. It should be possible to use NSOM-like nanoprobes to optically scan the surfaces of cells or organelles to <~1 nm resolution, mapping their topography and spectroscopic characteristics to depths of tens of nanometers without penetrating the surface. However, a proper membrane-sealing invasive light funnel ~20 nm in diameter might not seriously disrupt some cellular or organelle membranes and thus could be inserted into the interiors of these bodies or through the cytoskeletal interstices to permit deeper volumetric scanning. The thermal conductivity of water at 310 K is 0.623 watts/m-K and the energy per 500 nm photon is 400 zJ, so for 1 micron^3 of watery tissue the maximum scan rate is ~35 micron^3/sec-K, if e^{SNR} photons are used to image each 1 nm^3 voxel with SNR = 2. Thus a 1-sec volumetric optical scan of an aqueous ~1-micron^3 sample volume to 1 nm^3 resolution requires a ~1 nanowatt scanner running at ~GHz bit rates, raising sample volume temperature by ~0.03 K. NSOM permits the determination of five of the six degrees of freedom for each molecule, lacking only the optically inactive rotation around the dipole axis. In principle, it should be possible to produce <~1 nm resolution near-field optical scans of in situ protein molecules, since with atomically precise fabrication and single lines of atoms as conductors a minimum light guide (metal-dielectric-metal) is 3 atoms wide. Given a molecular laser and adequate collimation, photons passing through folded proteins will scatter according to the molecular structure. Detection of sufficient photons comprising these scattering patterns should allow the noninvasive determination of protein structure; polarized photons provide information on chirality. Absorption and fluorescent signals will be visible from phenyl rings, tryptophan, and bound cofactors such as ATP and adenine (which is fluorescent). Positions of monoclonal antibodies on virus surfaces are now identifiable experimentally using NSOM; it should be possible to map binding sites on virus and cell surfaces using fluorescently labelled antibodies. Single molecule detection has been proposed as a tool for rapid base-sequencing of DNA.

Other optically-based cellular imaging techniques must be distinguished. Optical Coherence Tomography (OCT) uses Michelson interferometry to achieve ~10 micron spatial resolutions over tissue depths of 2-3 mm in nontransparent tissue in the near infrared. However, OCT requires numerous physical components not easily implemented on a micron-size detector (e.g., beam splitter, lens-grating pair, galvanometer mirror, optical prism), femtosecond pulse shaping, and illumination power levels of ~10^7 watts/m^2 >> 100 watts/m^2 "safe" continuous limit in tissue. Bioluminescence techniques in which the light source is placed inside the tissue (e.g., a transgenic mouse with a luciferase gene in every cell of its body) has a spatial resolution limited to ~10 per cent of the depth, or ~10 microns resolution at a ~100 micron depth. Coherent anti-Stokes Raman scattering (CARS) already permits organelle imaging in living cells and can in theory create a point-by-point chemical map of a cell using two intersecting lasers. Three-dimensional observation of microscopic biological nonliving structures by means of X-ray holography requires a high degree of spatial coherence and good contrast between target and surroundings. Good contrast may be achieved in the wavelength range between the K absorption edges of carbon (l = 4.37 nm) and oxygen (l =2.33 nm), the "water window" where carbon-containing biological objects absorb radiation efficiently but water is relatively transparent. A 50-micron diameter emitter has been tested that uses near-IR 5-femtosecond laser pulses impinging upon a helium gas target to create a well-collimated (<1 milliradian) beam of coherent soft X-rays at a 1 KHz repetition rate producing a brightness of 5×10^8 photons/mm^2-milliradian^2-sec, a peak X-ray intensity of $>10^{10}$ watts/m^2 on the propagation axis behind the He target.

Many cellular parameters need not be measured directly in order to be detected by a medical nanodevice. Cell volume sensing is a case in point. An intracellular nanodevice can indirectly monitor changes in the volume of the cell in which it resides by one of two methods. First, measurements of the mechanical deformation of the cellular membrane, stretch-activated channels, or cytoskeletal strains and structural changes are quite sensitive to alterations in total cell volume. Second, concentration or dilution of the cytoplasmic environment through cell shrinkage or swelling leads to the activation of various volume-regulation responses which may be detected by the nanorobot. Changes in the concentration of soluble cytosolic proteins may nonspecifically affect enzyme activity via "macro- molecular crowding". Minor changes in cell volume can cause severalfold changes in ion transport. Cellular signalling entities that have been linked to the transduction and amplification of the primary volume signal include Ca^{++} transients, phosphoinositide turnover, eicosanoid metabolism, kinase/phosphatase systems such as JNK and p38, cAMP, and G-proteins. These signal amplification pathways may be monitored using chemical concentration sensors aboard the medical nanodevice, allowing the nanorobot to eavesdrop on the natural sensory channel traffic of the cell.

In many ways the neuron is the most nanomedically important class of cell in the human body. Nanomedical applications regarding neurons and the brain are addressed in. The following is a brief summary of noninvasive (i.e., no axonal membrane penetration) nanotechnological methods for monitoring the electrical traffic of individual neurons. Noninvasive measurement of axonal traffic within nerve bundles will require multiple sensors and greater sensitivity to compensate for shielding by the perineurium, a tight resistive sheath enclosing the bundle ~20 microns thick with resistivity ~4000 ohm-cm.

The "typical" ~20 micron human neuron discharges 5-100 times per second, moving from -60 mV potential to +30 mV potential in ~10^{-3} sec. Thus the variation in electric field at the axonal surface is ± 4500 volts/m. Since electric field sensors can detect 100 volt/m fields up to GHz frequencies, an electric sensor attached by circumaxonal cuff or pressed against the axonal surface (possibly at the node of Ranvier) should readily detect each action potential discharge. (The diameter of human nerve axons is 0.1-20 microns.) By 1998 silicon-to-neuron extracellular junctions already permitted direct stimulation of individual nerve cells in vitro without killing the cells, and extracellular electrodes were commonly used to detect neuronal electrical activity noninvasively, both in vivo and in vitro. Artificial electric fields may also be employed to trigger or moderate neural signals. Cell membrane capacitance is typically ~0.01 picofarads/micron^2, and varies with the state of the health of the cell.

The magnetic flux density caused by a single action potential discharge is ~0.1 microtesla at the axonal surface, which may be detected by a "compass oscillator" type magnetosensor with a ~KHz maximum sampling rate. It might be possible for artificial magnetic fields to directly influence neural transmissions. Even a static 65 millitesla field has been shown to reduce frog skin Na^+ transport by 10-30 per cent. Each neuronal discharge develops an electrical energy of ~20 picojoule (~10^{10} kT), far smaller than the magnetic energy stored in a B = 1.4 tesla field of a permanent micromagnet traversing an L^3 = (20 micron)^3 volume which is $B^2 L^3/2 m_0$ ~ 6000 pJ. If properly manipulated, such a field may be sufficient to enhance, modulate, or extinguish a passing neural signal.

While neurons come in many shapes and sizes our "exemplar" ~14,000-micron^3 neuron discharging ~90 mV into an input impedance of ~500 Kohms produces ~0.2 microampere current per pulse and generates a continuous (average) 100-300 pW of waste heat as measured experimentally. The discharge rate of 5-100 Hz can produce brief surges up to ~2000 pW during a high-frequency train, but the duty cycle of such trains is far less than 100 per cent, reducing time-averaged dissipation to the observed 100-300 pW range. Single impulses are measured experimentally to produce 2-7 microkelvin temperature spikes in cold or room-temperature mammalian non-myelinated nerve fibers and ~23 microkelvins in non-myelinated garfish olfactory nerve fibers at an energy density ranging from 270-1670 joules/m^3-impulse from 0-20°C. In non-myelinated fibers the initial heat occurs in two temperature-dependent phases: a burst of positive heat, followed by rapid heat reabsorption (called the negative heat). The positive heat derives from the dissipation of free energy stored in the membrane capacity, and from the decrease in entropy of the membrane dielectric with depolarization. An L ~ 20-micron neuron in good thermal contact with an aqueous heat sink at 310 K has thermal conductance L K_t ~ 10^{-5} watts/K, so trains of 5-100 Hz impulses lasting 1 second should raise cellular temperature by 10-30 microkelvins; up to ~200 microkelvin thermal spikes from such trains have been observed experimentally. These events are easily detectable by nanoscale thermal sensors capable of ~1 microkelvin sensitivity up to ~1 KHz. The ~microkelvin heat signature of individual impulses or very short pulse trains can probably be temporally resolved because the minimum pulse repetition time is ~10 millisec (at 100 Hz) which is much longer than the thermal time constant for an L ~ 20-micron neuron (thermal conductivity K_t = 0.528 watts/m-K and heat capacity C_V = 3.86 × 10^6 joules/m^3-K for brain tissue; which is t_{EQ} ~ L^2 C_V/K_t ~ 3 millisec.

The synaptic cleft between the axonal presynaptic terminal and the dendritic postsynaptic membrane is 10-20 nm in most synapses, although in the vertebrate myoneural junction it may be as large as 100 nm. Contact area per bouton is ~1 micron^2, giving a total gap volume of ~10^7-10^8 nm^3. The density of acetylcholine receptors is highest in muscles along the crests and upper thirds of the junctional folds (~10,000/micron^2), and is lowest in the extrasynaptic regions (~5/micron^2). Each action potential discharge triggers the release of ~10^4-10^5 molecules of acetylcholine into the gap volume of an active neuromuscular junction (diffusion time ~1 microsec), raising c_{ligand} from near zero to ~3 × 10^{-4} molecules/nm^3 (~0.0005 M) in ~1 millisec, followed by near-complete hydrolyzation by acetylcholinesterase during the 1-2 millisec refractory period. Into the gap volume may easily be inserted a ~10^5 nm^3 neurotransmitter concentration sensor able to measure ~100 acetylcholine molecules in ~1 millisec, thus detecting pulses at the fastest discharge rate. A similar device could be used to precisely regulate neuro-transmitter concentration at the junction, and hence the neural signal itself, under nanodevice control. Simple electrochemical and mechanochemical artificial synapses have been demonstrated.

Many other noninvasive neurosensing techniques are readily conceivable. For instance, counting rotors or sodium magnetic resonance imaging could detect changes in the ionic composition (e.g., Na^+, K^+) of the periaxonal fluid before, during, and after discharge.

Starting in 1968, H. Frohlich, observing that millivolt electrical potentials maintained across cell membranes ~10 nm thick give rise to huge fields ~10^7 volts/m possibly producing an electret state, theorized that membrane molecules must be highly electrically polarized and thus could interact to produce coherent surface acoustic vibrational modes in the 10-100 GHz (microwave) frequency range; the longest wavelength is about twice the membrane thickness. Interestingly, this frequency span is very close to the maximum trigger/reset frequency for bioelectronic molecules). Note that (~100 mV) (1.6×10^{-19} coul) ~ 4 kT, so a membrane molecule with a single charge on either end should be reliably reoriented by a depolarization wave, coupling pressure waves and electrostatic field fluctuations. However, the direct detection of 10-100 GHz millimeter radiation by non-nanotechnological means is experimentally difficult and controversial because the tests must be performed in vivo in close proximity to an actively metabolizing cell in water—and water strongly absorbs microwaves over macroscale ranges (e.g., ~99 per cent absorption in 3 mm at 100 GHz).

Nevertheless, active cells have shown enhanced Raman anti-Stokes scattering, an effect ascribed to the converse of the Frohlich oscillations. In one study, the normalized growth rate of yeast cultures was enhanced or inhibited when irradiated by CW microwave fields of ~30 watts/m^2 of various frequencies; growth rate data spanning 62 separate runs revealed a repeatable frequency-dependent spectral fine structure with six distinct peaks of width ~10 MHz near 42 GHz. Investigations of related phenomena are ongoing and voluminous; the interested reader should peruse *Bioelectromagnetics*, the archival journal of this field.

100 GHz waves attenuate only ~1 per cent after passing through ~3 microns of soft tissue. A single electron injected into an integral membrane protein could act as an oscillating dipole, making a 300 volt/m signal 10 nm from the protein antenna with an energy transfer of ~0.004 kT per cycle; ~1000 oscillating electrons could produce a measurable field. A 20-micron diameter cell modeled as a nonuniform spherical dipole layer with transmembrane dipoles located 10 nm apart and embedded in a dissipative medium could produce 10^2-10^5 volt/m microwave fields 1-10 microns from the cell surface.

Thus, a variety of rf and microwave electromagnetic emanations may in theory be detectable both within and nearby living cells which could prove diagnostic of numerous internal states. Such states may include cytoskeletal dynamics, metabolic rates, plasmon-type excitations due to the collective motion of ions freed in chemical reactions, positional, rotational or conformational changes in biological macromolecules and membranes, internal movements of organelles and nerve traffic conduction, cellular pinocytosis, cellular reproduction events, cell membrane identity (e.g., distinguishing erythrocyte, Gram-positive and Gram-negative cell coat conductivities at 10 KHz), and cell-cell interactions.

Macrosensing is the detection of global somatic states (inside the human body) and extrasomatic states (sensory data originating outside the human body). While the treatment here is necessarily incomplete, the discussion nevertheless gives a good feel for the kinds of environmental variables that internally-situated nanodevices could sense. Not all capabilities outlined here need be available on every nanorobot, since injection of a cocktail of numerous distinct but mutually cooperative machine species allows designers to take full advantage of the benefits of functional specialization. In many cases, a given environmental variable can be measured by several different classes of sensor device. However, since these devices are microscopic it is in theory possible to operationalize almost all of the macrosensing capabilities described below in one patient using just a billion devices (~10 mm^{-3} whole-body deployment density), a total volume of ~1 mm^3 of nanorobots or ~0.1 per cent of the typical ~1 cm^3 therapeutic dose.

Can sounds generated by a single cell be detected, and thus be useful for diagnosis? Probably not, given that low frequency acoustic radiators are notoriously inefficient. For example, mitochondrion organelles of the giant amoeba *Reticulomyxa* are shuttled back and forth by 1-4 cytoplasmic dynein motors while riding on the outside of a bundle of 1-6 microtubules. Each dynein motor generates 2.6 pN of force and drives the mitochondria at up to ~10 micron/sec, developing $0.3\text{-}1.0 \times 10^{-16}$ watts of mechanical power within each 320-nm diameter organelle. Taking each organelle as a cylindrical acoustic radiator with a mechanical input power of $P_{in} \sim 10^{-16}$ watts at n ~ 1000 Hz, the output acoustic pressure at the organelle surface is only ~10^{-9} atm an acoustic power intensity $I \sim 10^{-15}$ watts/m^2 which is not detectable by micron-sized nanorobots. Nevertheless, cells and intracellular elements are capable of vibrating in a dynamic manner with complex harmonics that can be altered by growth factors and by the process of carcinogenesis, so the possibility cannot be completely ruled out.

Blood pressure ranges from 0.1-0.2 atm in the arteries to as low as 0.005 atm in the veins. The systolic/diastolic differential ranges from 0.05-0.07 atm in the aorta and 0.01-0.02 atm in the pulmonary artery, falling to 0.001-0.003 atm in the microvessels, or 0.003-0.005 atm if the precapillary sphincter is dilated. In venous vessels, pulse fluctuations are 0.002-0.010 atm in the superior vena cava, 0.004-0.006 atm in the subclavian vein, ~0.004 atm in venules generally, and ~0.0005 atm in the brachial vein. There is also a ~0.05 Hz random fluctuation in the microvessels with amplitude on the order of 0.004-0.007 atm. Both blood pressure and pulse rate can be reliably monitored by a medical nanodevice virtually anywhere in the vascular system using a

$(68\ nm)^3$ pressure sensor with ~0.001 atm sensitivity. Pulse propagation through body tissue is somewhat muted due to absorption in compressible fatty membranes, but most cells lie within 1-3 cell-widths of a capillary so the cardiac acoustic signal should still be measurable using more sensitive detectors. The time-averaged interstitial pressure in subcutaneous tissue is 0.001-0.004 atm. Arterial pulse waves (vascular oscillations) carry subtle messages about the health of internal organs and the arterial tree. The idea of using pulse waves for diagnosis has a long history dating back 2000 years in China. For example, in the *Book on Pulse Waves* by Wang Shu-He (201-285 AD), waves detected by manual probing are classified using such subjective and qualitative descriptors as floating, deep, hidden, rapid, slow, moderate, feeble, replete, full, thready, faint, weak, soft, slippery, hesitant, hollow, firm, long, short, swift, running, intermittent, uneven, taut, string-tight, gigantic, or tremulous. Abnormal waves were empirically related to disease states. Wave data gathered by nanodevices could make possible a theoretically sound, quantitative system of noninvasive observation, classification, and diagnosis as a supplement to other nanomedical tools.

The variation of mechanical pressure over a complete respiratory cycle is ~0.003 atm in the pleura, ~0.002 atm at the alveoli, detectable by nanomedical pressure sensors positioned in the vicinity of the respiratory organs. Holding a deep breath further stretches the pulmonary elastic tissue, up to 0.02 atm. However, turbulent flows at Reynolds number $N_R > 2300$ in the trachea, main bronchus and lobar bronchus produce a whooshing noise that may be the loudest noncardiac sound in the human torso during conventional auscultation.

Many other mechanical body noises should be globally audible to properly instrumented medical nanodevices. If normal chewing motions (of hard foods) release 1-10 milliwatts in a ~100 cm^3 oral volume with a ~1 sec jawstroke, power density is ~10 watts/m^3 or ~10^{-4} atm of tooth-crunching noise. A stomach growl registering 45 dB (vs. 30 dB whisper, 60 dB normal conversation) at 2 meters has a source power of 160 milliwatts; released from a 10 cm^3 gastric sphincter volume gives a ~2×10^{-6} atm acoustic wave, detectable throughout the body. Walking and running releases 20-100 joules/footfall for a 70 kg man; assuming the energy is absorbed within a ~1 cm thickness or within ~1 second by the sole of the foot, an upward-moving planar compression wave of 0.4-2.0 atm, easily detectable by acoustically instrumented nanodevices body-wide. (Shoe insoles dissipate energy and alter the shock wave pulse shape.) Hand-clapping generates 0.02-0.2 atm pulses, also easily detectable. Lesser noises including ~30 millisec hiccups at (4-60)/min, intestinal and ureteral peristalsis, sloshing of liquid stomach contents, heart murmurs, a tap on the shoulder by a friend, nasal sniffling and swallowing, clicks from picking or drumming fingernails, crepitations, manustuprations and ejaculations, the rustling noise of clothing against the skin, flapping eyelids, anal towelling, bruits (including murmurs and thrills) due to vascular lesions, dermal impact of water while showering, copulatory noises, urethral flow turbulence during urination, transmitted vibrations from musical instruments, creaking joints, and squeaking muscles can be detected locally if not globally. Implantation of significant interconnected in vivo diamondoid structures may produce increased sensitivity to internal noises, due to the extremely low acoustic absorption coefficient of diamond.

Average source power for conversational speech in air is ~10 microwatts at the vocal cords (60 dB), up to ~1000 microwatts for shouting (90 dB) and as little as 0.1 microwatts (30 dB) for whispering. Vocal cord surface area ~1 cm^2, giving an acoustic intensity I ~ 0.001-10 watts/m^2. (Using the decibel notation, dB = 10 $\log_{10}$ (I/I_0), where $I_0 \sim 5 \times 10^{-13}$ watts/m^2 in air, $I_0 \sim 1 \times 10^{-16}$ watts/m^2 in water.) For water at 310 K, r = 993.4 kg/m^3 and v_{sound} = 1500 m/sec, therefore A_p = 0.0005-0.05 atm for speech, detectable by nanodevices throughout the body due to minimal attenuation at audible frequencies. Other easily detectable vocalizations include whistling, humming, coughing, sneezing, rales, wheezing, expectorating, eructations, flatus, vomiting, hawking and noseblowing. In the case of spherical waves diverging from an omnidirectional transmitter of radius r at a distance X from the transmitter, average intensity declines inversely as the square of the distance and so.

Of course, an ex vivo acoustic nanosensor may receive sound that has passed through no interface, hence may detect pressure waves ~3 orders of magnitude lower in amplitude. Assuming $A_{transmit} = A_{incident}$, $P_R \sim 600$ microwatts, so ex vivo nanorobots with a 0.3 micron^3 sensor could hear people shouting at x_R = 2 meters. To hear talking (~10 microwatt source) requires a 2.4 micron^3 ex vivo pressure sensor (limit ~10^{-7} atm). Optimally positioned and calibrated nanomedical pressure sensors could directly measure changes in the ambient barometric pressure to within ± 10^{-6} atm. Normal atmospheric variation due to weather ranges from 0.94-1.05 atm; such slow moving changes are readily monitored. Very near the Earth's surface, the air pressure P at altitude h above sea level is approximated by $P = e^{-k_p h}$ (atm), where $k_p = 1.16 \times 10^{-4}\ m^{-1}$ at 20°C; at sea level, a 10^{-6} atm change in pressure reflects a change in altitude of only ~1 cm. However, the opening or closing of a door inside a (~5 m)^3 room that displaces >125 cm^3 of air also causes a minimally detectable >10^{-6} atm pressure pulse. Other sources of environmental pressure variation such as infrasonic (~0.2 Hz) microbaroms from offshore ocean storms, wind entering through open windows, forced-air currents from central heating or A/C systems, or even the movements of nearby people and pets may be detectable and thus, may further confuse the measurement, reducing absolute accuracy unless suitable corrections are made.

Using a navigational transponder network, a population of ~10^{11} nanodevices each spaced an average ~100 microns apart throughout the body tissues can determine relative location to a positional accuracy of ~3 microns and an angular accuracy of ~2 milliradian, with the data updatable once every millisecond. Thus, dispersed, the network can continuously monitor and record the relative positions of all limbs with worst-case (cumulative error) ~0.3 mm accuracy over a 2-meter span. These devices can prepare high-resolution dynamic maps of body position, velocity, acceleration and rotation, allowing precise real-time digital kinesthesia during sports or artistic activities such as gymnastics, pole vaulting, or ballet; during transportative activities such as driving cars around hairpin turns, roller-coaster rides, military aircraft maneuvers and space launches; during precision tool-using such as needle-threading, antique watch repair, or while using the fingers as measurement calipers; during self-defense activities requiring complex motions such as karate or judo; and during emergency situations such as automobile crashes and tumbling motions during falls from great heights. All such sensory data is readily outmessaged to the patient or user in real time. Nanorobots could assist in path integration or the reconstruction of experienced limb trajectories. Given the ability to determine rotation rates and to measure applied forces and torques, the network should also be able to forecast the anticipated future positions of limbs. By comparing these projections to actual results, the network can then infer the viscosity of the medium in which the activity is taking place (e.g., air or water), whether the environment is stable or is translating or rotating in some direction, and whether the human user is physically supported or in free-fall. Internal nanodevices can directly measure if a body is sitting, standing, laying prone, inverted, falling, floating or diving, and then communicate that information directly to the human user. Detection of patient activity states—e.g., the patient is sitting, standing, or walking—can be used to control nanorobot behaviours, activate or deactivate outmessaging displays, and so forth. A simple macroscale wearable tactile "compass belt" that would convey desired geographical directions has already been proposed.

Since nanopendular sensors can reliably determine the direction of the local gravity field vector to within ~2 milliradians in ~10^{-4} sec, the navigational network can poll its members and arrive at an accurate consensus on which direction is up. Nanodevices affixed to relatively stable hard body parts will exhibit more consistent orientational readings. Measurement of the gravity vector allows the vertical orientation of the rest of the body to be precisely fixed in space, especially useful for gymnasts, trapeze artists, underwater divers in murky lakes, and vestibular-impaired individuals, to whom this information may be outmessaged directly. The gravitational vector may also be indirectly measured, albeit more slowly and less accurately, by monitoring the body's natural reactions to changes in the axis of gravitational loading. For instance, the variation in head-to-toe hydrostatic pressure for a 1.7 m tall, 70 kg human standing in a 1-g gravity field is ~0.17 atm; the ability to measure a systematic cross-body differential of 10^{-6} atm allows the detection of a ~10^{-5} g change along the lengthwise aspect, or ~10^{-4} g change along the transverse aspect. Thus a continuously updated whole-body barostatic map allows the human body to serve as a three-dimensional gravity/orientation sensor.

A network of nanodevices can inventory the volume and density of each of the body's ~10^8 $(1 \text{ mm})^3$ voxels using a combination of acoustic ranging, somatic mapping, and flowmetry. Each voxel is identified as fat, muscle tissue, bone mass, interstitial fluid, and so forth. These measurements allow the body's weight to be precisely computed as volume times density, rather than the usual method of determining weight using gravitational force data. The existence of a natural physiological "ponderostat," a crudely analogous humoral body-mass detector, has been proposed; indeed, the insulin-leptin system has been found to serve a related purpose, and is easily eavesdropped by medical nanorobots.

To achieve such phenomenal positional accuracies, the nanodevice must be able to computationally resolve several complicating factors. First, localized mass concentrations representing nonuniformities in crustal density produce residuals of up to ±0.0006 m/ sec^2, which may be removed from the data using a standard map of known terrestrial isostatic variations and anomalies. Indeed, matching observations to such a map could provide useful longitudinal information as well. Another complication is the variation in gravity due to tidal forces amounting to ~3 × 10^{-7} g's, twice daily, which lies at the limits of detectability for a 20-micron gravimeter. Other minor geodesic and terrain-related corrections, too complicated for discussion here, may also need to be applied in certain circumstances. Note that the presence of nearby heavy objects does not influence measurement accuracy: a 100-ton building 10 meters away adds a lateral acceleration of only 7 × 10^{-9} g's to a human body. One final complication is that patient movements create kinematic accelerations that must be distinguished from the gravitational accelerations. Gravity readings can be corrected by taking derivatives of the signals from kinesthetic monitoring to give gravity in the reference frame of the patient's room, and many individual measurements may be averaged to improve accuracy since the gravity vector normally changes only very slowly.

In vivo studies of the electric properties of blood vessels shows that the walls of veins and arteries present a specific electric resistance ~200 times greater than the conductive medium of blood (plasma)—roughly 200 ohm-m vs. 0.7 ohm-m. Thus blood vessels are properly regarded as relatively insulated conducting cables which can electrically connect an injured tissue with surrounding noninjured tissue. Capillaries form an electric junction with the interstitial fluid, so the electric gradient can be canceled by ionic transport. Cell membranes are insulating dielectrics with resistance and capacitance, penetrated by ionic channels or gates for ionic transport. Additionally, in 1941 Szent-Gyorgyi suggested the possibility of semi-conduction in proteins, a theory which has since been elaborated by many investigators. B. Nordenstrom has proposed that the above schema represents a system for selective electrogenous mass transport of material between blood and tissue which he calls Vascular-Interstitial Closed (electric) Circuits (VICCs)—in effect, an additional electro-circulatory system operating in parallel with the well-known diffusive, osmotic, and hydrodynamic mechanisms of regular bloodstream materials transport. As long as ions can leak through open pores and migrate through ion channels, long distance transport cannot take place and the VICC system remains primarily a local (e.g., within tissues), not global, electrical circuit. Connections also exist with conductive media including cerebrospinal fluid, bile, and urine. Nanorobots capable of monitoring the status of local VICCs may rapidly and efficiently acquire a wealth of systemic information without the need for direct inspection of the affected tissues. For example, when a working muscle produces catabolic products such as lactic acid, an electrochemical potential gradient develops between the muscle and surrounding tissue in a known manner. Injured tissues become polarized in relation to surrounding tissue, as initial catabolic degradation products acidify the tissue. Malignant neoplasms, benign neoplasms, and internally necrotic granulomas all may polarize electrically in relation to surrounding tissue; a vascular thrombus also contains ionized material. Large deflections in the diffusion potential of blood occur if blood is deoxygenated or is infected with Gram-negative bacteria. Blood changes its electric potential from +500 mV to +1000 mV during spontaneous coagulation, and the spontaneous electric potential at the site of a crush injury of the iliac crest in rats oscillated several times between +190 mV and -60 mV over a four day experimental trial.

Nordenstrom proposes that in vivo electrophoresis via the VICC system may also play a role in mediating leukotaxis. When tissue is artificially polarized by electrodes, accumulation of granulocytes with margination and development of pseudopods is the result. In one experiment, exposure of a mesenterial membrane to a 1 microampere 1 volt field for 30 minutes stimulated diapedetic bleeding near the anode. At higher power, the arterio-capillaries contracted, narrowed, and emptied of blood cells, while the veno-capillaries and venules widened and filled with granulocytes. Monitoring such actions of local VICCs experiencing natural voltage fluctuations could alert medical nanorobots to the presence of local injuries, tissue changes, or pathologies that otherwise might go unnoticed for a considerable time.

Detection of electric fields to ~100 volts/m, the normal vertical atmospheric gradient, in theory might allow determination of aboveground altitude to ~1 m accuracy if the time between recalibrations is very brief. However, a human body (which is a good conductor), standing on level ground, acquires a slight negative surface charge and becomes "part of the ground" electrically. This distorts upward the equipotentials that usually run parallel to the ground, thoroughly corrupting an altitude measurement. Nearby lightning hits and storms are detectable, which is useful. However, medical nanodevices with appropriate magnetometers can measure variations in magnetic field to ~0.1 microtesla using a (660 nm)^3 permanent magnet sensor, in a measurement time t_{meas} ~ 0.6 millisec. Geomagnetic field maps of the Earth's surface are highly nonisotropic in both latitudinal and longitudinal directions; knowledge of the field plus the absolute direction of true north (e.g., using a nanogyroscope) gives longitude information from the separation of the planetary magnetic and spin poles. This implies that properly equipped in vivo nanodevices can establish their latitude and longitude on Earth's surface ~1000 times every second by this means. In particular, the horizontal component of the geomagnetic field vector ranges from 0-41 microtesla from magnetic pole to magnetic equator; the independent vertical field component ranges from 0-70 microtesla. Thus a 0.1-microtesla sensor should resolve each component to an accuracy of ~11 arcmin or ~20 km in both latitude or longitude. Terrestrial surface magnetic anomalies (e.g., iron deposits) produce local variations of 0.03-30 microtesla and are thus detectable. Additional complications to these measurements include artificial magnetic field sources, the 0.01-0.1 microtesla solar daily variation, occasional magnetic storms producing erratic geomagnetic fluctuations of 0.01-5 microtesla, sudden-commencement ionospheric electrojets up to 0.3 microtesla, and various long-term secular variations in the geomagnetic field. Radio, television, and direct broadcast satellite signals probably are not detectable by individual nanodevices, but such detection may be indirectly achieved in vivo using nanorobot-accessible dedicated macroscale antennas implanted in the body.

The piezoelectric effect is "the production of electrical polarization in a material by the application of mechanical stress". Many materials in the human body are piezoelectric, including tendon and elastin, dentin and bone. This polarization, or surface charge, varies as the physical stress imposed on the material changes over time. Thus, measurement of the piezoelectric effect in bone or tendon can provide information on the physical loads that are being carried by these materials. For example, a shear stress applied along the long axis of a bone alters the polarization voltage that appears on a surface at right angles to the axis. In another experiment, piezoelectric surface charges on a loaded human femur were measured to range from -131 to +207 picocoulombs/cm^2 (-8 to +13 charges/micron^2) from one end of the bone to the other end, depending upon position along the shaft. Real-time monitoring of this data would permit specific inferences as to the amount of load the bone was carrying and from what direction, including bending, shearing, and twisting forces, from which whole-body activity states could subsequently be inferred. Knowledge of these surface charge variations might also be exploited in osteographic or functional navigation.

In vivo medical nanodevices can gather little useful optical information from the external environment. Here's why. In biological soft tissues, scattering dominates absorption except in the pigmented layers of the epidermis and stratum corneum. Thus the propagation of light in tissues may be regarded as occurring in two steps.

Over the optical band from 400-700 nm, incident intensity I_0 = 100-400 watts/m^2 when standing in direct sunlight; artificial lighting in homes and offices is typically 0.1-10 watts/m^2; moonlight provides only $I_0 \sim 10^{-4}$ watts/m^2; and the absolute threshold for human vision is ~10^{-8} watts/m^2. For $s_d \sim 900$ m^{-1}, the intensity ratio of transmitted/incident visible light falls to $I_d/I_0 = 0.1$ at z = 2.5 mm depth (~eyelids), ~10^{-4} at z = 1 cm depth, and ~10^{-11} at z = 2.8 cm—a depth at which the tissue would be completely dark to the human eye even under direct sunlight illumination at the outermost skin surface. Let us assume that a patient is standing in $I_0 \sim 400$ watt/m^2 direct unfiltered sunlight, and that t_{meas} = 1 sec, SNR = 2, N_{sensor} = 25,000 elements giving a nanorobot eyespot of $A_e N_{sensor}$ = 0.025 micron^2, and that s_d = 900 m^{-1}. The maximum tissue depth for optical photon detection is $z_{max} \sim 17$ mm ($I_d \sim 10^{-4}$ watt/m^2) which includes a volume of tissue comprising up to ~30 per cent of total body volume. Changes in illumination at the minimum indoor artificial level of ~0.1 watts/m^2 are visible to $z_{max} \sim 7$ mm depth. This eyespot, if exposed to air on the outermost surface of the skin, is just sensitive enough to detect full moonlight. Our general conclusion is that variations in normal indoor lighting may be directly measurable by nanorobots stationed within the outermost ~1 cm of body tissues. Imaging, as opposed to the simple illumination detection system described here, is a much more difficult design challenge

The ability to detect individual neural cell electrical discharges noninvasively in many different ways, coupled with the abilities (A) to recognize and identify specific desired target nerve cells and (B) to pool data gathered independently by spatially separated nanodevices in real time, offers the possibility of indirect neural macrosensing of complex environmental stimuli by eavesdropping on the body's own regular sensory signal traffic. For example, facultatively mobile nanodevices may swim into the spiral artery of the ear and down through its bifurcations to reach the cochlear canal, then position themselves as neural monitors in the vicinity of the spiral nerve fibers and the nerves entering the epithelium of the organ of Corti (cochlear or auditory nerves) within the spiral ganglion. These monitors can detect, record, or rebroadcast to other nanodevices in the communications network all auditory neural traffic perceived by the human ear. Advanced speech recognition systems may permit recovery of spoken words and identification of individual speakers (including vocalizations by the user), recognition of background noises, reception and validation of cues or commands spoken directly to in vivo nanosystems by authorized medical personnel, and so forth. Since properly configured monitors can also modulate or stimulate nerve impulses, these devices may add audible signals to the audio traffic, thus may be employed as hearing aids (using feedback loops), real-time language translation mechanisms, continuous vocalization/audition recorders, voice-stress analyzers, or nanodevice-user communications links. Nanomonitors positioned at the afferent nerve endings emanating from hair cells located in the otolithic membrane of the utricle and saccule and in the cristae ampullaris of the semicircular canals allow medical nanodevices to directly record, amplify, attenuate, or modulate the body's own sensations of gravity, rotation, and acceleration, although kinesthetic sensory management might also be required for complete control. Motor neurons likewise can be monitored to keep track of limb motions and positions, or specific muscle activities, and even to exert control. Neuron-resident nanorobots may detect auditory effects induced in the brain by pulsed microwaves from external sources. Feline cochlear neurons sensitive to audible frequencies of >300 Hz respond to single microwave pulses at a threshold-specific absorption rate of 6-11 watts/kg-pulse, while in humans the threshold for effect, which depends on energy per pulse, may be as low as ~0.02 joule/m^2-pulse for people with low hearing threshold. Olfactory and gustatory sensory neural traffic similarly may be eavesdropped by nanosensory instruments. Nerve taps in the medulla oblongata or at the phrenic nerve that drives the diaphragmatic muscles allow direct monitoring of respiratory activity. Pain signals may be recorded or modified as required, as can mechanical and temperature nerve impulses from other receptors located in the skin. Even psychological variables such as emotionality, vigilance, and mental workload may be directly monitored by measuring ANS activity in sympathetic efferent fibers outside of the brain. The most complex and difficult challenge in neural macrosensing will be optic nerve taps. The retina is thoroughly vascularized, permitting ready access to both photoreceptor (rod, cone, bipolar and ganglion) and integrator (horizontal, amacrine, and centrifugal bipolar) neurons. However, the optic nerve bundle itself has ~10^6 tightly bunched individual nerve fibers, a 10-100 MHz signal bandwidth, and significant natural data compression techniques which all must be untangled in real time. Developing algorithms capable of interpreting raw optical nerve traffic, say, to recognize a specific human face or a specific scene in the vision field, would prove a significant research challenge. (Direct monitoring of photo-receptors or retinal membrane potentials and other techniques may simplify untangling of the signal compression.) Rapid visual field identification using artificial neural nets is also a subject of much current research interest. Eyeball rotations (e.g., via resident intradevice nanogyroscopes, eyelid position, pupil aperture, and lens accommodation under ciliary muscle control must be monitored to supplement the vision field analysis.

We have only scratched the surface of the potential for nanorobot macrosensing. Medical nanodevices can quantitatively monitor variables not normally accessible to human consciousness, such as:

1. Hormone and neurotransmitter levels.
2. Gastroelectric oscillations and skin conductivity.
3. Pupil dilations.
4. Drug and alcohol (and breakdown product) concentrations in the blood.
5. Internal organ damage or malfunction.
6. Digital real-time performance data on vital organs or limbs (e.g., continuous kidney volume throughput, pancreatic insulin output, cholesterol metabolism in the liver, or lactic acid production in specific muscles during exercise).
7. Continuous mapping of thermoregulatory isotherms from which the temperature and heat capacity of the external medium (e.g., swimming pool water, cool night air) and the presence of sunlight or shade on the skin (via dermal thermal differentials) may be inferred.

Various chemical substances cyclically increase and decrease in serum concentration with time of day, permitting crude temporal macrosensing. Other substances significantly vary in serum concentration depending upon whether the patient is supine or standing, thus permitting limited biochemical postural macrosensing. If nanorobots can leave and re-enter the body, then macrosensing may include direct sampling of the external environment upon demand and the possibilities for remote data acquisition become virtually limitless.

Sensors pervade many aspects of modern living. They are built into many consumer electronic devices, cars, medical devices, security and safety devices, and systems for monitoring pollution and environmental conditions. Many applications demand miniaturisation to reduce power consumption for integration into portable devices. Affordable mass production is also a prerequisite for sensors for consumer products, and for disposable devices such as sensors for medical diagnostics and pollution monitoring. Sensors support applications across the economy—industrial processes, and those in construction, extractive industries, agriculture, health care and so on—and can be incorporated into new or existing products.

Sensors can model various parameters: physical parameters such as temperature, displacement, acceleration, flow and so on; and chemical and biochemical parameters, such as concentrations of gases, ions or molecules, and molecular interactions)

The application of nanotechnology to sensors should allow improvements in functionality. In particular, new biosensor technology combined with micro and nanofabrication technology can deliver a huge range of applications. They should also lead to much decreased size, enabling the integration of 'nanosensors' into many other devices.

We can also expect to see actuators that control movement on the nanoscale. Sensor/actuator combinations will deliver 'smart' and precise functions in products and processes. For example, nanofabrication and inspection tools require sensors and actuator systems that can position objects with nanometre accuracy. In this way, sensors and actuators constitute another enabling technology.

Gas Sensors

MS Modeling's DMol^3 has been used to investigate the nitrogen dioxide sensing mechanism of tin dioxide nanoribbons. Understanding the sensing mechanism will enable the efficient design and manufacture of nanoscale chemical sensors—an important commercial application of nanotechnology.

Nanowires and Nanoribbons

Practical challenges with carbon nanotubes involving cost of synthesis, control of chirality and diameter, separation from bundles, and attachment of functional groups have prompted researchers to explore other types of one-dimensional nanostructures. Two of the most promising candidates in this regard are nanowires and nanoribbons. As the name suggests, nanoribbons are solid objects (unlike nanotubes, which are hollow) with a near-uniform rectangular cross-section.

So far, nanoribbons have primarily been synthesized from the oxides of metals and semiconductors. In particular, SnO_2 and ZnO nanoribbons have been materials systems of great current interest because of potential applications as catalysts, in optoelectronic devices, and as chemical sensors for pollutant gas species and biomolecules. Although they grow to tens of microns long, the nanoribbons are remarkably single-crystalline and essentially free of dislocations. Thus they provide an ideal model for the systematic study of electrical, thermal, optical, and transport processes in one-dimensional semiconducting nanostructures, and their response to various external process conditions.

Tin Dioxide Nanoribbons as Chemical Sensors

Recent experiments with SnO_2 nanoribbons indicate that these are highly effective in detecting even very small amounts of harmful gases like NO_2. Upon adsorption of these gases, the electrical conductance of the sample decreases by more than an order of magnitude. More interestingly, it is possible to get rid of the adsorbates by shining UV light, and the electrical conductance is completely restored to its original value. Such single-crystalline sensing elements have several advantages over conventional thin-film oxide sensors: low operating temperatures, no ill-defined coarse grain boundaries, and high active surface-to-volume ratio.

Investigating the NO_2 Sensing Mechanism of SnO_2

To be able to fully commercialize their potential, it is important to better understand the sensing mechanism of such systems. With this goal in mind, researchers at the Brookhaven National Laboratory, Lawrence Berkeley National Laboratory, and Accelrys used MS Modeling's DMol[3] to investigate the nitrogen dioxide sensing mechanism of SnO_2 nanoribbons. Reporting in *Nano Letters*, the researchers examined the NO_2-sensing mechanism of SnO_2 nanoribbons with exposed (1 0'1) and (0 1 0) surfaces.

The density functional theory (DFT) calculations revealed that:

- The most stable adsorbed species involved an unexpected NO_3 group doubly bonded to Sn centers
- An orders-of-magnitude drop in electrical conductance can be explained by significant electron transfer to the adatoms
- Computed binding energies were consistent with adsorbate stability up to 700 K, with X-ray absorption spectroscopy indicating predominantly NO_3 species on the nanoribbon surface.

Investigating the O_2 and CO Sensing Mechanism of Nanoribbons

The ability of the nanoribbons to sense O_2 and CO was also investigated. In the case of O_2, the response of the nanoribbon was highly sensitive to the concentration of O-vacancies on the surface. Thus, in the absence of any surface vacancies, the calculation predicted negligible charge transfer. However, when surface vacancies were present, an O_2 molecule can adsorb as a peroxide bridge, and withdraw a significant amount of electronic charge from the nanoribbon surface, thereby decreasing its electrical conductance. In the case of CO adsorption, there was a net electron transfer from the

CO to the nanoribbon surface. Thus the calculation predicted an increase in nanoribbon electrical conductance upon CO-adsorption, in agreement with experimental results.

Redesigning Platinum at the Nanoscale

Researchers from the Department of Energy's Sandia National Laboratories and the University of New Mexico (UNM) have developed a new way of mimicking photosynthetic proteins to manipulate platinum at the nanoscale. The method has the potential of changing the metal's properties and benefiting emerging technologies. While we are in the early stages of research, we see the possibility of manipulating the nanoscale structure of platinum so that we can have control over the size, porosity, composition, surface species, solubility, stability and other functional properties of these metal nanostructures," says John Shelnutt, the Sandia scientist leading the research effort. "Such control means that the redesigned platinum could be used in many new applications, including catalysis, sensors and optoelectronic and magnetic devices." He adds that while research groups have reported a few platinum nanostructures — including nanoparticles, nanowires, nanosheets and others — the addition of new types of nanostructures is "highly desirable and potentially technologically important." Working with Shelnutt in the research are Frank van Swol from Sandia, UNM graduate student Yujiang Song and Eulalia Pereira from the University of Porto in Portugal. The new method of manipulating platinum was detailed in a paper in the *Journal of American Chemical Society* published in December. The idea for the technique is similar to photosynthesis, in which plants use the energy from sunlight to produce sugar. But instead of manufacturing sugar, the new method changes a platinum ion to the neutral metal atoms. The photosynthetic protein mimicks this repeatedly, allowing metal to be deposited as desired at the nanoscale.

The method involves putting porphyrins — the active part of photosynthetic proteins — along with the platinum salt in an aqueous solution of ascorbic acid at room temperature. The porphyrins are placed in specific locations in the solution where it is intended that metal should be deposited. For example, the porphyrins may be confined to micelles, or liposomes. Micelles are spherical assemblies of detergent molecules in which the heads are exposed to the water and the tails stick together in the interior. Liposomes are similar structures, but they are larger and have water on the inside and outside separated by a closed membrane — sort of like a cell. The membrane is composed of two layers of detergent molecules, with the heads on the inner and outer surface facing the water and the tails forming the interior of the membrane. When light is shined on the porphyrins located in these detergent structures, the porphyrins excite, becoming catalysts for platinum reduction and deposition. As this occurs, the metal grows onto the surfaces of the surfactant structures as a thin sheet or in other ways. In the case of micelles, the platinum grows into balls that look like the common toy "Koosh" ball. The ball size can be controlled by the amount of porphyrins and platinum in the solution, the amount of light illuminating the solution, and the amount of time the light is on.

For the metals platinum and palladium that form these nanostructures, it is enough for the porphyrin molecule to grow only a small metal "seed" particle composed of about 500 atoms. When it reaches this size, the seed starts to catalyze its own rapid growth (by oxidation of ascorbic acid), budding off arms in all directions and creating the Koosh-ball-like nanostructures. The porphyrin provides a convenient method of making these seeds at the location and time desired, leading to a uniform and selectable nanostructure size The platinum nanostructures take on a different form when they are prepared under different conditions. When the porphyrin is in a micelle, the platinum nanostructures produced look like Koosh balls. When the porphyrin is in the bilayer membrane of a liposome, the platinum grows in 2-nanometer thick sheet or platinum lace on the outer surface of the membrane, giving circular sheets — sort of like two-dimensional Koosh balls. Under solution conditions for which the liposomes aggregate, growth can occur along the interfaces of the liposomes to give platinum foamlike materials and foam nanoballs. The type of nanostructure is mainly determined by the type of surfactant assembly upon which the platinum grows and the extent of growth from the individual seed nanoparticles. Since the porphyrin remains attached to the platinum nanostructure and active in the presence of light, it can also perform other functions besides growing itself. For example, when illuminated with light, the platinum nanostructure evolves hydrogen from water. This reaction is similar to one of interest to car manufacturers looking for new ways to build automobiles powered by hydrogen fuel cells. Shelnutt says that in addition to structuring the platinum, the process also happens very fast. A few minutes in light will create many seeds, which then grow into the mature nanostructures in tens of minutes. And the process is easy to do.

5

Nano-optics: Applications & Utilities

Nano-Optics: An Introduction

Today we encounter a strong trend towards nanoscience and nanotechnology. This trend is motivated by the fact that as we move to smaller and smaller scales the underlying physical laws change from macroscopic to microscopic. The exploitation of quantum effects for technological applications is the most obvious driving force behind the current miniaturization. The recent rapid advances are due in large part to our newly acquired ability to measure and manipulate individual structures on the nanoscale (scanning probe techniques, optical tweezers, and high-resolution electron microscopes). The increasing trend towards nanoscience and nanotechnology makes it necessary to address the key issues of optics on the nanometer scale. Since the diffraction limit does not allow us to focus light to dimensions smaller than roughly half a wavelength, traditionally it was not possible to interact selectively with nanoscale features. In recent years several new approaches have been put forth to 'shrink' the diffraction limit (confocal microscopy) or to even overcome it (near-field microscopy). For example, with our tip-enhancement technique we are able to do Raman spectrocopy and multiphoton fluorescence imaging with a spatial resolution of less than 20nm. To date, this is the highest optical resolution of a spectroscopic measurement. The reason for the hard effort to advance the field of optics to the nanometer scale is the fact that the energy of light lies in the range of electronic and vibrational transitions in matter. Therefore, the interaction of light with matter renders unique information about the structural and dynamical properties of matter. These unique spectroscopic capabilities are of great importance for the study of biological and solid-state nanostructures. We are applying near-field optical techniques to probe complex semiconductor nanostructures as well as individual protein molecules. We plan to explore the possibility to optically interact with semiconductor nanostructures on length scales smaller than the extent of their quantum wavefunctions. Probing and manipulating these wavefunctions might open up various exciting applications such as data storage and optical switching based on quantum logic. Nano-optics addresses the broad spectrum of optics on the nanometer scale covering technology and basic science. On the technological side, we find topics like nanolithography and high density optical data storage. On the basic sciences end, we might mention atom-photon interactions in the optical near-field and their potential applications for atom trapping and manipulation experiments. Compared with free propagating light the optical near-field is enriched by so-called virtual photons. These are the same sort of particles responsible for molecular binding (van der Waals/ Casimir forces) and are therefore promising for selective probing of atomic structures. The consideration of virtual photons in the field of quantum optics will enlarge the range of fundamental experiments and applications.

Nano-optics deals with the controlled manipulation of evanescent optical fields and their exploitation in investigations of nanometer scale structures. The basic idea is to detect the evanescent field distribution, which is significant in the near-field region that extends to only a few tens of nanometers from the sample surface. The detection of the evanescent waves is typically done with a nanoprobe. The probe acts as a local scattering center transforming some of the evanescent field components into homogeneous (propagating) waves that can be detected in the far zone. By scanning the nanoprobe over the sample surface, the evanescent field distribution can be detected with a resolution limited mainly by the size of the nanoprobe and the scanning height. In practice, the resolution achieved in scanning near-field optical microscopy (SNOM) is on the order of a few tens of nanometers, i.e., the technique offers an order of magnitude improvement to the best conventional far-zone measurements. Our research has concentrated mainly on three specific areas. The main efforts have been focused on developing the shear-force feedback system of our SNOM apparatus, on the near-field investigation of a broad-area semiconductor laser, and on the theoretical modeling of the near-field-matter interaction. Insight into the origin of the high spatial resolution in a SNOM measurement may be gained by investigating the contributions of the homogeneous and evanescent electromagnetic waves separately in some selected geometries. The simplest model for the sample-probe system in SNOM can be obtained by considering the probe tip and the sample structure as being composed of a collection of point-dipoles. Of fundamental importance in this approach is to understand how the electromagnetic field of a single dipole is decomposed into its evanescent and homogeneous parts. Knowing this one can then proceed to generalize these results to systems containing several point-dipoles by including the couplings between the dipoles in a self-consistent calculation.

The point-dipole model has been widely employed as an aid in discussing the properties of near-field microscopy. Caution should, however, be exercised when interpreting the results. The point dipole does not offer an adequate model for describing mesoscopic structures in near-field investigations, since strong evanescent field gradients exist close to the sources. By applying the point-dipole model, however, we can investigate how the nanoprobe affects the near field and how the resolution depends on the scanning height of the probe over the sample in the different operation modes of SNOM. Furthermore, we can study the effects of different polarization directions of the nanoprobe and the external polarizing field on the resolution of a SNOM measurement. In particular, we can see the central role of the evanescent field components in breaking the classical diffraction barrier. The homogeneous, evanescent and total field distribution at the probe site when a z-polarized nanoprobe is scanned at the height of 40 nm over the sample. The probe dipole locally illuminates the sample and the same probe is also used to collect the scattered light. As a result of the local interaction between the probe and the sample dipoles the scanned homogeneous field distribution now reveals the same information about the sub-wavelength scale structures of the sample as the evanescent field, although in lower magnitude. Nanooptics, i.e. optical spectroscopy with a spatial resolution of <100 nm, opens up fascinating insights into the physical processes occuring on a nanometer scale. In our group, we apply this technique to investigate *single* semiconductor quantum dots, sometimes refered to as 'artificial atoms' because of their d-like density of states.

"Embossing" is an ancient technical term in English traceable to the period 1350 to 1400, and used by Chaucer in its present sense. The technical meaning, creating a raised design or pattern by pressing a die having the raised negative image of the design against the material to be patterned, has not changed during the ensuing 650 years. What has changed are the applications, materials used, and the scale of the design. Today's embossed patterns may have features measured in nanometers, bringing new capabilities to integrated optical systems. Nanoembossing, often referred to as nano-imprint lithography (NIL), can rapidly pattern relatively large area substrates with feature sizes below 100 nm. Creating patterns at this scale previously required far slower and more costly techniques such as electron-beam writing. NIL, used in combination with related technologies, can create an entire wafer of integrated optical structures with nanometer features in minutes. An array of replicated lenses formed on the surface of a 4-in. silicon wafer. Wafer-level NIL promises reductions in the size, cost and complexity of integrated optics, while improving performance. Integrating embossed components on a wafer offers major size reductions. Producing truly "monolithic" optical systems could reduce centimeter-sized systems to millimeter-sized die. Optical performance is improved by locating components physically closer together, reducing the optical path length and the attendant losses. For example, optical elements such as lenses can be fabricated directly onto optical sources, such as lasers.

Surprising to the non-expert, the small physical size of the nano-structures also creates optical elements that differ significantly in some characteristics from normal macro-optics. Optical elements with feature sizes smaller than the wavelength of the incident light, sometimes referred to as sub-wavelength elements, interact differently with the light than optical elements with larger feature sizes. More specifically, light interacts in different ways with periodic structures having spacing smaller than its wavelength. The altered interactions create a variety of devices with characteristics that differ from similar but larger structures. The sub-wavelength optical elements offer some new, desirable characteristics. Sub-wavelength grating structures, for example, may have a wider acceptance angle than conventional gratings. Wider acceptance angles relieve manufacturing tolerances and may permit automated assembly, further reducing costs. Sub-wavelength gratings also perform more consistently over a broader range of wavelengths than conventional gratings. NIL has demonstrated a growing variety of nano-optical components, including gratings, filters, anti-reflective coatings, refractive and diffractive components, fresnel lenses, waveguides, and lens arrays. With all of these potential products and proven benefits, why has nano-optics languished so long in the laboratories? Because formidable difficulties must be surmounted to replicate these tiny dimensions in volume manufacturing processes.

To understand the manufacturing challenges that are now being overcome, look at the basic processing techniques that have been developed. NIL in its various forms has five processing steps:

- Design and fabricate a master stamp with features patterned by high-resolution techniques.
- Impress the master stamp on a layer of suitably prepared material.
- Maintain the master stamp pressure while the material first flows to duplicate the stamp, and then solidifies to preserve the duplicate pattern.
- Separate the patterned replica from the master stamp, without damaging nanoscale features.
- Post-process the replica further for its desired optical function.

The starting point for NIL is the creation of a master stamp or mold. The ultimate resolution capability of the embossing depends solely on the master stamp. The design and fabrication of the master stamp is critical to minimum feature size, uniformity, fluid flow, and ease of separation. Designing a stamp to allow quick and complete fluid flow of the softened material is challenging. Once the design is satisfactory, it must be transferred to the master. Fine-line approaches, such as electron beam, are used to create nanoscale features on the master stamp. Electron beams have fine-line capability, but direct e-beam writing has a working area only a few millimeters square, and is relatively slow. Stamps used for hot embossing or for injection molding are commonly made from nickel or steel. Stamps for UV embossing must be of UV-transparent materials. While the completed master stamp could be directly used to produce many replicas, wear or deterioration through repeated use would require creating a new master stamp. In practice, to preserve its pristine quality, the completed master stamp generally is cloned, using the replication techniques described here. Cloned sub-masters are used to produce the nanoscale features in volume production, either simultaneously over a whole substrate, or repetitively, at chip size, onto a substrate. The substrates may be silicon, quartz, glass, or other materials. Replication layers are generally coatings of heat-curable or UV curable polymers, with carefully controlled coating uniformity and thickness. While polymers are common materials for embossing, they may not offer the dimensional precision or the long-term stability under challenging environmental conditions demanded of some system components. This need has been filled by the development of special materials, such as of sol-gel organic/inorganic composite materials. These composites can be UV cured or thermally cured like polymers, and after curing display optical and mechanical properties close to those of glass. Embossing and curing the pattern on the substrate varies with different processes. Once the coating is cured, the stamp must be separated from the cured material, without damaging the replicated pattern. As feature size becomes smaller, clean separations are more difficult. A special coating may be applied to the stamp before molding, to facilitate separation.

The embossed wafer may be post-processed by etching, deposition, or by repeating further lithographic processes. The cured patterned material can be the etch resist, for reactive ion etching of the underlying substrate. Alternatively, the patterned material may be used as a resist for metal deposition and lift-off. Post-processing might also include additional lithography, further depositions, and the addition of antireflective or protective coatings. Hot embossing of wafers uses the cloned master die as the wafer stamp. Wafers are first coated with a thermoplastic polymer, that is, a polymer that softens when heated. The wafer is positioned in a controlled-atmosphere stamping chamber. After heating the wafer, the stamp is brought down with pre-determined pressure and temperature to soften the polymer and simultaneously form the image on the whole wafer. After allowing time for polymer flow during heating, the stamp and wafer are cooled to harden the image. To maintain the pattern, pressure must continue to be applied until both stamp and substrate are cooled below the thermoplastic softening point. The hot-embossed wafer may be used for its embossed features, or it may be post-processed as a resist for etching, material deposition, and lift-off patterning. For resist use, curing may be followed by a light etch to remove the remaining traces of materials from the clear areas.

An alternative hot embossing approach is step-and-repeat stamping, using a smaller than wafer size master stamp. Step-and-stamp imitates the action of step-and-repeat optical steppers used in semiconductor fabrication. The smaller master stamp is simpler to design, produce, and control. It is also less costly than a full wafer stamp. The step-and-repeat approach mitigates the dimensional, planarity, flow, and separation anxieties inherent in wafer embossing, by literally reducing the problem. The tradeoffs include the added time required for multiple repetitions of the stamping process, the need to confine heating to the stamp area, and the cost and complexity of a precision step-and-repeat apparatus.

Cold embossing uses UV light to polymerize and solidify materials after forming by the stamp. The starting materials are thin films of UV-curable materials. Typical materials may be monomers, or inorganic-organic hybrid polymers, which are polymerized by UV to polymers with desirable optical properties. The films may be deposited onto silicon, gallium arsenide, or similar wafers. The UV-transmissive patterned master template is pressed against the film, imprinting the pattern, which is UV cured before pressure is released. UV embossing allows accurately aligning and overlaying multiple levels of features. UV embossing has a clear advantage over hot embossing in allowing multi-level applications, since the first level is polymerized after exposure, making it irrevocably formed and dimensionally stable, regardless of subsequent layer applications and exposures. UV embossing is a low pressure, room temperature process, with little shrinkage of the pattern. It also allows partial UV casting, by limiting the illumination to maintain clear areas on the wafer. Cold embossing may be done on one or both sides of the wafer. Double-sided embossing allows micromolding. The laser lens array was formed by cold embossing. As with hot embossing, UV embossing may be used with a single wafer size stamp, or with a smaller stamp in a "step and flash" procedure similar to the "step and stamp" described above.

Injection molding of optical components offers throughput and cost advantages for high-volume production. Injection molding can produce 3-D objects and integrated multicomponent systems. To move beyond conventional injection molding and capture nanofeatures, electroformed metal inserts are nanopatterned and included as part of the master mold. Double-sided injection molding at wafer scale has been used to create optical structures on both sides of a supporting substrate. These structures may be the same on both sides of the substrate, or they may be different; for example, combining a diffractive lens on one side with a refractive lens on the other. The two sides may be processed separately, or simultaneously. In either case, precise alignment must be made between the features through the intervening substrate. Injection molding has been used to produce optical components in hundreds of thousands. However, the high tooling cost can only be justified for these large quantities. Injection molding draws upon a wider range of materials than direct embossing. Acrylics or polycarbonates may be used for all-plastic systems in low-cost, high-volume injection molding, producing inexpensive commercial products in the hundreds of thousands.

A summary comparison of the three embossing processes. Hot embossing, at wafer level or by step-and-stamp, lends itself to single side patterning on thermoplastic materials. Process temperatures are typically 90°C above the glassivation temperature. Throughput is limited by the time required to heat, hold, and cool the die and substrate. Step and stamp avoids the high cost and tolerances of a wafer-size stamp, at a penalty in multiplying the time delays. Hot embossing has some well established and perhaps more profitable commercial applications. Music CDs are commonly replicated in volume by embossing. Roll-to-roll hot embossing has been used to add interesting color effects to wrapping paper, and to produce material with difficult to counterfeit optical patterns for security documents, such as identification cards.

Cold (UV) embossing may be preferred for critically dimensioned materials. It can be a room temperature process, eliminating the heating and cooling times and thermal expansions/contractions of hot embossing. It allows single or repeated multiple patterns to be layered on one or both sides of a substrate with accurate alignments. Both sides of the substrate may be embossed with identical or different patterns. For applications with more critical planarity, dimensional tolerances, or environmental requirements, cold embossing of photosensitive organic-inorganic sol-gel materials gives results close to those of glass.

Injection molding produces 3-D molding, allowing entire systems to be molded as a unit. It is not a room temperature process; processing time and temperature depend on the materials used. These may be inexpensive commercial plastics, or highly accurate specialty materials. Patterning can be single- or double-sided, in 3-D, although not generally multilayered as in cold embossing. The high tooling cost makes injection molding practical for high-volume (e.g. 100,000 units and up) products.

A mixture of techniques and equipment from the semiconductor and micropackaging industries has been adapted to nano-optics fabrication. Much of the development work reported above used one-of-a-kind laboratory equipment, built for that purpose, or modified equipment, originally developed for other purposes. One report referred to a modified semiconductor mask aligner. Another recent paper reported demonstrating hot embossing using a standard flip chip aligner-bonder. One nano-optics manufacturer has developed customized robotic embossing systems specifically for volume production. Only recently have major equipment manufacturers announced production equipment intended to meet the challenges of nanometer optics. Dedicated aligner-steppers with sub-20-nm embossing resolution are now available. Product applications based upon NIL include wavelength monitors, beam splitters, pattern generators, polarization filters, switches, and many others.

Also commercially available now are replicated components that are high performance building blocks for fabricating optical assemblies. These offer manufacturing advantages in comprising a common family of similar products, allowing standardization of manufacturing operations and equipment. A typical component size. Embossing has come a long way since Chaucer. After decades of laboratory development, nanoembossing promises to open new optical frontiers with truly monolithic production systems. The growing commercialization of nano-optic systems will subject them to the scrutiny and judgment of the marketplace. Success in surmounting the manufacturing challenges will determine whether nanoembossed optical systems remain a technology niche, or become a broad new technology opportunity.

Near-Field Optical Microscopy

Near field scanning optical microscopy (NSOM) was developed in the mid 1980s as a means to break the diffraction limit on spatial resolution attainable with optical measurement. Traditional NSOM utilizes a tapered, metal-coated optical fiber with a small aperture as either an excitation source or collection device. The spatial resolution attainable with this technique is roughly defined by the size of the aperture itself. Unfortunately, rough metallic coatings on glass fibers, coupled with the finite skin depth of metals at optical frequencies, place a limit on the spatial resolution attainable with fiber probes. Typical spatial resolution for the fiber approach to NSOM is ~50-100 nm, though careful probe and sample preparation have yielded ~30 nm resolution. Research groups have long been interested in mapping out the spatial distribution of the proteins in photosynthetic membranes. These proteins are approximately 10 nm in diameter, and electron microscopy indicates that they are closley packed in their native configuration. The density of this arrangement precludes optical mapping with either far-field measurement or traditional aperture NSOM.

In the mid-1990's, a number of groups began working on what would prove to be the solution to the fundamental resolution limitations in aperture NSOM. The idea was to spatially map an optical signal from the surface of a sample by scattering that signal with a solid probe. The choice of probe ranged from tungsten STM probes to metallized AFM tips to silicon probes. While optical image interpretation was complicated by topographic coupling, the spatial resolution in the apertureless technique was improved to ~10's of nm. In 1999 the Xie group began work on a type of apertureless NSOM that takes advantage of the phenomenon of field enhancement by noble metal probes. With the appropriate combination of excitation polarization and probe geometry, a gold or silver probe is capable of strongly enhancing the magnitude of the incident illumination. This strong, localized field can be used as an excitation source for a spectroscopic transition of interest, such as fluorescence or Raman scattering. By locally exciting a spectroscopic transition in the sample and detecting the Stokes shifted light in the far field, the interpretation of apertureless NSOM images is greatly simplified. They utilize the nonlinear response of the sample by exciting two-photon fluorescence in the sample. This tip enhanced nonlinear optical microscopy (TENOM) amplifies the contrast. The enhancements achievable with specifically fabricated gold probes are sufficient to yield near-field two photon fluorescence signal above the background generated in the diffraction-limited far-field region. Furthermore, TENOM has improved spatial resolution to the sub-20 nm regime. This high resolution and strong signal generation should be capable of, for example, mapping the spatial distribution of PSI and PSII in the thylakoid membrane. While TENOM has presented a solution to issues of high resolution optical imaging, it does have several intrinsic drawbacks. These drawbacks include the quenching of non-energy-transferred fluorescence, the necessity to carefully tailor the probe geometry to yield high field enhancement, and the practical requirement of a nonlinear signal source.

Currently exploration is on to get a variety of approaches to eliminate the few drawbacks of TENOM and to expand the capabilities of apertureless NSOM in general. The first concern has been to address the problem of fluorescence quenching by the metal probe. To this end, an ion- and electron-beam assited deposition (IBAD and EBAD, respectively) of dielectrics has been applied to the apertureless probes. The growth of glass with these techniques allows us to build a spacer layer between the probe and the sample, reducing quenching while still maintaining high field enhancement at the sample. The theoretical treatment of fluorescence quenching by probes of complex geometries is currently being explored through the use of finite difference time domain (FDTD) simulation and Poynting vector integration. Due to the complex nature of light-metal interactions, electromagnetic simulation is vital in the TENOM and apertureless NSOM research. The FDTD technique is a flexible, powerful method for fully three dimensional simulation of the response of metal probes to incident radiation. The power of the FDTD technique lies largely in its conceptual simplicity. A geometry of interest is defined by a number of cells, called Yee cells. The complex permittivities of each cell are assigned. After a geometry has been defined, an incident optical field is simulated, and the Maxwell's equations are integrated in a leapfrog fashion. The result is a method that can calculate the resultant fields for any arbitrary geometry illuminated with optical sources. FDTD has been applied to the study of NSOM in the past. Unfortunately, the large computational burden associated with this technique made three dimensional simulations quite difficult. Two dimensional calculations have generated insight into the aperture method of NSOM. In this research a commercial FDTD package (RemCom's XFDTD) was utilized to simulate apertureless probes in three dimensions, working to adapt this microwave-focused package to the complex permittivities intrinsic to the optical regime behavior of real metals. This commercial software, coupled with the recent tremendous improvement in computational power, has provided a general method for simulating complex geometries, such as groups of spheres. In order to expand apertureless NSOM to linear signal sources, intensity enhancements on the order of 10,000 were required in these experiments. In order to develop a physical intuition for the nature of strong field enhancement, and to rationally design probes for high field enhancement, FDTD was used to model apertureless NSOM. The simulations indicated that probes with finite size on the scale of the excitation light were preferable to the "quasi-infinite" tips used in the past. Furthermore, by properly tuning the size of a right trigonal pyramid to the excitation light, field enhancements sufficient for exciting linear signal sources and yielding signal above the far field background were predicted. The simulated high field enhancement has been experimentally verified. A focused ion beam instrument was used to make apertureless NSOM probes. This instrument (FEI Strata DB-235) allowed to mill materials with ~10 nm resolution, image with ~5 nm resolution, and deposit glass or platinum with ~20 nm resolution. This powerful tool for nano-optics and near field microscopy applications, including surface enhanced Raman scattering (SERS) and chemical sensors was utilized.

Electromagnetic coherence theory for optical near fields refers to the electromagnetic field at the immediate vicinity of the sample where the evanescent or non-propagating waves are strong. The near field is a primary example of a highly non-paraxial field and, in general, it is composed of three orthogonal electric field components. The conventional tools of electromagnetic coherence theory are, however, valid mainly for beam-like fields. We have put forward the definitions for the degree of polarization of general three-dimensional fields and for the electromagnetic degree of coherence. We have shown that in the limit of complete coherence, the electromagnetic coherence tensors, both in the space-time and space-frequency domains, behave analogously to the corresponding scalar functions. Furthermore, we have rigorously formulated the electromagnetic theory of coherent modes, and constructed the mode expansion for a class of statistically homogeneous and isotropic electromagnetic fields. Currently, we are investigating the optical coherence in a near-field detection system, and tightly focused electromagnetic fields.

We study the state of partial polarization of tightly focused optical fields. In the vicinity of such a focus the electromagnetic field displays a complicated three dimensional structure. For example, the light is fully polarized on concentric rings around the optical axis even when the incident light is unpolarized in the 2D sense. The distribution of the degree of polarization in the image plane of an imaging system of high numerical aperture. The rings of full polarization are clearly resolved in the image.

We have modeled the electromagnetic field in various nanostructures using a boundary-integral method in two dimensions. In the boundary-integral method the electromagnetic scattering problem is casted into a form of an integral equation involving integrations only over the boundaries of the system. We have applied the method to analyze changes in the degree of polarization in light transmission through a tip aperture of a scanning near-field optical microscope. Furthermore, we have carried out spectral analysis of light transmission through sub-wavelength slits in metal films. We study experimentally the detection and spectroscopy of very small gold nanoparticles. Such particles are an interesting model system to study the transition from individual atoms to the optical properties of bulk matter. Nanoparticles also have important applications in chemical sensing and as labels in biology. In our first experiments we used supercontinuum light produced in a photonic crystal fibre to detect sub 10-nm gold particles. A confocal microscope image of 15 nm gold particles. The cross section shows that the supercontinuum light can be focused to almost a diffraction limited spot. The spectrum of the light scattered by an individual particle. The scattering spectrum displays a clear resonance which can be used to distinguish gold nanoparticles from other scatterers in a biological medium. This work was performed in the nano-optics group of professor Vahid Sandoghdar at ETH Zürich. We are currently continuing this work at HUT with experiments on the absorption spectroscopy of individual metallic nanoparticles.

Single walled carbon nanotubes (SWNTs) are tubular graphitic molecules that can be microns long and less than one nanometer wide. The combination of the unique properties of the carbon–carbon bond and their interesting molecular structure have given SWNTs unique physical properties and provided the potential for revolutionary applications. SWNT structure is characterized by two integers (n,m) that can be easily related to both its diameter and chirality. SWNTs have an unusual electronic configuration: each SWNT with a distinct (n,m), where n–m divisible by three is metallic, while the rest are semiconducting Further, SWNT electronic properties exhibit a strong structural dependence which helps drive potential applications. Simple one electron models indicate that the density of states (DOS) of a nanotube is characterized by singularities, with each singularity corresponding to a quantum subband. Recent studies of individual SWNT electronic structure using scanning tunneling spectroscopy (STS) agree with these general predictions. Simple one electron models also imply that the dipole-allowed optical transitions occur between mirror image peaks in the DOS. Absorption spectra of SWNT bundles contain three broad absorption regions corresponding to the first and second allowed transitions for semiconductor SWNTs, and the first allowed transition for metallic SWNTs, respectively. However, these spectra do not exhibit any features identified with a particular nanotube structure, since strong electronic coupling mixes the energy states from the different SWNTs in the bundle. This strong intertube coupling, combined with the large inhomogeneity in nanotube structures present in an ensemble, completely obscures any fine structure in the absorption spectra. Isolation of individual nanotubes is critical for decoupling the optical response of one nanotube from another, and was recently demonstrated by O'Connell et al. for nanotubes manufactured by the high-pressure CO chemical vapor deposition (HiPCo) process (H-SWNTs). Absorption spectra of an ensemble of isolated SWNTs (Mic-SWNTs) show many more highly resolved features than the corresponding corresponding absorption spectrum from nanotube bundles. These distinct features correspond to dipole-allowed optical transitions from SWNTs with different (n,m) values and/or from higher interband transitions for a nanotube with a given (n,m) value. The general isolation procedure can be applied using numerous surfactants, as well as nanotubes produced by various methods such as the laser oven (L-SWNTs), and arc discharge (A-SWNTs) processes. Bundled SWNTs do not fluoresce since any metallic tubes present in the bundle provide an efficient nonradiative pathway for photoexcited electrons. However, for HiPCo SWNTs, Mic-SWNT samples show strong, dipole allowed fluorescence at the bandedge. The measured Stokes shift between the absorbing and emitting lines is quite small (_40 cm_1). SWNTs are poor emitters; the fluorescence quantum yield (QY) of SWNTs is on the order of QY _10_4. The dependence of the SWNT electronic structure on diameter and chirality [(n,m) values] is fundamentally important and is also of technical interest. Simple one electron models predict that the energy of the dipole-allowed optical transitions depends linearly on the nanotube diameter. However, the observation of structurally dependent photoluminescence from SWNTs represented a major breakthrough, as the dependence of electronic transition energies was shown to depend on (n,m) values explicitly and not simply on the SWNT diameter. In particular, the magnitude and sign of deviations from the predicted linear dependence were observed to depend on the value of (n-m) mod 3, suggesting that trigonal warping effects were playing a significant role. Empirical fitting parameters have been introduced modifying the simple one-electron picture to compensate for these effects, thus allowing for a precise prediction of the transition energies for different tube (n,m) structures. However, despite the observation of fluorescence spectral features attributable to specific nanotube (n,m) structures, it is not fully understood why the transition frequencies differ significantly from simple theoretical predictions, although many-particle interactions and/or trigonal warping effects likely play a role in this discrepancy.

Notwithstanding these recent advances in SWNT separation and isolation, the optical spectra clearly suffer from severe inhomogeneous broadening due to the distribution of different nanotube structures in the ensemble. Thus, early optical studies could only determine an averaged overall behavior that obscured crucial electronic characteristics of the individual nanotube. For example, since spectra from SWNTs with different (n,m) values will overlap, it is difficult to make definite conclusions regarding important emission features such as the spectral line width or details of the band shape, or to observe evidence of coupling to vibrations. Therefore, new approaches are needed to further advance the understanding of SWNT optical properties and accelerate the de velopment of novel SWNT photonic applications, such as nanometer scale integrated electroluminescent devices.

Spin-casting the Mic-SWNT suspension onto a cover slip allows for isolation of individual nanotubes by standard confocal microscopy. A typical fluorescence image from the spun Mic-SWNT sample is shown, with a spatial resolution of 300 nm. The image was obtained by collecting an optical spectrum at every pixel and then binning regions of the individual spectra where strong fluorescence signals are expected according to the ensemble spectrum. Since atomic force microscopy (AFM) measurements reveal predominantly short SWNTs (length 200–300 nm) lying on top of residual surfactant patches, distinct tubular features in the Mic-SWNT fluorescence images are typically unresolved. Alternatively, by dispersing SWNTs directly in organic solvents such as dichloroethylene, followed by spin coating, occasionally individual nanotubes that fluoresce can be isolated, both the fluorescence and Raman images clearly show the characteristic one-dimensional features of the SWNT thus proving that the fluorescence in fact results from the single nanotube. Fluorescence images from the isolated Mic-SWNTs show distinct, bright spots at different positions for different emission wavelengths. Displays representative spectra detected from one of these bright spots. Each spectrum now exhibits a single emission band with a smooth line shape; all traces of contributions to the fluorescence spectra from the different SWNT (n,m) structures have been eliminated. The fluorescence maximum for all three spectra matches three transitions observed in the ensemble spectrum. In so much as the ensemble fluorescence spectrum is a weighted superposition of spectra from individual SWNTs. Single nanotube spectroscopy clearly eliminates the overlapping spectral features associated with ensemble averaging and thus allows for the potential discovery of crucial SWNT optical and electronic properties. For example, SWNT fluorescence spectra are described by a single, featureless, symmetrical band corresponding to a single Lorentzian line shape function

The fluorescence from individual semiconductor quantum dots and from small molecules exhibit an emission intermittency or on-off "blinking" behavior for all excitation intensities, all temperatures, and on time-scales spanning many orders of magnitude. In fact, fluorescence intermittency is commonly used as the conclusive indicator of single chromophore optical emission. Surprisingly, the fluorescence intensity from a single SWNT at 300 K recorded using moderate excitation intensities (<70 kWcm_2) is constant over many seconds with no indication of fluctuations on the 40-ms to 100-second time scale. At extremely high excitation powers (>100 kWcm_2), fluorescence intensity fluctuations can be observed, likely due to laser-induced sample heating. The use of single-photon sources is extremely desirable for applications in quantum optics, such as quantum cryptography, since the presence of even two photons in a signal pulse is an opportunity for an eavesdropper to steal or manipulate information. Thus, single molecules or single nanoparticles are attractive as single-photon "on demand" sources in quantum optics. However, their unavoidable fluorescence intermittency drastically hampers their potential utility. For example, in quantum cryptography it is difficult for the receiver to determine whether an "off" period is related to the information being transferred, or just the result of a random blinking event. Some stable sources of single photons do exist, such as nitrogen vacancies in diamond nanoparticles. However, these sources have a very broad linewidth typically over 100 nm, thus limiting the bandwidth of any potential devices.

SWNTs have many advantages for possible use as single-photon sources for quantum optics applications. On the millisecond to minutes time scale, and at room temperature, SWNT fluorescence is stable showing no signs of intensity or spectral blinking. SWNTs also have an extremely narrow linewidth, and can be pumped electrically which allows for the exciting possibility of future nanometer scale integrated single-photon photonic devices. SWNTs also are resistant to photobleaching, as their fluorescence intensity does not change despite hours of intense photoillumination. Finally, unlike dye molecules or impurities in diamond, SWNTs have an easily size-tunable emission that is in the infrared at technologically relevant wavelengths (1.3–1.5 mm). Single molecule sources are also highly desirable for applications in biology and biotechnology. For example, the use of single fluorophores to track individual molecules inside a cell is a compelling method for studying intracellular processes. However, photobleaching of organic dye labels severely restricts the single molecule processes that can be studied in cells. Detection of a single analyte as reported through changes in the optical emission of a single fluorophore is also a promising approach to achieving the ultimate in detection sensitivity. On the other hand the fluorescence intermittency of single molecules makes the fluorometric detection of single analytes extremely difficult, since the modulation of fluorescence upon binding of the analyte cannot be distinguished from a random blinking event. SWNTs hold great promise for possible use as single-molecule fluorophores. First, SWNT fluorescence does not photobleach at 300 K even after hours of continuous laser excitation. Thus, extremely long-lived biological processes on the singlemolecule level can be followed much more easily. Second, the fact that SWNTs do not exhibit fluorescence blinking allows them to be used as fluorometric reporters of a single biological agent. SWNT photonic devices show strong near-infrared (NIR) electroluminescence, allowing for easy "on-chip" integration of single

SWNT sensors with silicon technology. Finally, SWNTs are optically active at wavelengths in the range of 800–1100 nm, which allows for NIR excitation and detection. This NIR optical activity has the tremendous advantage that background fluorescence from the sample can be dramatically reduced allowing for much improved sensitivity on the single fluorophore level. For some nanoparticles the processes that contribute to fluorescence blinking events slow down as the temperature is lowered. For example, in CdSe semiconductor nanocrystals a higher fraction of nanoparticles are "on" (or bright) at low temperatures versus at room temperature. Interestingly, at low temperatures (T=1.8 K) a significant fraction (about 2/3) of SWNTs show an intermittency in their fluorescence intensity, and also a spectral wandering of their fluorescence energy as shown. Spectral diffusion cannot be observed at room temperature, since the fluorescence linewidth is much larger than the change in energy during a spectral jump. However, it is currently unclear whether the processes that cause blinking "turn on" at low temperatures, or whether SWNTs blink inherently on timescales much faster than 10 ms. With the time resolution of past experiments, fluctuations in the sub-millisecond regime were not observable.

Combined fluorescence and Raman scattering measurements on the single molecule level open up new ways to characterize SWNTs. Single nanotube studies enable the connection between fluorescence energy, (n,m) structural parameters, and RBM frequency of a particular SWNT with a single measurement, and without any complicated fitting procedures necessitated by working with macroscopic samples. Single nanotube studies also reveal new insights into nanotube optical properties, such as a variation of the fluorescence energies of nanotubes having the same structure. Unlike for single molecules, for SWNTs the photoluminescence unexpectedly does not show any intensity or spectral fluctuations at room temperature. The fact that SWNTs show no emission intensity blinking or bleaching demonstrates that SWNTs have the po tential to provide a stable, single molecule infrared photon source, with an extremely narrow linewidth, which would have a significant impact for future nanometer scale integrated photonic devices and which shows promise for applications in quantum optics and biological sensing.

Optical Processing

A new class of devices is smashing the size, manufacturing, cost and reliability hang-ups that have hampered optical elements to date. The commercial introduction of wafer-based, nanofabrication techniques now makes it possible to create a new class of optical components called subwavelength optical elements (SOEs). With physical structures far smaller than the wavelength of light, the physics of the interaction of these fine-scale surface structures with light yields new arrangements of optical-processing functions. These arrangements have greater density, more robust performance and greater levels of integration when compared with many existing technologies and could fundamentally alter approaches to optical system design. However, as with any new technology, the designer must clearly understand the general concepts underlying the technology as well as its potential impact (in this case, on optical component design), as well as implications for the evolution of the technology.

Integrated optical component and module design has achieved a high degree of sophistication in the combination of discrete optical elements. The properties of optical materials and their combinations, clever optics design, and advances in alignment and assembly are applied to increase component and module density, reduce costs and increase reliability. But there are limits. The natural properties of many discrete optical elements are relatively fixed, limiting the flexibility of the optical component designer and reducing transferability from design to design. This is where SOEs come in. SOEs are the realization of nanotechnology as applied to optical elements. As such, they offer excellent optical properties while being easy to integrate with other optical materials in a broad range of configurations. In addition, they self-integrate, allowing the combination of optical functions to reduce parts count and increase reliability, and to allow increased flexibility in optical component design. A broad range of optical effects can be achieved by the manipulation of nanostructures in the optical beam path.

The interaction of light with subwavelength grating structures-gratings with one or more dimensions one or more orders of magnitude smaller than the wavelength of the incident light-produces a range of controllable effects. To explain those effects, a simple subwavelength grating structure. When light is normally incident on the structure, the transmitted portion of the light-that which passes through the grating structure-is affected. The effect on the passed light can be varied by adjusting the size, shape and spacing of the grating structure. In particular, for the one-dimensional grating structure, appropriate selection of dimensions will result in behavior as a polarizer, waveplate or polarization-dependent filter. More sophisticated effects, such as polarization-independent filters, can be achieved by two-dimensional gratings. The general class of these structures is referred to as SOEs. Because the diameter of the incident light beam is often much larger than the structures of the grating, the effect on the transmitted light is really the statistical sum of numerous local interactions between the light and the grating. For example, if a light beam measuring 300 microns in diameter is incident on a two-dimensional grating such as that shown, where the structure size is roughly 100 x 100 nanometers, more than a million nanostructures will be illuminated. Therefore, varying the grating dimensions spatially across the incident beam front allows additional control over optical processing effects. The physics of the SOE depends on rigorous application of the boundary conditions of Maxwell's equations to describe the interaction of light with the structures. At the wavelengths of light used in telecommunications-980 through 1,800 nm-the structures required to achieve those effects have some dimensions on the order of tens to a few hundred nanometers. At the lower end of the scale, single electron or quantum effects may also be observed. Whereas the principles of reflection, refraction, diffraction and interference describe the behavior of traditional optical elements, with SOEs the equations describing conventional optical behavior do not fully cover the resultant phenomena because quantum mechanical effects now come into play.

In many applications, subwavelength structures act as a nanoscale diffraction grating whose interaction with incident light can be modeled by rigorous application of diffraction-grating theory and the abovementioned boundary conditions of Maxwell's equations. Consider refraction, a key property exploited in optical components. Traditionally, different materials must be used to obtain different refractive indices. In SOEs, however, different refractive indices can be obtained with the same material by adjusting the physical structure. As an example, SOE structures can be used to create an "artificial" birefringence effect. If a is the grating period and t is the grating width. With that sort of flexibility, why are SOEs just now becoming available for telecommunications applications? The primary reason has been manufacturability. While these optical effects have been researched for at least 20 years, cost-effective manufacturing of the optical elements has not been available. Building subwavelength grating structures in a research environment has generally required high-energy techniques, such as electron-beam lithography, or extremely precise process control, for instance, to grow nanostructures via "self-alignment." Second, there must be a consistent process for creating a broad range of nanostructure patterns. Many techniques for creating nanostructure patterns are limited in the range of possible patterns. Nanoimprint lithography overcomes those limits and has four key steps: making a mold inscribed with the complement of the desired nanostructure, impressing this mold into a resist-coated wafer, separating the mold and selectively removing the resist with reactive ion etching to transfer the nanopattern to the target material. After that, postimprint processing may be used to add metal layers to enhance performance and to provide a protective overcoat for handling in a standard manufacturing environment. Testing and dicing follow. Because nanoimprint lithography forms the nanostructures of SOEs by a direct, physical process rather than by beams of energy, the wave diffraction, scattering and interference in the resist do not limit fabrication resolution. Silicon dioxide is commonly used to make a mold patterned with the desired nanostructures. Various techniques, including electron beam lithography, can be used to create the desired negative image of a nanostructure. Because the mold can be replicated and reused, complex multistep and multiprocess methods can be used to create the desired nanostructure. Since those initial tooling steps do not need to be repeated for each production wafer, they are amortized over the full production run of a particular SOE. Using different molds with different nanostructure patterns allows the same manufacturing process to create the full range of SOEs. Unpackaged for use in free-space applications, the final SOE is an optical chip with a subwavelength grating on one side of a substrate. The overall thickness of the element is dependent on the thickness of the substrate. The (x,y) dimensions depend on application requirements.

The practical consequences of using SOEs can be explored by looking at a specific realization—an SOE polarization beam splitter/combiner, or PBS/C. By appropriate selection of the dimensions of a one-dimensional subwavelength grating structure, one polarization will be transmitted, while the orthogonal polarization will be reflected, as illustrated. At that scale, the subwavelength-diffraction grating exhibits only zero-order diffraction, resulting in a number of useful element characteristics. These include consistent performance across a broad range of wavelengths, the grating structure performs the same for wavelengths from 980 nm to 1,800 nm, and wide variations in the angle of incidence of the light beam that are up to 20 degrees of deviation from normal. Typical performance parameters for an SOE PBS reveal insertion loss of less than 0.13 dB on both the reflected and the transmitted beams. Extinction ratios are greater than 40 dB and 20 dB for the transmitted and reflected beams, respectively. The flexibility of SOEs allows trade-offs around those parameters. The benefits of a PBS/C SOE can be realized in two areas. First, there are architectural benefits in optical component design. Because of their small form factor, they allow more compact component designs. Reduced insertion losses are also realized because SOEs can be abutted to other elements with minimum free-space separation. There are differences in the way that the elements handle light, for instance, the polarization beam splitter is a reflective device with 180 degrees of effective separation obtained in less than 1 micron. Because of those differences, the beam path can be laid out to simplify device design and possibly eliminate other optical elements. Further, the reflective nature of the SOE PBS lets it support high-energy applications associated with laser transmitters and fiber amplifiers. Second, there are advantages for manufacturing. The wide acceptance angle simplifies the alignment process, reducing manufacturing time and cost. It is also possible to use automated "pick-and-place" manufacturing technologies. The SOEs are robust: with proper selection of materials they can be subjected to temperature ranges of –200°C to 400°C. Consequently, they can be subjected to a broad range of manufacturing processes. Finally, reduced size may both simplify packaging and reduce packaging costs. Applications of the SOE PBS/C have been explored for fiber amplifiers, circulators and isolators, interleavers, optical switches and variable optical attenuators. Here, their size and low power consumption are attractive relative to traditional technologies.

Current devices are the initial applications of SOEs, but the possibilities are expanding in several directions. SOE-based building-block functionality will be introduced both as chips and as packaged devices. Demonstrated SOE functionality spans polarizers, polarization beam splitters/combiners, filters, photodetectors and photonic bandgap devices. Dynamic control for switching, attenuation and tuning is also possible. Monolithic integration of SOEs can be achieved by stacking SOE layers to create aggregated optical effects. Nanoimprint lithography permits direct layering on of SOEs without resorting to lamination techniques. Combining SOEs with optically active layers allows optical control circuits to be built, resulting in complex optical components "on-a-chip." Multilayer SOE integration has already been demonstrated by combining filters with photodetector arrays to create dynamic optical feedback loops. Because SOEs are are self-compatible and fabricated via wafer-scale manufacturing with relatively minor differences from device to device, the means exist to implement them.

Optoelectronic Space Applications

Diffractive optical elements, optoelectronic transducers and photonic components, which play an important role in optical data communication, can be substantially improved by lateral nanostructures. With the development of lateral optoelectronic nanostructures the way to controllable diffractive optics is paved. For this, elements with specific interference structures are necessary, which act as specific and possibly controllable transmission or reflection filters. Nanostructured optoelectronic components (e.g. quantum well or quantum dot lasers, photonic crystals) offer large market potentials in the future, e.g. for optical data communication or in the range of consumer electronics (for example laser television).

Nanostructured optoelectronic components offer promising application in space in the fields of optical satellite telecommunications or sensor technology (infrared sensors, high resolution charge-coupled devices, etc). With optical wireless data links (OWL) for intrasatellite communication as well as optical intersatellite links, significantly smaller and lighter devices a higher bandwidth could be realized in comparison to conventional microwave communications. Optical intersatellite links were demonstrated in the frame of the ARTEMIS mission of the ESA. For the data transmission, extreme frequency-stable solid state lasers (Nd:YAG lasers) are used, which are pumped with diode lasers.

The German company, Tesat, is a leading manufacturer of laser terminals for optical intersatellite communication. Such laser terminals are also interesting for scientific applications; for example, as an injection seeder for a satellite-based Doppler-Lidar (ALADDIN), as a satellite-based measuring device for gravitation wave detectors (LISA, with SMART as demonstration mission), or as a frequency normal for a satellite-based FT-spectrometer (POISON).

Semiconductor quantum dots, which have been manufactured in high quality by means of self-organization for about the last 5 years, offer a new degree of freedom in selecting the working wavelength of photonic elements. They allow manufacturers to cover almost completely the entire spectral region from the ultraviolet to the far infrared, with a small number of substrate materials. Further advantages of quantum dot lasers are a small energy consumption through low threshold current densities, a high modulation range for high-speed applications as well as an improved temperature stability. Beyond that, an improved radiation hardness has already been proven by quantum dot lasers compared with quantum well lasers. First commercial uses of quantum dot lasers are expected in 2003. Due to their radiation hardness and the low energy consumption, quantum dot lasers in principle are relevant for space applications, e.g. as pump lasers for solid state lasers, which are needed for different applications. In order to realize the potential of quantum dot lasers in space applications, appropriate measures have to be accomplished by the space industry for the specification, system integration and space qualification.

Photonic crystals are a further example of nano-optoelectronic components with application potential in optical data communication. Photonic crystals exhibit a periodic refractive index and possess an analogy to semiconductors in electronics, a "photonic band gap" for certain frequency in the visible and IR wavelength ranges. The lattice constant of photonic crystals lies in the range of half the wavelength of the light in the medium. For visible light this means that for the production of photonic crystals, a precision within the range of 10 nm is necessary. Two-dimensional structures today can be routinely manufactured with high precision.

At present, intensified efforts are made for the development of three-dimensional photonic crystals, e.g. with utilization of lithography and self-organization procedures, in which nanoscale colloids (e.g. from polymers or silicates) arrange spontaneously to a cubic lattice. These lattices are used as templates for lattices from more interesting materials such as metals and metal oxides. Three-dimensional photonic crystals would open up new possibilities in optical data communication (light could be guided and branched to arbitrary directions) and offer in principle the potential for the realization of purely optical circuits (optical computing). Such photonic transistors are however at present still very far from realization. In the long run, photonic crystals will find applications in optical satellite communications.

IR sensors offer a multiplicity of application potential in space, e.g. for the satellite-based earth observation and atmosphere research, for astronomy, as navigation aid for space systems or for optical data communication. Approaches for the miniaturization and further improvement of infrared sensors are based among other things on the application of two- (quantum well), one- (quantum wire) or zero-dimensional (quantum dot) nanostructures. With the help of quantum well or quantum dot structures the detection characteristics of IR sensors can be adjusted selectively to the relevant spectral region (band gap engineering). Quantum well IR sensors, based on GaAs are developed, for example, by the Center for Space Microelectronics Technology of NASA for special space applications. This quantum well infrared photodetector (QWIP) consists of a GaAs-layer, which is embedded, sandwich-like, in two $Al_xGa_{1-x}As$ layers. The characteristics of the quantum wells can be adjusted by varying the thickness of the GaAs layer and the composition of the barrier layer. By means of molecular beam epitaxy, nm-thick layers can be produced on large areas with atomic precision. Ga-As QWIP can also be used for long-wave IR radiation > 6 μm.

The Technical University of Munich pursues a different approach in the frame of a BMBF (Bundesministerium für Bildung und Forschung—Federal Ministry of Education and Research) joint project on the self organization of Si/Ge islands on silicon. The research activities are focused on controlling the characteristics of epitactical, defect free Si/Ge islands on silicon substrates, which can be produced with the self-organizing, parallel Stranski-Krastanow procedure in the material system Si/SiGe. The objective here is to develop coupled systems with several layers of quantum dots in a homogeneous layer system, which exhibit new functions by charge transfer and electrostatic coupling and can be used as optical detectors particularly in the mid IR (MIR) range. With combined QD/QW structures a 50-fold increased photoresponse can be obtained compared with sole quantum dot structures. Expected advantages of QD MIR (mid-infrared) detectors to be mentioned are the extended spectral range, a high durability and reproducibility, radiation hardness and a low dark current. For the year 2003, the realization of a prototype 2-colour IR detector in co-operation with DaimlerChrysler was done.

Integrated Optical Circuits

Nano-optics is a valuably novel class of technology that takes advantage of light's unique interaction with subwavelength, nanoscale patterned materials, and nanotechnology enabled fabrication methods to create a broadly applicable optical device and manufacturing platform. In optical circuits used in consumer electronics, communications, and security applications, nano-optic devices offer significant benefits by virtue of their novel behavior, small size, ease of integration, and lower overall costs. Through these virtues, nano-optics delivers compelling alternatives to conventional bulk optical solutions, and offers functionality and form factors that conventional optics cannot match. The following is a basic review of nano-optics theory and applications.

The unique optical properties of nanooptic devices derive from appropriately patterning material on the nanoscale with structures whose critical dimensions are sized to be several times less than the wavelength of light at which they operate, thereby creating optical materials with highly useful modified optical functions. Through a proper combination of materials and structures, nano-optic devices can perform any passive optical function including polarization filtering, phase retardation, spectral filtering, and propagation management (e.g., lenses and beam splitters). Various functions can be designed for both free-space- and waveguide- based applications. The former is far and away the most common in commercial applications. Nano-optic devices also can be designed to operate over any wavelength range. The basic methodology is applicable to UV, visible, and IR wavelengths with appropriate variations in structural dimensions and materials. Additionally, they can form the core of an optical system. Practical nano-optic applications require a complete optical system comprised of the nano-patterned material and adjacent materials including the optical substrate and thin-film coatings. The nano-optic structures are defined as a combination of material and physical attributes, as illustrated. Attributes can be divided in to three types:

- *Structural:* pattern (e.g., linear, mesh, and circular), dimensions (e.g., period, thickness, and duty cycle), and spatial variations (e.g., "chirping", arrays, and multiple layers).

- *Material:* nano-structure material (e.g., dielectrics, metals, and polymers), interstructure fill material, adjacent thin film materials, and substrate material (e.g., glass, plastics, dielectrics, and crystalline material).
- *System:* relationship of the nano-optic structures with the optical beam path and any actuator materials and structures (e.g., liquid crystal, solid state materials, electro-optic polymers, and MEMS).

Subwavelength structure size is a key to nano-optics' optical and physical advantages. Nano-optic elements employ microstructures one or more orders of magnitude smaller than the wavelengths of the incident light—with dimensions typically on the order of 10s to a few 100s of nanometers. At such dimensions, these structures may interact with light both according to the principles of classical optics and also in a realm of comparatively small-scale local interactions; rigorous application of the boundary conditions of Maxwell's equation describes the interactions of light with nano-optic structures. For very small structures, single electron or quantum effects may be observed. As a consequence, many nano-optic devices exhibit unique optical behaviors—such as high polarization discrimination in a micron-thin space, achromatic phase retardance, and single layer combinations of optical functions like polarization and spectral filtering—either by coincidence or intent. These nano-optics provide useful alternatives for optical circuit design; others can simply be designed as drop-in replacements for classical bulk optics. In addition to unique behaviors, the devices also achieve their optical effects over relatively short distances—nano-optic devices are generally very thin, usually one micron or less in thickness. This has strong positive repercussions for integrated optics since nano-optic structures can often be applied, similar to thin films, as a "coating layer" during a complex manufacturing process. Single electron and quantum effects also enable optical performance to be readily customized. Modifying material properties by adding nano-structures results in a continuous spectrum of optical functionality—essentially this combination creates hybrid, "artificial" optical materials—that allows performance specifications and optical wavelengths to be incrementally and accurately varied to meet specific application requirements.

Nano-optic structures can be realized through a variety of manufacturing methods. For commercial usefulness, an optical device is not optimally practical unless it can be manufactured in high volumes, in multiple application specific variations, and readily integrated with other relevant technologies. The first criterion is necessary to achieve economies of scale; the second, to ensure that the relevant market opportunity is broad enough to be interesting; and the third, to facilitate easy integration of nano-optic elements into more complex systems. The available fabrication approaches (self-assembly and nano-lithography) achieve these objectives to varying degrees. Some materials can be caused to form regular, nanoscale structures under appropriate, controlled conditions—a.k.a. self-assembly. The difficulty with this approach is lack of flexibility in achievable structures and usable materials, which limits the functions that can be realized. Nanolithography, which is akin to standard semi-conductor manufacturing, generally uses some method to create an image in a polymer resist layer; this image is then used as an etching mask to transfer the nanoscale pattern into the target material. Nanolithography methods include interference lithography, e-beam (electron beam) lithography, and nanopattern replication. Overall, each of these methods is optimal for some set of nanostructure patterns, materials, and volume requirements. As a manufacturing method, nanolithography supports a range of integration modes for optical circuits:

- Optimization of traditional multi-element, multi-technology optical circuit architectures is possible by taking advantage of the unique optical behavior of nano-optics.
- Monolithic "self-integration" can be achieved by layering one atop the other to create aggregate optical effects.
- Spatial integration can be accomplished by organizing varying optical functions into an array structure via nanopattern replication.
- Hybrid integration is achieved by adding a nano-optic layer(s) to functional optical materials.
- Integration of nano-optics with optically active layers, such as liquid crystal or MEMS structures, can create electronically controllable optical devices.

Integrated optical devices and subassemblies based on nano-optic elements realize a range of benefits. Compared to conventional bulk optics, their size and optical properties make alignment in assembly more forgiving, less labor-intensive, and less costly. When physically combined with other nano-optics or other materials in manufacturing to create monolithic integrated optics, this eliminates the complication and cost of multidevice lamination, while improving reliability. Also, arrayed nano-optic devices can be applied in multi-beam or multi-path optical circuits, eliminating the need to individually align discrete optics. In general, the process by which nano-optic elements are fabricated is flexible and robust, allowing for cost-saving manufacturing process integration in creating hybrid and monolithic optics; again, the resultant optics are often smaller, more robust, more functional, and easier to assemble. Because nano-optic devices are manufactured utilizing wafer-based processes proven in semi-conductor fabrication, this enables flexible sharing of manufacturing capacity and supports high-volume manufacturing throughput. Nano-optics draws on a growing "design vocabulary" that is functionally equivalent to the design libraries used in semiconductor applications. Multiple design elements can be used with complex, multi-step, and multi-process methods to create particularly sophisticated nano-patterns and multi-layer devices. The broad range of different nano-structure patterns, materials, and modes of integration means that there are no inherent limits in the functionality and operational wavelength of nanooptic devices. These attributes are solely driven by the needs of the application and the optical circuit. To date, the commercialization of nano-optics has demonstrated the portability of the manufacturing platform and its products across multiple markets including optical data storage, digital imaging, projection display, and telecommunications. As nano-optic technology continues to propagate, it promises to further expand both in applications and in benefits realized by users of nano-optic based devices.

Interference Lithography: Readily creates interference patterns with useful dimensions using UV light sources. The benefit of this method is simplicity; the difficulty is in creating complex shapes and arrays.

E-Beam Lithography: Writes arbitrary, complex patterns using a focused beam of electrons. The benefit is that almost any pattern can be created; a drawback for commercial production is that individual wafer processing times can be fairly long (tens to thousands of hours).

Nano-Pattern Replication: Relies on other methods, e.g., the two above—to create a master plate capturing the inverse of the desired pattern. A printing-like transfer method then patterns the polymer resist. The benefit is high-fidelity reproduction allowing extremely small feature size, complex patterns, effective material independence during repetitive production, and for different structures in the same production line. It is not applicable for limited runs or when only a fixed interference pattern is required.

A nano-optic polarization beam splitter/combiner (PBS/C) provides 180 degrees of effective separation—meaning that one polarization is transmitted and the other is reflected—via a submicron-thick nanoustructural layer. This device's beam separation and thin form allow adjacent components to be closer coupled compared with conventional optics, reducing existing optical path lengths by up to one-half. In turn, shorter optical path lengths reduce beam divergence, improving coupling efficiency and reducing power loss. Many optical circuit designs also incorporate a wave plate adjacent to the PBS/C.

Digital Imaging, Communications and Sensors: Physically, a nano-optic enabled variable optical attenuator (VOA) consists of a crossed pair of polarizing nano-structures sandwiching a liquid crystal cell; applying an electrical field across the liquid crystal will rotate the liquid crystal molecules and control the fraction of light that is passed through the structure. This tunable optic addresses the frequent need to control and block the output power of polarized light sources, such as optical transceivers, in order to dynamically optimize their signals and to facilitate maintenance. Because the nano-optic polarizer is directly integrated into the liquid crystal cell construction, this allows the entire component to be less than 1 mm thick.

REFERENCES

Averback, R.S., J. Bernholc, and D.L. Nelson. 1991. *MRS Symposium Proceedings* Vol. 206.

Baker, R.T.K. Synthesis, properties and applications of graphite nanofibers. In *R&D status and trends*, ed. Siegel et al.

Bate, R., Frazier, G., Frensley, W., Reed., M., "An Overview of Nanoelectronics," *Texas Instruments Technical Journal*, July-August 1989, pp. 13-20.

Beck, J.S., J.C. Vartuli, W.J. Roth, M.E. Leonowicz, C.T. Kresge, K.D. Schmitt, C.T.-W. Chu, D.H. Olsen, E.W. Shepard, S.B. McCullen, J.B. Higgins, and J.L. Schlenker. 1992. *J. Am. Chem. Soc.* 114:10834.

Bensaude-Vincent, B.: 2001, 'The Construction of a Discipline: Materials Science in the United States', *Historical Studies in the Physical and Biological Sciences*, 31, 223-248.

Berube, D.M.: 2004, 'The Rhetoric of Nanotechnology', in: D. Baird, A. Nordmann & J. Schummer (eds.), *Discovering the Nanoscale*, Amsterdam: IOS Press, pp. 173-192.

Bowes, C.L., A. Malek, and G.A. Ozin. 1996. *Chem. Vap. Deposition* 2:97.

Braun, T.; Schubert, A. & Zsindely, S.: 1997, 'Nanoscience and nanotechnology on the balance', *Scientometrics*, 38, 321-325.

Brinker, C.J. 1996. *Curr. Opin. Solid State Mater. Sci.* 1:798.

Brown, N.: 2004, 'Needle on the Real: Technoscience and Poetry at the Limits of Fabrication', in: N.K. Hayles (ed.), *Nanoculture: Implications of the New Technoscience*, Bristol, UK: Intellect Books, pp. 173-190.

Bueno, O.: 2004, 'The Drexler-Smalley Debated on Nanotechnology: Incommensurability at Work?', *Hyle: International Journal for Philosophy of Chemistry*, 10(2), 83-98.

Bueno, O.: 2004, 'Von Neumann, Self-Reproduction and the Constitution of Nanophenomena', in: D. Baird, A. Nordmann & J. Schummer (eds.), *Discovering the Nanoscale*, Amsterdam: IOS Press, pp. 101-115.

Carroll, J.S.: 2001, 'Social Science Research Methods for Assessing Societal Implications of Nanotechnology', in: M.C. Roco and W.S. Bainbridge (eds.), *Societal Implications of Nanoscience and Nanotechnology*, Dordrecht: Kluwer, pp. 188-192.

Chen, C-Y., S.L. Burkett, H.-X. Li, and M.E. Davis. 1993. *Microporous Mater*. 2:27.

Chianelli, R.R. 1998. Synthesis, fundamental properties and applications of nanocrystals, sheets, and fullerenes based on layered transition metal chalcogenides. In *R&D status and trends*, ed. Siegel et al.

Chianelli, R.R., M. Daage, and M.J. Ledoux. 1994. *Advances in Catalysis* 40:177.

Claeson, T., and Likharev, K.," Single Electronics," *Scientific American*, June 1992, pp. 80-85.

Coenen, C.: 2004, 'Nanofuturismus: Anmerkungen zu seiner Relevanz, Analyse und Bewertung', *Technikfolgenabschätzung ' Theorie und Praxis*, 13 (2), 78-85.

Crandall, B.C. and Lewis, J., *Nanotechnology: Research and Perspectives*, MIT Press, Cambridge, 1992.

Crow, M. & Sarewitz, D.: 2001, 'Nanotechnology and Societal Transformation', in: M.C. Roco and W.S. Bainbridge (eds.), *Societal Implications of Nanoscience and Nanotechnology*, Dordrecht: Kluwer, pp. 45-54.

Decker, M.; Fiedeler, U.; Fleischer, Th.: 2004, 'Ich sehe was, was Du nicht siehst... zur Definition von Nanotechnologie', *Technikfolgenabschätzung ' Theorie und Praxis*, 13 (2), 10-16.

Dresselhaus, M., and G. Dresselhaus. 1995. *Ann. Rev. Mat. Sci*. 25:487.

Dresselhaus, M.S. 1998. Carbon-based nanostructures. In *R&D status and trends*, ed. Siegel et al.

Dresselhaus, M.S., G. Dresselhaus, and P. Eklund. 1996. *Science of fullerenes and carbon nanotubes*. San Diego: Academic Press.

Drexler, K.E.: 2004, 'Nanotechnology: From Feynman to Funding', *Bulletin of Science, Technology & Society*, 24, 21-27.

Dupuy, J.-P.: 2004, 'Complexity and Uncertainty: A Prudential Approach To Nanotechnology', in: European Commission (Community Health and Consumer Protection): *Nanotechnologies: A Preliminary Risk Analysis on the Basis of a Workshop, Brussels, 1-2 March 2004*, pp. 71-93 (www.europa.eu.int/comm/health/ ph_risk/documents/ev_20040301_en.pdf).

Dupuy, J.-P.: 2004, 'Pour une évaluation normative du programme nanotechnologique', *Annales des Mines*, (February 2004), 27-32.

Einsiedel, E.F.; Goldenberg, L.: 2004, 'Dwarfing the Social? Nanotechnology Lessons from the Biotechnology Front', *Bulletin of Science, Technology & Society*, 24, 28-33.

ETC: 2003a, *The Big Down: Atom Tech ' Technologies Converging at the Atomic Scale*. Winnipeg, Canada: Action Group on Erosion, Technology and Concentration [www.etcgroup.org/documents/TheBigDown.pdf].

ETC: 2003b, 'No Small Matter II: The Case for a Global Moratorium ' Size Matters!' *Occasional Paper Series* 7(1) [www.etcgroup.org/documents/Occ.Paper_Nanosafety.pdf].

Etzkowitz, H., 2001, 'Nano-Science and Society: Finding a Social Basis for Science Policy', in: M.C. Roco and W.S. Bainbridge (eds.), *Societal Implications of Nanoscience and Nanotechnology*, Dordrecht: Kluwer, pp. 121-128.

European Comission: *European Workshop on Social and Economic Research on Nanotechnologies and Nanosciences, Brussels, 14-15 April 2004* [www.stage-research.net/STAGE/PAGES/Nano.html].

European Commission (Community Health and Consumer Protection): 2004, *Nanotechnologies: A Preliminary Risk Analysis on theBasis of a Workshop, Brussels, 1-2 March 2004* [www.europa.eu.int/comm/health/ ph_risk/documents/ev_20040301_en.pdf].

European Commission: 2004, *Converging Technologies: Shaping the Future of European Societies (Report of the High Level Expert Group "Foresighting the New Technology Wave")*, Brussels: European Commission Research [http://europa.eu.int/comm/research/conferences/2004/ntw/index_en.html].

Feynman, R., "There's Plenty of Room at the Bottom: An invitation to Enter a New Field of Physics," Talk at the Annual Meeting of the American Physical Society, 29 December 1959.

Fielder, F.A.; Reynolds, G.H.: 1994, 'Legal Problems of Nanotechnology: An Overview', *Southern California Interdisciplinary Law Journal*, 3, 593-629.

Fleischer, Th.; Decker, M. & Fiedeler, U. (eds.): 2004, *Große Aufmerksamkeit für kleine Welten '*

Nanotechnologie und ihre Folgen, special issue of *Technikfolgenabschätzung ' Theorie und Praxis*, 13 (2), 5-85 [www.itas.fzk.de/tatup/042/inhalt.htm].

Fogelberg, H. & Glimell, H.: 2003, 'Molecular Matters: In Search of the Real Stuff', in: H. Fogelberg & H. Glimell, *Bringing Visibility to the Invisible: Towards A Social Understanding of Nanotechnology*, Göteborg: Göteborg University, pp. 5-32 [www.sts.gu.se/publications/STS_report_6.pdf].

Fogelberg, H. & Glimell, H.: 2003, *Bringing Visibility to the Invisible: Towards A Social Understanding of Nanotechnology*, Göteborg: Göteborg University [www.sts.gu.se/publications/STS_report_6.pdf].

Fogelberg, H.: 2003, 'The Grand Politics of Technoscience: Contextualizing Nanotechnology', in: H. Fogelberg & H. Glimell, *Bringing Visibility to the Invisible: Towards A Social Understanding of Nanotechnology*, Göteborg: Göteborg University, pp. 33-54 [www.sts.gu.se/publications/STS_report_6.pdf].

Frazier, G., "An Ideology For Nanoelectronics," in *Concurrent Computations: Algorithms, Architecture, and Technology*, Plenum Press, New York, 1988.

Glimell, H.: 2001, 'Challenging Limits ' Excerpts from an Emerging Ethnography of Nano Physicists', in: H. Glimell & O. Johlin (eds.), *The Social Production of Technology: On the Everyday Life with Things*, Göteburg, SE: BAS Publisher, chapter 7, pp. 111-131 (reprinted in: H. Fogelberg & H. Glimell, *Bringing Visibility to the Invisible: Towards A Social Understanding of Nanotechnology*, Göteborg: Göteborg University, pp. 115-139.

Glimell, H.: 2001, 'Dynamics of the Emerging Field of Nanoscience', in: M.C. Roco and W.S. Bainbridge (eds.), *Societal Implications of Nanoscience and Nanotechnology*, Dordrecht: Kluwer, pp. 156-160.

Glimell, H.: 2003, 'A Nano Narrative: The Mircopolitics of a New Generic Enabling Technology', in: H. Fogelberg & H. Glimell, *Bringing Visibility to the Invisible: Towards A Social Understanding of Nanotechnology*, Göteborg: Göteborg University, pp. 55-77 [www.sts.gu.se/publications/STS_report_6.pdf].

Glimell, H.: 2003, 'Dynamics of the Emerging Field of Nanoscience', in: H. Fogelberg & H. Glimell, *Bringing Visibility to the Invisible: Towards A Social Understanding of Nanotechnology*, Göteborg: Göteborg University, pp. 79-85 [www.sts.gu.se/publications/STS_report_6.pdf].

Glimell, H.: 2004, 'Grand Visions and Lilliput Politics: Staging the Exploration of the 'Endless Frontier'', in: D. Baird, A. Nordmann & J. Schummer (eds.), *Discovering the Nanoscale*, Amsterdam: IOS Press, pp. 231-246.

Goldhaber-Gordon, D., Montemerlo, M.S., Love, J.C., Opiteck, G.J., and Ellenbogen, J.C., "Overview of Nanoelectronic Devices," submitted to the *Proceedings of the IEEE*, February 1997. For more information, please send e-mail to nanotech@mitre.org.

Gorman, M.E.; Groves, J.F. & Shrager, J.: 2004, 'Societal Dimensions of Nanotechnology as a Trading Zone: Results from a Pilot Project', in: D. Baird, A. Nordmann & J. Schummer (eds.), *Discovering the Nanoscale*, Amsterdam: IOS Press, pp. 63-73.

Grinbaum, A. & Dupuy, J.-P.: 2004, 'Living with Uncertainty: Toward the Ongoing Normative Assessment of Nanotechnology', *Techne: Research in Philosophy and Technology*, 8(3) (forthcoming).

Grinbaum, A.: 2004, 'La condition de l'homme moderne et les nanotechnologies', in: G. Nivat (ed.),

Les limites de l´humain. 39èmes Rencontres Internationales de Genève, L´Age d´Homme, Genève, p. 141.

Grunwald, A.: 2004, 'Ethische Aspekte der Nanotechnologie. Eine Felderkundung', *Technikfolgenabschätzung ' Theorie und Praxis*, 13 (2), 71-78.

Gupta, V.K. & Pangannaya, N.B.: 2000, 'Carbon nanotubes: bibliometric analysis of patents', *World Patent Information*, 22, 185-189.

Hansson, S.O.: 2004, 'Great Uncertainty about Small Things', *Techne: Research in Philosophy and Technology*, 8(3) (forthcoming).

Haruta, M. 1997. *Catalyst surveys of Japan* 1:61 and references therein.

Hayles, N.K. (ed.): 2004, *Nanoculture: Implications of the New Technoscience*, Bristol, UK: Intellect Books.

Hayles, N.K.: 2004, 'Connecting the Quantum Dots: Nanotechscience and Culture', in: N.K. Hayles (ed.), *Nanoculture: Implications of the New Technoscience*, Bristol, UK: Intellect Books, pp. 11-26.

Hemerén, G.: 2004, 'Nano Ethics Primer', in: European Commission (Community Health and Consumer Protection): *Nanotechnologies: A Preliminary Risk Analysis on theBasis of a Workshop, Brussels, 1-2 March 2004* (www.europa.eu.int/comm/health/ ph_risk/documents/ ev_20040301_en.pdf), pp. 95-102.

Hennig, J.: 2004, 'Changes in the Design of Scanning Tunneling Microscopic Images from 1980 to 1990', *Techne: Research in Philosophy and Technology*, 8(3) (forthcoming).

Hessenbruch, A.: 2004, 'Nanotechnology and the Negotiation of Novelty', in: D. Baird, A. Nordmann & J. Schummer (eds.), *Discovering the Nanoscale*, Amsterdam: IOS Press, pp. 135-144.

Hullmann, A. & Meyer, M.: 2003, 'Publications and Patents in Nanotechnology. An overview of previous studies and the state of the art', *Scientometrics*, 58 (3) 507-527.

Huo, Q., D.I. Margolese, U. Ciesla, P. Feng, T.E. Gier, P. Sieger, R. Leon, P.M. Petroff, F. Schuth, and G.D. Stucky. 1994. *Nature* 368:317.

Jena, P., S.N. Khanna, and B.K. Rao. 1996. In *Science and technology of atomically engineered materials*, ed. P. Jena. River Edge, NJ: World Scientific.

Johansson, M.: 2003, 'Plenty of room at the bottom: Towards an anthropology of nanoscience', *Anthropology Today*, 19 (No. 6), 3-6.

Johnson, A.: 2004, 'The End of Pure Science: Science Policy from Bayh-Dole to the NNI', in: D. Baird, A. Nordmann & J. Schummer (eds.), *Discovering the Nanoscale*, Amsterdam: IOS Press, pp. 217-230.

Jones, C.W., and W.J. Koros. 1994. *Carbon* 32:1419.

Jounet, C., W.K. Maser, P. Bernier, A. Loiseau, M. Lamy de la Chapelle, S. Lefrant, P. Deniard, R. Lee, and J.E. Fischer. 1997. *Nature* 388:756.

K.E. Drexler, *Nanosystems: Molecular Machinery, Manufacturing, and Computation*, Wiley, New York, 1992.

Karger, J., and D.M. Ruthven. 1992. *Diffusion in zeolites*. New York: J. Wiley.

Keiper, A.: 2003, 'The Nanotechnology Revolution', *The New Atlantis*, 2, 17-34.

Khushf, G.: 2004, 'A Hierarchical Architecture for Nano-scale Science and Technology: Taking

Stock of the Claims About Science Made By Advocates of NBIC Convergence', in: D. Baird, A. Nordmann & J. Schummer (eds.), *Discovering the Nanoscale*, Amsterdam: IOS Press, pp. 21-33.

Khushf, G.: 2004, 'Systems Theory and the Ethics of Human Enhancement: A Framework for NBIC Convergence', *Annals of the New York Academy of Sciences*, 1013, 124-149.

Krätschmer, W., L.D. Lamb, K. Fostiropoulos, and D.R. Huffman. 1990. *Nature* 347:354.

Kresge, C.T., M.E. Leonowicz, W.J. Roth, J.C. Vartuli, and J.S. Beck. 1992. *Nature* 359:710.

Kupperman, A., S. Nadimi, S. Oliver, G. Ozin, J. Garcés, and M. Olken. 1993. *Nature* 365:239.

Kuusi, O.; Meyer, M.: 2002, 'Technological generalizations and leitbilder ' the anticipation of technological opportunities', *Technological Forecasting & Social Change*, 69, 625-639.

Landon, B.: 2004, 'Less is More: Much Less is Much More: The Insistent Allure of Nanotechnology Narratives in Science Fiction', in: N.K. Hayles (ed.), *Nanoculture: Implications of the New Technoscience*, Bristol, UK: Intellect Books, pp. 131-146.

Laszlo, P.: 2004, 'Is There Life After Partington?', *Hyle: International Journal for Philosophy of Chemistry*, 10(2), 169-178.

Lenhard, J.: 2004, 'Nanoscience and the Janus-Faced Character of Simulations', in: D. Baird, A. Nordmann & J. Schummer (eds.), *Discovering the Nanoscale*, Amsterdam: IOS Press, pp. 93-100.

Lent, C.S., Tougaw, P.D., Porod, W., Bernstein, G.H., "Quantum Cellular Automata," *Nanotechnology*, Vol. 4, p. 49, 1993.

Lewak, S.E.: 2004, 'What's the Buzz? Tell Me What's A-Happening: Wonder, Nanotechnology, and Alice's Adventures in Wonderland', in: N.K. Hayles (ed.), *Nanoculture: Implications of the New Technoscience*, Bristol, UK: Intellect Books, pp. 201-.

Lide, D.R., ed. 1993-1994. *CRC Handbook of Chemistry and Physics*, 74th ed.

Lin-Easton, P.C.: 2001, 'It's Time for Environmentalists to Think Small ' Real Small: A Call for the Involvement of Environmental Lawyers in Developing Precautionary Policies Molecular Nanotechnology', *Georgetown International Law Review*, 14, 106-134.

López, J.: 2004, 'Bridging the Gaps: Science Fiction in Nanotechnology', *Hyle: International Journal for Philosophy of Chemistry*, 10(2), 129-152.

Lösch, A.: 2004, 'Nanomedicine and Space: Discursive Orders of Mediating Innovations', in: D. Baird, A. Nordmann & J. Schummer (eds.), *Discovering the Nanoscale*, Amsterdam: IOS Press, pp. 193-202.

Marshall, K.: 2004, 'Atomizing the Risk Technology', in: N.K. Hayles (ed.), *Nanoculture: Implications of the New Technoscience*, Bristol, UK: Intellect Books, pp. 147-160.

Martin, T.P., N. Malinowski, U. Zimmerman, U. Naher, and H. Schaber. 1993. *J. Chem. Phys.* 99:4210.

Martin, T.P., U. Naher, H. Schaber, U. Zimmerman. 1993. *Phys. Rev. Lett.* 70:3079.

Mayer, S.: 2002, 'From genetic modification to nanotechnology: the dangers of 'sound science", in: T. Gilland (ed.), *Science: Can We Trust the Experts?*, London: Hodder and Stoughton, pp. 1-15.

Mehta, M.D.: 2002, 'Nanoscience and Nanotechnology: Assessing the Nature of Innovation in These Fields', *Bulletin of Science, Technology & Society*, 22(4), 269-273.

Mehta, M.D.: 2004, 'From Biotechnology to Nanotechnology: What Can We Learn from Earlier Technologies?, *Bulletin of Science, Technology & Society*, 24, 34-39.

Meyer, M. & Kuusi, O.: 2004, 'Nanotechnology: Generalizations in an Interdisciplinary Field of Science and Technology', *Hyle: International Journal for Philosophy of Chemistry*, 10(2), 153-168.

Meyer, M.: 2000a, 'Does science push technology? Patents citing scientific literature', *Research Policy*, 29, 409-434.

Meyer, M.: 2000b, 'Patent citations in a novel field of technology: What can they tell about interactions of emerging communities of science and technology?', *Scientometrics*, 48, 151-178.

Meyer, M.: 2001a, 'Patent citations in a novel field of technology: An exploration of nano-science and nano-technology', *Scientometrics*, 51, 163-183.

Meyer, M.: 2001b, 'Socio-economic Research on Nanoscale Science and Technology: A European Overview and Illustration', in: M.C. Roco and W.S. Bainbridge (eds.), *Societal Implications of Nanoscience and Nanotechnology*, Dordrecht: Kluwer, pp. 217-241.

Meyer, M.; Persson, O.: 1998, 'Nanotechnology ' interdisciplinarity, patterns of collaboration and differences in application', *Scientometrics*, 47, 195'205.

Milburn, C.: 2002, 'Nanotechnology in the age of post-human engineering: science fiction as science', *Configurations*, 10, 261-295 [reprinted in: N.K. Hayles (ed.), *Nanoculture: Implications of the New Technoscience*, Bristol, UK: Intellect Books, 2004, pp. 109-130].

Milburn, C.: 2004, 'Nano/Splatter: Disintegrating the Postbiological Body', *New Literary History*, 35 (in print).

Mnyusiwalla, A.; Abdallah, S.D.; Singer, P.A.: 2003, 'Mind the gap: science and ethics in nanotechnology', *Nanotechnology*, 14, R9-R13.

Mody, C.C.M.: 2004, 'How Probe Microscopists Became Nanotechnologists', in: D. Baird, A. Nordmann & J. Schummer (eds.), *Discovering the Nanoscale*, Amsterdam: IOS Press, pp. 119-133.

Mody, C.C.M.: 2004, 'Instruments in Training: The Growth of American Probe Microscopy in the 1980s', in: D. Kaiser (ed.), *Pedagogy and the Practice of Science: Producing Physical Scientists, 1800-2000*, Cambridge, MA: MIT Press (forthcoming).

Mody, C.C.M.: 2004, 'Small, but Determined: Technological Determinism in Nanoscience', *Hyle: International Journal for Philosophy of Chemistry*, 10(2), 99-128.

Mody, C.C.M.: 2004, *Crafting the Tools of Knowledge: The Invention, Spread, and Commercialization of Probe Microscopy, 1960-2000*, Ph.D. dissertation, Cornell University.

Montemerlo, M.S., Love, J.C., Opiteck, G.J., Goldhaber, D. J., and Ellenbogen, J.C., "Technologies and Designs for Electronic Nanocomputers," MITRE Technical Report 96W0000044, The MITRE Corporation, McLean, VA, July 1996. For more information, send e-mail to nanotech@mitre.org.

Moor, J.H. & Weckert, J.: 2004, 'Nanoethics: Assessing the Nanoscale From an Ethical Point of View', in: D. Baird, A. Nordmann & J. Schummer (eds.), *Discovering the Nanoscale*, Amsterdam: IOS Press, pp. 301-310.

Munn Sanchez, E.: 2004, 'The Expert's Role in Nanoscience and Technology', in: D. Baird, A. Nordmann & J. Schummer (eds.), *Discovering the Nanoscale*, Amsterdam: IOS Press, pp. 257-266.

Nordmann, A.: 2003, 'Shaping the World Atom by Atom: Eine nanowissenschaftliche WeltBildanalyse', in: A. Grunwald (ed.), *Technikgestaltung zwischen Wunsch und Wirklichkeit*, Berlin: Springer, pp. 191-199.

Nordmann, A.: 2004, 'Molecular Disjunctions: Staking Claims at the Nanoscale', in: D. Baird, A. Nordmann & J. Schummer (eds.), *Discovering the Nanoscale*, Amsterdam: IOS Press, pp. 51-62.

Nordmann, A.: 2004, 'Nanotechnology: Convergence and Integration', Presentation at *EuroNanoForum, Trieste, December 10, 2003* (Proceedings in preparation).

Nordmann, A.: 2004, 'Nanotechnology's WorldView: New Space for Old Cosmologies', *IEEE Technology and Society Magazine*, 23 (forthcoming).

Nordmann, A.: 2004, 'Social Imagination for Nanotechnology', in: European Commission (Community Health and Consumer Protection): *Nanotechnologies: A Preliminary Risk Analysis on the Basis of a Workshop, Brussels, 1-2 March 2004*, pp. 111-113 [www.europa.eu.int/comm/health/ph_risk/documents/ev_20040301_en.pdf].

Nordmann, A.: 2004, 'Was ist TechnoWissenschaft? ' Zum Wandel der Wissenschaftskultur am Beispiel von Nanoforschung und Bionik', in: T. Rossmann & C. Tropea (eds.), *Bionik ' Neue Forschungsergebnisse aus Natur-, Ingenieur- und Geisteswissenschaften*, Berlin: Springer, 2004 (forthcoming)

Paschen, H.; Coenen, C.; Fleischer, T.; Grünwald, R.; Oertel, D. & Revermann, C.: 2004, *Nanotechnologie: Forschung, Entwicklung, Anwendung*, Berlin: Springer (*Nanotechnologie, TAB-Arbeitsbericht 92*, Berlin: Büro für Technikfolgen-Abschätzung beim Deutschen Bundestag).

Pitt, J.C.: 2004, 'The Epistemology of the Very Small', in: D. Baird, A. Nordmann & J. Schummer (eds.), *Discovering the Nanoscale*, Amsterdam: IOS Press, pp. 157-163.

Pressman, J.: 2004, 'Nano Narrative: A Parable from Electronic Literature', in: N.K. Hayles (ed.), *Nanoculture: Implications of the New Technoscience*, Bristol, UK: Intellect Books, pp. 191-200.

Prigogine, I., and S. Rice. 1988. *Advances in chemical physics*, Vol. 70, Parts 1 & 2. New York: J. Wiley.

R. T. Bate, "Nanoelectronics," *Nanotechnology*, Vol. 1, pp. 1-7, 1990.

Rademann, K., B. Kaiser, U. Even, F. Hensel. 1987. *Phys. Rev. Lett.* 59:2319.

Rao, C.N.R., B.C. Satishkumar, and A. Govindaraj. 1997. *Chem. Commun.* 1581.

Rao, M.B., and S. Sircar. 1993. *Gas Separation and Purification* 7:279.

Reed, M.A., "Quantum Dots," *Scientific American,* January 1993, pp. 118-123.

Roberts, J.A.: 2004, 'Deciding the Future of Nanotechnologies: Legal Perspectives on Issues of Democracy and Technology', in: D. Baird, A. Nordmann & J. Schummer (eds.), *Discovering the Nanoscale*, Amsterdam: IOS Press, pp. 247-255.

Robinson, C.: 2004, 'Images in NanoScience/Technology', in: D. Baird, A. Nordmann & J. Schummer (eds.), *Discovering the Nanoscale*, Amsterdam: IOS Press, pp. 165-169.

Robison, W.L.: 2004, 'Nano-Ethics', in: D. Baird, A. Nordmann & J. Schummer (eds.), *Discovering the Nanoscale*, Amsterdam: IOS Press, pp. 285-300.

Roco, M.C. & Bainbridge, W.S. (eds.): 2001, *Societal implications of nanoscience and nanotechnology*,

(Proceedings of a workshop organized by the National Science Foundation, September 28-29, 2000), Kluwer: Dordrecht [available online at http://itri.loyola.edu/nano/societalimpact/nanosi.pdf]

Roco, M.C. & Tomellini, R. (eds.): 2002, *Nanotechnology: Revolutionary Opportunities and Societal Implications, Workshop, Lecce (Italy), 31 January - 1 February 2002*, Luxemburg: Office for Official Publications of the European Communities, [200 pp.]

Roco, M.C.; Bainbridge, W.S. (eds.): 2002, *Converging Technologies for Improving Human Performance: Nanotechnology, Biotechnology, Information Technology and the Cognitive Science*, Arlington, VA: National Science Foundation.

Rohlfing, E.A., D.M. Cox, and A. Kaldor. 1984. *J. Chem. Phys.* 81:3846.

Rosen, A. 1998. A periodic table in three dimensions: A sightseeing tour in the nanometer world. In *Advances in quantum chemistry*. In press.

Ruthven, D.M., S. Farooq, K.S. Knaebel. 1994. *Pressure swing adsorption*. New York: VCH Publishers.

Sarewitz, D.; Woodhouse, E.: 2003, 'Small is Powerful', in: A. Lightman, D. Sarewitz & Chr. Desser, (eds.), *Living with the Genie: Essays on Technology and the Quest for Human Mastery*, Washington, DC: Island Press, pp. 63-83.

Sayari, A. 1996. *Chem. Mater*. 8:1840.

Schiemann, G.: 2004, 'Dissolution of the Nature-Technology Dichotomy? Perspectives on Nanotechnology from an Everyday Understanding of Nature', in: D. Baird, A. Nordmann & J. Schummer (eds.), *Discovering the Nanoscale*, Amsterdam: IOS Press, pp. 209-213.

Schmidt, J.C.: 2004, 'Unbounded Technologies: Working Through Technological Reductionism of Nanotechnology', in: D. Baird, A. Nordmann & J. Schummer (eds.), *Discovering the Nanoscale*, Amsterdam: IOS Press, pp. 35-50.

Schuler, E.: 2004, 'Perception of Risks and Nanotechnology', in: D. Baird, A. Nordmann & J. Schummer (eds.), *Discovering the Nanoscale*, Amsterdam: IOS Press, pp. 279-284.

Schummer, J.: 2001, 'Ethics of Chemical Synthesis', *Hyle: International Journal for Philosophy of Chemistry*, 7, 103-124 [online: www.hyle.org/journal/issues/7/schummer.htm].

Schummer, J.: 2003, 'Aesthetics of Chemical Products: Materials, Molecules, and Molecular Models', *Hyle: International Journal for Philosophy of Chemistry*, 9, 73-104 [online: www.hyle.org/journal/issues/9-1/schummer.htm].

Schummer, J.: 2003, 'The Notion of Nature in Chemistry', *Studies in History and Philosophy of Science*, 34 (2003), 705-736.

Schummer, J.: 2004, 'Interdisciplinary Issues in Nanoscale Research', in: D. Baird, A. Nordmann & J. Schummer (eds.), *Discovering the Nanoscale*, Amsterdam: IOS Press, pp. 9-20.

Schummer, J.: 2004, 'Multidisciplinarity, Interdisciplinarity, and Patterns of Research Collaboration in Nanoscience and Nanotechnology', *Scientometrics*, 59, 425-465.

Schummer, J.: 2004, 'Naturverhältnisse in der modernen Wirkstoff-Forschung', in: K. Kornwachs (ed.), *Technik ' System ' Verantwortung*, Münster: LIT, pp. 629-638.

Schummer, J.: 2004, 'Why do Chemists Perform Experiments?', in: D. Sobczynska, P. Zeidler, E. Zielonacka-Lis (eds.), *Chemistry in the Philosophical Melting Pot*, Peter Lang (Frankfurt/M.), 2004, pp. 395-410.

Schummer, J.: 2005, 'Reading Nano: The Public Interest in Nanotechnology as Reflected in Book Purchase Patterns", *Public Understanding of Science*, 14 (2) (forthcoming).

Schummer, J.; Baird, D. (eds.): 2004-5, *Nanotech Challenges*, joint special issue of *Hyle: International Journal for Philosophy of Chemistry & Techne: Research in Philosophy and Technology* (forthcoming).

Schwarz, A.E.: 2004, 'Shrinking the 'Ecological Footprint' with NanoTechnoScience?', in: D. Baird, A. Nordmann & J. Schummer (eds.), *Discovering the Nanoscale*, Amsterdam: IOS Press, pp. 203-208.

Schwarz, R.B. 1998. Storage of hydrogen in powders with nanosized crystalline domains. In *R&D status and trends*, ed. Siegel et al.

Siegel, R.W. 1991. *Ann. Rev. Mater. Sci.* 21:559.

Siegel, R.W., E. Hu, and M.C. Roco, eds. 1998. *R&D status and trends in nanoparticles, nanostructured materials, and nanodevices in the United States*. Baltimore: Loyola College, International Technology Research Institute. NTIS #PB98-117914.

Soffer, A., J.E. Koresh, S. Saggy. 1987. United States Patent 4,685,940.

Stix Gary., "Toward 'Point One'," *Scientific American*, February 1995, pp. 90-95.

Stuckless, J.T, D.E. Starr, D.J. Bald, C.T. Campbell. 1997. *J. Chem. Phys.* 107:5547.

Suchman, M.: 2001, 'Envisioning Life on the Nano-Frontier', in: M.C. Roco and W.S. Bainbridge (eds.), *Societal Implications of Nanoscience and Nanotechnology*, Dordrecht: Kluwer, pp. 211-216.

Suchman, M.: 2002, 'Social Science and Nanotechnologies', M. Roco & R. Tomellini (eds.), *Nanotechnology ' Revolutionary Opportunities and Societal Implications*, Luxembourg: European Communities, pp. 95-99.

Sweeney, A.E.; Seal, S. & Vaidyanathan, P.: 2003, 'The promises and perils of nanoscience and nanotechnology: Exploring emerging social and ethical issues', *Bulletin of Science, Technology & Society*, 23 (4), 236-245.

Swiss Re: 2004, *Nanotechnology: Small Matter, Many Unknowns*, Zurich: Swiss Re [www.swissre.com].

Tanev, P.T., M. Chibwe, and T.J. Pinnavaia. 1994. *Nature* 368:321.

Taylor, K.J., C.L. Pettiette-Hall, O. Cheshnovsky, and R.J. Smalley. 1992. *J. Chem. Phys.* 96:3319.

ten Wolde, A. (ed.): 1998, *Nanotechnology: Towards a Molecular Construction Kit*, The Hague: The Netherlands Study Centre for Technology Trends (STT).

Tenner, E.: 2001, 'Nanotechnology and Unintended Consequences', in: M.C. Roco and W.S. Bainbridge (eds.), *Societal Implications of Nanoscience and Nanotechnology*, Dordrecht: Kluwer, pp. 241-246.

The Royal Society & The Royal Academy of Engineering: 2004, *Nanoscience and nanotechnologies: opportunities and uncertainties*, London [www.nanotec.org.uk/finalReport.htm].

Toumey, C.: 2004, 'Anticipating Public Reactions to Nanotechnology', *Techne: Research in Philosophy and Technology*, 8(3) (forthcoming).

Trudeau, M.L., and J.Y. Ying. 1996. *Nanostructured Materials* 7:245.

Tschope, A.S., W. Liu, M. Flyzani-Stephanopoulos, and J.Y. Ying 1995. *J. Catal.* 157:42.

Turton, R., *The Quantum Dot*, Oxford University Press, 1995.

Verbeek, A. & Callaert, J.: 2002, *Detailed analysis of the science-technology interaction in the field of Nanotechnology*, Luxemburg: Office for Official Publications of the European Communities (Linking Science to Technology. Bibliographic References in Patents, Vol. 9) [79 pp.].

Vermaas, P.E.: 2004, 'Nanoscale Technology: A Two-Sided Challenge for Interpretations of Quantum Mechanics', in: D. Baird, A. Nordmann & J. Schummer (eds.), *Discovering the Nanoscale*, Amsterdam: IOS Press, pp. 77-91.

Wald, C.A.: 2004, 'Working Boundaries on the NANO Exhibition', in: N.K. Hayles (ed.), *Nanoculture: Implications of the New Technoscience*, Bristol, UK: Intellect Books, pp. 83-108.

Weckert, J.: 2001, 'The Control of Scientific Research: The Case of Nanotechnology', *Australian Journal of Professional and Applied Ethics*, 3, 29-44.

Weil, V.: 2001, 'Ethical Issues in Nanotechnology', in: M.C. Roco and W.S. Bainbridge (eds.), *Societal Implications of Nanoscience and Nanotechnology*, Dordrecht: Kluwer, pp. 193-198

Glossary

Affinity: A thermodynamic measurement of the strength of binding between molecules, say between an antibody and antigen. Each antibody/antigen pair has an association constant, Ka, expressed in L/mol.

Anisotropic: Exhibiting different values of a property in different crystallographic directions.

At the Core: A Non-Heavy Metal Quantum Dot. Research at Evident, which is focused on advancing product development, led to the first commercially available nanocrystals that comprise a III-V semiconductor material: Indium, Gallium and Phosphide core quantum dots, with a unique molecular metallic platting compound to form the shell or InGaP/ZnS EviDots-MP™.

Avogadro's number: The number of atoms in exactly 12 grams of pure ^{12}C, equal to 6.022×10^{23}.

Bogosity Filter: A mechanism for detecting bogus ideas and propositions.

Bone substitutes: Synthetic or natural materials for the replacement of bones or bone tissue. They include hard tissue replacement polymers, natural coral, hydroxyapatite, beta- tricalcium phosphate, and various other biomaterials. The bone substitutes as inert materials can be incorporated into surrounding tissue or gradually replaced by original tissue.

Breakdown: Failure of a material resulting from an electrical overload. The resulting damage may be in the form of thermal damage (melting or burning) or electrical damage (loss of polarization in piezoelectric materials). The failing of a physical principle or unit of measurement as the scale is changed.

Brownian Assembly: Brownian motion in a fluid brings molecules together in various position and orientations. If molecules have suitable complementary surfaces, they can bind, assembling to form a specific structure. Brownian assembly is a less paradoxical name for self-assembly (how can a structure assemble itself, or do anything, when it does not yet exist?).

Brownian Motion: Motion of a particle in a fluid owing to thermal agitation, observed in 1827 by Robert Brown. (Originally thought to be caused by vital force, Brownian motion in fact plays a vital role in the assembly and activity of the molecular structures of life).

CPU: The central processing unit of a computer, responsible for executing instructions to process information.

Creep: The time-dependent permanent deformation that occurs under stress; for most materials it is important only at elevated temperatures.

Crosslinked polymer: A polymer in which adjacent linear molecular chains are joined at various positions by covalent bonds.

Disruptive techonlogies: In his book, *The Innovator's Dilemma*, Harvard professor Clayton

Christensen used the term "disruptive technologies" to describe developments capable of displacing established market leaders.

Dissolution: Deterioration in an organism such that its original structure cannot be determined from its current state.

Enzyme immunoassay/EIA: In an enzyme immunoassay (EIA), an enzyme-labeled antibody or antigen is used for the detection and quantification of the antigen-antibody reaction. In an electrochemical EIA, the enzyme-catalyzed reaction is monitored electrochemically (amperometric, potentiometric, voltametric or conductometric). In EIA, the antibody-antigen reaction furnishes the needed specificity. The enzyme label provides the sensitivity via chemical amplification. A large molecule, usually a protein, that catalyzes biological reactions. A protein that acts as a catalyst in a biochemical reaction. Protein that catalyzes a specific chemical reaction.

Enzymes: Molecular machines found in nature, made of protein, which can catalyze (speed up) chemical reactions.

Fail-stop: Describes a component or subsystem that, in the event of a failure, produces no output (e.g., of material or data) rather than producing a damaged or incorrect output.

Fertilization: Fusion of a male and a female gamete (both haploid) to form a diploid zygote, which develops into a new individual.

FET: Field-Effect Transistor — semiconductor device whose insulated gate electrode controls current flow.

Fiber-optic: Relates to transmission of information as modulated light in tiny transparent fibers instead of copper wires.

Fick's first law: The diffusion flux is proportional to the concentration gradient. This relationship is employed for steady-state diffusion situations. The time rate of change of concentration is proportional to the second derivative of concentration. This relationship is employed in non-steady-state diffusion situations.

Field-Effect Transistor (FET): Field-Effect Transistor — semiconductor device whose insulated gate electrode controls current flow.

Fluorescent dye: Molecule that absorbs light at one wavelength and responds by emitting light at another wavelength; the emitted light is of longer wavelength (and hence of lower energy) than the light absorbed.

Free-energy change (DG): Change in the free energy during a reaction: the free energy of the product molecules minus the free energy of the starting molecules. A large negative value of DG indicates that the reaction has a strong tendency to occur.

Frequency response: Two relations between sets of inputs and outputs. One relates frequencies to the output-input amplitude ratio; the other relates frequencies to the phase difference between the output and input.

G1 phase: Gap 1 phase of the eucaryotic cell-division cycle, between the end of cytokinesis and the start of DNA synthesis.

G2 phase: Gap 2 phase of the eucaryotic cell-division cycle, between the end of DNA synthesis and the beginning of mitosis.

GAG (glycosaminoglycan): Long, linear, highly charged polysaccharide composed of a repeating pair of sugars, one of which is always an amino sugar. Mainly found covalently linked to a protein core in extracellular matrix proteoglycans. Examples include chondroitin sulfate, hyaluronic acid, and heparin.

Gain: The ratio of the amplitude of an output to input signal.

Gastrula: Animal embryo at an early stage of development where cells are invaginating to form the rudiment of a gut cavity. (From Greek *gaster*, belly.)

Gate: Circuit whose logical output variables are determined by its inputs. In digital logic, a component that can switch the state of an output dependent on the states of one or more inputs.

Grain boundary: The interface separating two adjoining grains having different crystallographic orientations.

Grain growth: The increase in average grain size of a polycrystalline material: for most materials, an elevated temperature heat treatment is necessary.

Gray goo: The name given to free-range self-replicating miniature machines that could, in theory, run out of control and cause severe damage to the biosphere. The actual threat is generally overrated, as we explain here.

Green ceramic body: A ceramic piece, formed as a particulate aggregate, that has been dried but not fired.

Group: A set of linked atoms in a molecule; a defined substructure. Typically, a set that is usefully regarded as a unit in chemical reactions of interest.

Harmonic oscillator: A system in which a mass is subject to a linear restoring force, like an ideal spring. A harmonic oscillator vibrates at a fixed frequency, independent of amplitude.

Heat capacity (C_v at constant volume and C_p at constant pressure): The quantity of heat required to produce a unit temperature rise per mole of material. The ratio of the heat input to the temperature increase in a system. Note that this definition does not imply that a system *contains* heat, despite the name *heat capacity*.

Heat shock protein (stress-response protein): Protein synthesized in response to an elevated temperature or other stressful treatment; usually helps the cell to survive the stress.

Heat: As defined in thermodynamics, heat is the energy that flows between two systems as a result of temperature differences (a system *contains* neither heat nor work, but can *produce* heat or *do* work). Heat thus differs from thermal energy.

Heisenberg Uncertainty Principle: A quantum-mechanical principle with the consequence that the position and momentum of an object cannot be precisely determined. The Heisenberg principle helps determine the size of electron clouds, and hence the size of atoms.

HeLa cell: Line of human epithelial cells that grows vigorously in culture. Derived from a human cervical carcinoma.

Heme: An iron complex. Cyclic organic molecule containing an iron atom that carries oxygen in hemoglobin and carries an electron in cytochromes.

Hemidesmosome: Specialized cell junction between an epithelial cell and the underlying basal lamina.

Hemoglobin: A biomolecule composed of four myoglobine-like units (proteins plus heme) that can bind and transport four oxygen molecules in the blood.

The major protein in red blood cells that associates with O_2 in the lungs by means of a bound heme group.

Immune Machines: Medical nanomachines designed for internal use, especially in the bloodstream and digestive tract, able to identify and disable intruders such as bacteria and viruses.

Immune response: Response made by the immune system of a vertebrate when a foreign substance or microorganism enters its body.

Immune system: Population of lymphocytes and other white blood cells in the vertebrate body that defends it against infection.

Immunoglobulin (Ig): An antibody molecule. Higher vertebrates have five classes of immunoglobulin - IgA, IgD, IgE, IgG, and IgM - each with a different role in the immune response.

Immunoglobulin like (Ig-like) domain: Characteristic protein domain of about 100 amino acids that is found in antibody molecules and in many other proteins that form the Ig superfamily.

IMP: Electronic implant, especially in the brain. [Ron Hale Evans]

Impedance: The complex ratio of a force-like quantity (force, pressure, voltage, temperature, or electric field) to a corresponding related velocity-like quantity (velocity, volume velocity, current, heat flow, or magnetic field strength).

***In situ* hybridization:** Technique in which a single-stranded RNA or DNA probe is used to locate a gene or an mRNA molecule in a cell or tissue.

***In vitro*:** Term used by biochemists to describe a process taking place in an isolated cell-free extract. Also used by cell biologists to refer to cells growing in culture (*in vitro*), as opposed to in an organism (*in vivo*). (Latin for "in glass.")

***In vivo*:** In an intact cell or organism. (Latin for "in life.")

Index of refraction (n): The ratio of the velocity of light in a vacuum to the velocity in some medium.

Inductance [in Henry, H]: That property of an electric circuit which tends to oppose change in current in the circuit. One henry (H) is the inductance of a closed circuit in which an electromotive force of 1 volt is produced when the electric current in the circuit varies uniformly at the rate of 1 ampere per second.

Induction (embryonic): Change in the developmental fate of one tissue caused by an interaction with another tissue.

Inductor: Energy storage circuit component consisting of a coil of wire and possibly a magnetic material.

Inflammatory response: Local response of a tissue to injury or infection. Caused by invasion of white blood cells, which release various local mediators such as histamine.

Infrared: Invisible electromagnetic radiation having a longer wavelength, and lower frequency, than visible red light.

Inhibitor: A chemical substance that, when added in relatively low concentrations, slows down a chemical reaction.

Inline Universities: (As opposed to online universities), nanocomputer implants serving to increase intelligence and education of their owners, essentially turning them into walking universities.

Inmessaging: In medical nanorobotics, conveyance of information from a source external to the human body, or external to working nanodevices, to a receiver located inside the human body.

Insertion point (in lithography context): Adaptation of a new lithography technique is referred to as the insertion point of that technique.

Insulator: A substance, object or material that does not conduct electricity. Material that conducts electricity very poorly.

Integrated Circuit (IC): An electronic circuit consisting of many interconnected devices on one piece of semiconductor, typically into 10 millimeters on a side. ICs are the major building blocks of today's computers. Semiconductor circuit, typically on a very small silicon chip, containing microfabricated transistors, diodes, resistors, capacitors, etc.

Intelligent Agent: Aka "software agent". Software that can do things without supervision, because it knows your patterns, history, preferences, likes, dislikes, and so forth. You want to take a vacation - it knows that you really enjoyed that trip to Hawaii, and that you prefer to fly at night, 1st class. It also knows that the bungalow you rented last time was marked as being 5-star, and worth a re-visit. Your IA then collates all your parameters, searches the internet for flights, car rentals, restaurant reservations, and lodgings, and schedules everything for you, with options on the side. No more travel agent - you have a software agent to handle things! Many experts agree that by 2010 we will each have one, and that they will greatly reduce our daily load of trivial and redundant tasks.

Interagency Working Group on Nanoscience, Engineering and Technology IWGN, National Science and Technology Council NSTC, US

Interface: In physical terms, an interface is the boundry between two phases, for instance between a solid and liquid or between a liquid and gas.

Intermolecular: Between more than one molecule. Describes an interaction (e.g., a chemical reaction) between different molecules.

Internal energy: The sum of the kinetic and potential energies (including electromagnetic field energies) of the particles that make up a system.

Internet: Worldwide digital communication network in which packets of information travel between senders and recipients.

Interstitial diffusion: a diffusion mechanism that causes atomic motion from interstitial site to interstitial site.

Intramolecular: Describes an interaction (e.g., a chemical reaction) within a single molecule. Intramolecular interactions between widely separated parts of a molecule resemble intermolecular interactions in most respects. (2) Within a single molecule.

Intrinsic: Characterizes pure undoped semiconductor; electrical conductivity depends only on temperature and the band gap energy.

Ion: An atom or molecule with a net charge. **(2)** An atom with more or fewer electrons than those needed to cancel the electronic charge of the nucleus. An ion is an atom with a net electric charge.

Ionic bond: A chemical bond resulting chiefly from the electrostatic attraction between positive and negative ions. A coulombic interatomic bond existing between two adjacent and oppositely charged ions.

Kilobyte (kB): 2^{10} (= 1024, or about one thousand) bytes of information.

Kilocalorie (kcal): Unit of heat energy equal to 1000 calories. Often used to express the energy content of food or molecules: bond strengths, for example, are measured in kcal/mole. An alternative unit in wide use is the kilojoule, equal to 0.24 kcal.

Kilohertz (kHz): One thousand cycles per second.

Kilojoule: Standard unit of energy equal to 1000 joules, or 0.24 kilocalories.

Kinesin: One type of motor protein that uses the energy of ATP hydrolysis to move along a microtubule.

Kinetic energy: Energy resulting from the motion of masses.

Kinetic molecular theory: A model that assumes that an ideal gas is composed of tiny particles (molecules) in constant motion.

Kinetochore: Complex structure formed from proteins on a mitotic chromosome to which microtubules attach and which plays an active part in the movement of chromosomes to the pole. The kinetochore forms on the part of the chromosome known as the centromere.

Knowbots: Knowledge robots, first developed Vinton G. Cref and Robert E. Kahn for National Research Initiatives. Knowbots are programmed by users to scan networks for various kinds of related information, regardless of the language or form in which it expressed. "Knowbots support parallel computations at different sites. They communicate with one another, and with various servers in the network and with users."

Label: Chemical group or radioactive atom added to a molecule in order to follow its progress through a biochemical reaction or to locate it spatially. Also, as a verb, to add such a group or atom to a cell or molecule.

Lagging strand: One of the two newly made strands of DNA found at a replication fork. The lagging strand is made in discontinuous lengths that are later joined covalently.

Lambda bacteriophage (l bacteriophage): Virus that infects *E. coli;* widely used as a DNA cloning vector.

Lamin: Extracellular matrix protein found in basal laminae.

Lamins: Intermediate filament proteins that form the fibrous matrix (nuclear lamina) on the inner surface of the nuclear envelope.

Langmuir-Blodgett (LB) Film: A film of organic material (often surfactant molecules) assembled at the liquid-gas interface and subsequently transferred onto a solid substrate.

Langmuir-Blodgett: The name of a nanofabrication technique used to create ultrathin films (monolayers and isolated molecular layers), the end result of which is called a "Langmuir-Blodgett film".

Laser trimming: A method for adjusting the value of thin- or thick-film resistors by using a computer-controlled laser system.

Laser tweezers: Laser tweezers and related approaches are important for manipulation and isolation of subcellular *organelles* and structural components. [National Center for Research Resources "Integrated Genomics Technologies Workshop Report" Jan 1999]

Laser: Light Amplification by the Stimulated Emission of Radiation — quantum device that produces coherent light.

LCD (Liquid Crystal Display): LCD is the predominant technology used in flat panel displays. The principle that makes the display work is this: A crystalís alignment can be altered with an electric current. If the crystal is lined up one way ñ it will allow the light waves to pass through a polarized filter, but if the electric current alters the crystalís alignment, it will guide light so that the polarized filter blocks the light. By densely packing red, blue and green light emitting crystals next to each other on a sheet (ìcalled a substrateî), one can create a full color display. The great thing about LCD is that the crystals can be packed together closely, allowing for a higher-resolution, finer-detail display. The con is that LCDs are somewhat fragile, require a lot of power and are relatively less bright. Liquid Crystal Display — display device employing light source and electrically alterable optically active thin film.

Leading strand: One of the two newly made strands of DNA found at a replication fork. The leading strand is made by continuous synthesis in the 5´-to-3´ direction.

Leakage: The loss of all or parts of a useful agent, as of the electric current that flows through an insulator or the magnetic flux that passes outside useful flux circuits.

Lectin: Protein that binds tightly to a specific sugar. Abundant lectins derived from plant seeds are often used as affinity reagents to purify glycoproteins or to detect them on the surface of cells.

LED: Light-Emitting diode — semiconducting diode that produces visible or infrared radiation.

LEDs (Light Emitting Diodes): LEDs work on a completely different concept. Traditionally LEDs are created from two semiconductors. By running current in one direction across the semiconductor the LED emits light of a particular frequency (hence a particular color) depending on the physical characteristics of the semiconductor used. The semiconductor is covered with a piece of plastic that focuses the light and increases the brightness. These semiconductors are very durable, there is no filament, they donít require much power, theyíre brighter and they last a long time. By densely packing red, blue and green LEDs next to each other on a substrate one can create a display. The disadvantage of LEDs is that they are much larger ñ therefore the resolution is not nearly as good as LCD displays. Thatís why most LED displays are large, outdoor displays, not smaller devices, like monitors.

Leucine zipper: Structural motif seen in many DNA-binding proteins in which two a helices from separate proteins are joined together in a coiled-coil, forming a protein dimer.

Life (lifetime): The length of time the sensor can be used before its performance changes.

Life Science Applications of Deep Red T2-MP EviTag Quantum Dots: The new T2-MP EviTags are available in the deep red wavelengths used in life science research – Macoun Red (660nm) and Jonamac Red (680nm) – with carboxyl, amine, biotin and non-functionalized surfaces. Having a biologically compatible quantum dot label like the T2-MP EviTags allows further exploration and development of in vivo assays, following cells inside living organism to get a new understanding of life. In particular, it offers the possibility to create new classes of assays for use where current testing materials could produce cytotoxicity or cell death. Potential applications include live cell and even whole animal imaging, blood cancer assays, and numerous other applications.

Ligand: In protein chemistry, a small molecule that is (or can be) bound by a larger molecule is termed a ligand. In organometallic chemistry, a moiety bonded to a central metal atom is also termed a ligand; the latter definition is more common in general chemistry.

Luminescence: Defined as the mission of light from a substance in an electronically excited state.

Depending on whether the excited state is singlet or triplet, the emission is called fluorescence (less than one second decay) or phosphorescence (longer than 1 second decay). Depending on the source, molecules get the needed extra energy from different types of luminescence are distinguished: radioluminescence, photoluminescence (in the same category are fluorescence and phosphorescence), chemiluminescence and bioluminescence, electrochemiluminescence, sonochemi-luminescence and thermoluminescence.

Machine-phase chemistry: The chemistry of systems in which all potentially reactive moieties follow controlled trajectories (e.g., guided by molecular machines working in vacuum).

Macroscale: Larger than nanoscale; often implies a design that humans can directly interact with; too large to be built by a single assembler (one cubic micron of diamond contains 176 billion atoms).

Macrosensing: In medical nanorobotics, the detection of global somatic states (inside the human body) and extrasomatic states (sensory data originating outside of the human body) by in vivo nanorobots.

Magnetic field strength (designated by H) [A/m]: Magnetic field produced by a current, independent of the presence of magnetic material. The units of H are ampere-turns per meter, or just amperes per meter.

Magnetic flux density or magnetic induction (designated by B): The magnetic field produced in a substance by an external magnetic field. The units of B are tesla (T). One tesla is the magnetic flux density given by a magnetic flux of 1 weber per square meter. One weber is a magnetic flux that, linking a circuit of 1 turn, would produce in it an electromotive force of 1 volt if it were reduced to zero at a uniform rate in 1 second. Both B and H are field vectors. One henry (H) is the inductance of a closed circuit in which an electromotive force of 1 volt is produced when the electric current in the circuit varies uniformly at the rate of 1 ampere per second. The magnetic field strength and flux density are related according to: $B = \mu H$, where μ is the permeability (see under permeability).

Magnetic Force Microscopy (MFM): A method for observing local magnetic fields near a surface by scanning the surface with a magnetic probe.

Magnetic susceptibility ($\div_m$): The proportionality constant between the magnetization M (see under"magnetization") and the magnetic field strength H. The magnetic susceptibility is unitless.

Magnetization (M): The total magnetic moment per unit volume of material. Also, a measure of the contribution to the magnetic flux by some material within an H field. The magnitude of M is proportional to the applied field as: $M = \div_m \times H$, with $\div_m$ the magnetic susceptibility.

Magnetostrictive material: A material that changes dimension in the presence of a magnetic field or generates a magnetic field when mechanically deformed.

Manufacturing: The ability to make products, in this case ranging from clothing, to electronics, to medical devices, to books, to building materials, and much more.

Martensite: A metastable iron phase supersaturated in carbon that is the product of a diffusionless (athermal) transformation from austenite.

Mask: Pattern on glass, like a photographic negative, for producing integrated-circuit elements on semiconductor wafer.

Mass to Charge Ratio: A number defining how a particle will respond to an electric or magnetic field that can be calculated by dividing the mass of a particle by its charge.

Massometer: In medical nanorobotics, a nanosensor device for measuring the mass of individual molecules or small physical objects to single-proton resolution.

Materials science: Science of ceramics, glass, metals, plastics, semiconductors.

MatML Materials Markup Language: Materials property data distributed on the World Wide Web in documents using hypertext markup language.

Matter as Software: "Autonomous, motile microdevices clearly are on the horizon. They may be regarded as the first step in the evolution of a technology for "programming" the structure and properties of material objects at the microscopic and the submicroscopic levels. As this evolution progresses, *the physical and economic properties of such programmable matter are likely to become much like those of present day software.*" [MITRE Corporation]

MCM: MultiChip Module; the interconnection of two or more semiconductor chips in a semiconductor-type package.

Measurand: A physical quantity, condition, or property that is to be measured.

Meat Machine: AKA Cabinet Beast. A box containing assemblers and raw material, within which is formed meat [or whatever else it was programmed to make].

Mechanical: Pertaining to the positions and motions of atoms, as defined by the positions of their nuclei; see electronic. A purely mechanical device can be described in terms of atomic positions and motions without reference to electronic properties, save through their effect on the potential energy function.

Mechanochemistry: Chemistry accomplished by mechanical systems directly controlling the reactant molecules; the formation or breaking of chemical bonds under direct mechanical control. In this volume, the chemistry of processes in which mechanical systems operating with atomic-scale precision either guide, drive, or are driven by chemical transformations. In general usage, the chemistry of processes in which energy is converted from mechanical to chemical form, or vice versa.

Mechanosynthesis: Chemical synthesis controlled by mechanical systems operating with atomic-scale precision, enabling direct positional selection of reaction sites; synthetic applications of mechanochemistry. Suitable mechanical systems include AFM mechanisms, molecular manipulators, and molecular mill systems. Processes that fall outside the intended scope of this definition include reactions guided by the incorporation of reactive moieties into a shared covalent framework (i.e., conventional intramolecular reactions), or by the binding of reagents to enzymes or enzymelike catalysts. Molecular tools with chemically specific tip structures can be used, sequentially, to modify a work piece and build a wide range of molecular structures.

Mechatronics: The study of the melding of AI and electromechanical machines to make machines that are greater than the sum of their parts. The synergistic combination of precision mechanical engineering with electronic control.

Megabyte (MB): 2^{20} (= 1,048,576, or about one million) bytes of information.

Megahertz (Mhz): One million cycles per second

Meme: An idea that replicates through a society as it is propagated through person-to-person interaction, both direct and indirect. Memetics is a field of study that focuses on memes' role in the evolution of a culture. Self-reproducing idea or other information pattern which is propagated in ways similar to that of a gene. [Richard Dawkins, 1976]

MEMS MicroElectro Mechanical Systems: MEMS is less an application area in itself than a manufacturing or fabrication technique that enables other application areas. Many authors use MEMS as shorthand to imply a number of particular application areas. As it is used here, MEMS is a "top- down" fabrication technology that is especially useful for integrating mechanical and electrical systems together on the same chip. It is grouped in the category of integrated microsystems because these same MEMS techniques can be extended in the future to also help integrate biological and chemical components on the same chip, as discussed below. Thus far, MEMS techniques have been used to make some functional commercial devices such as sensors and single- chip measurement devices. Many researchers have used MEMS technologies as analytical tools in other areas of nanotechnology [Central Intelligence Agency, US The Global Technology Revolution, Chapter Two Technology Trends, Genomics, 2001] Stood originally for Micro-ElectroMechanical System — microscopic mechanical elements, fabricated on silicon chips by techniques similar to those used in integrated circuit manufacture, for use as sensors, actuators, and other devices. Today almost any miniaturized device (based on Si technology or traditional precision engineering, chemical or mechanical) is referred to as a MEMS device generic term to describe micron scale electrical/mechanical devices.

MEMS: Stood originally for Micro-ElectroMechanical System — microscopic mechanical elements, fabricated on silicon chips by techniques similar to those used in integrated circuit manufacture, for use as sensors, actuators, and other devices. Today almost any miniaturized device (based on Si technology or traditional precision engineering, chemical or mechanical) is referred to as a MEMS device.

Mesoscale: A device or structure larger than the nanoscale (10^-9 m) and smaller than the megascale; the exact size depends heavily on the context and usually ranges between very large nanodevices (10^-7 m) and the human scale (1 m). [AS]

Messenger molecule: A chemically recognizable molecule which can convey information after it is received and decoded by an appropriate chemical sensor.

Metal nanoshells: A new type of *nanoparticle* composed of a *semiconductor* or dielectric core coated with an ultrathin conductive layer.. By adjusting the relative core and shell thicknesses, metal nanoshells can be fabricated that will absorb or scatter light at any wavelength across the entire visible and infrared range of the electromagnetic spectrum.

Metamorphic: In medical nanorobotics, capable of adopting multiple physical configurations via smooth changes from one configuration to another.

Metastable: A classical system is metastable if it is above its minimum-energy state, but requires an energy input before it can reach a lower-energy state; accordingly, a metastable system can act like a stable system, provided that energy inputs (e.g., thermal fluctuations) remain below some threshold. Systems with strong metastability are commonly described as stable. Quantum mechanical effects can permit metastable states to reach lower energies by tunneling, without an energy input; an associated, broader definition of *metastable* embraces all systems that have a long lifetime (by some standard) in a state above the minimum-energy state.

Micellar nanocontainers: Block copolymer micelles are water- soluble biocompatible nanocontainers with great potential for delivering hydrophobic drugs. An understanding of their cellular distribution is essential to achieving selective delivery of drugs at the subcellular level. R. Savic, L. Luo, A. Eisenberg, D. Maysinger, Micellar nanocontainers distribute to defined cytoplasmic organelles Science 300 (5619): 615- 618, Apr. 25, 2003

Micelles: Micelles are small, spherical structures composed of molecules that attract one another to reduce surface tension. The head of the molecule is hydrophilic, meaning it likes water, while the interior portion is hydrophobic, meaning it avoids water.

Microbiotagraphics: Mapping the microbiotic populations present in the human body.

Microbubbles: Very small encapsulated gas bubbles (diameters of micrometers) that can be used in diagnostic and therapeutic applications. Upon exposure to sufficiently intense ultrasound, microbubbles will cavitate, rupture, disappear, release gas content, etc. Such characteristics of the microbubbles can be used to enhance diagnostic tests, dissolve blood clots, and deliver drugs or genes for therapy. MeSH 2004

Microchemistry: The development and use of techniques and equipment to study or perform chemical reactions, with small quantities of materials, frequently less than a milligram or a milliliter. MeSH 2003

Microdevices: Narrower terms: integrated microdevices, nanodevices, SED Single Electron Devices. Related terms microelectronics, microfluidics, micro- TAS.

Micro-Electromechanical Systems (MEMS): "The fabrication or micro-machining of materials to make stationary and moving structures, devices and systems of a nominal size scale from a few centimeters to a few micrometers."

Microelectronics: Narrower terms: MEMS, nanoelectronics, optoelectronics, SED Single Electron Devices. Related terms: molecular electronics, semiconductors

Microencapsulation: Individually encapsulated small particles.

Microfabrication: This technology includes techniques used to manufacture integrated circuits (ICs), discrete microelectronic devices, MEMS devices such as sensors and actuators, and various electro-optic devices. [University of Louisville Microfabrication Lab] is currently serving as a center for research activity in the areas of micromachined sensors and actuators, electro- optic devices, special- purpose microelectronic devices, planar waveguides, chemical transducers, microstrip and microgap radiation detectors, micromachined nozzles, and micromachined ink- jet printheads. Microfabrication holds the key to developing monolithically integrated devices capable of extremely high throughput. ... Microfabrication of surface characteristics provides a means of controlling the cell- environment interaction. Microfabricated structures also present new opportunities for cell, protein and nucleic acid separations and analysis- including high throughput single protein molecule sequencing. New approaches to analysis utilizing conductance techniques or tunneling probe microscopy merely emphasize the breadth of the potential of nanofabrication for future technology development relevant to biological issues. [National Center for Research Resources "Integrated Genomics Technologies Workshop Report" Jan 1999]

Microinjection: The insertion of a substance into a cell through a microelectrode. Typical applications include the injection of drugs, histochemical markers (such as horseradish peroxidase or lucifer yellow) and RNA or DNA in molecular biological studies. To extrude the substances through the very fine electrode tips, either hydrostatic pressure (pressure injection) or electric currents (ionophoresis) is employed. [OMD] A technique for introducing a solution of DNA, protein, or other soluble material into a cell using a fine microcapillary pipet. [Life Sciences Dictionary]

MNT: An abbreviation for molecular nanotechnology; refers to the concept of building complicated machines out of precisely designed molecules.

Mobility (electron, and hole): The proportionality constant between the carrier drift velocity and applied electric field.

Modulus of elasticity (E) : The ratio of stress to strain when deformation is totally elastic. Also the Young's modulus.

MOEMS MicroOpticalElectroMechanical systems: In addition to mechanical and electrical components, integrate waveguides or other optical features into the body of the silicon chip. Intellisense Corp. MEMS The Next Small Thing Google = about 10 Aug. 8, 2002; about 12 June 23, 2004; about 10 May 2, 2005

Moiety: A portion of a molecular structure having some property of interest.

Molality: The molality or molal concentration (symbol m) is the amount of substance per unit mass of solvent or mol kg^{-1}. Concentration in a liquid solution (symbol c), in terms of the number of moles of a solute dissolved in 10^6 mm^3 (10^3 cm^3) of solution in mol l^{-1}.

Molding (plastics): Shaping a plastic material by forcing it, under pressure at a high temperature, into a mold cavity.

Mole: A number of instances of something (typically a molecular species) equaling ~6.022 1023. *Mole* ordinarily means gram-mole; a kilogram-mole is ~6.022 1026.

Molecular - Nanotechnology: The technology of precisely- constructed molecular-scale machines; from nanometer: a billionth of a meter. For a good introduction, see What is Nanotechnology? or Ralph C. Merkle's nanotechnology page.

Molecular assembler: A general-purpose device for molecular manufacturing, able to guide chemical reactions by positioning individual molecules to atomic accuracy (e.g. mechanosynthesis) and to construct a wide range of useful and stable molecular structures according to precise specifications. Also known as an assembler, a molecular assembler is a molecular machine that can build a molecular structure from its component building blocks. [ZY]

Molecular Beam Epitaxy: [MBE] Process used to make compound (multi-layer) semiconductors. Consists of depositing alternating layers of materials, layer by layer, one type after another (such as the semiconductors gallium arsenide and aluminum gallium arsenide).

Molecular Integrated Microsystems (MIMS): microsystems in which functions found in biological and nanoscale systems are combined with manufacturable materials. See Molecular Integrated Microsystems

Molecular machine: A mechanical device that performs a useful function using components of nanometer scale and defined molecular structure; includes both artificial nanomachines and naturally occurring devices found in biological systems. Any machine with atomically precise parts of nanometer dimensions; can be used to describe molecular devices found in nature.

Molecular Manipulator: A device combining a proximal probe mechanism for atomically precise positioning with a molecule binding site on the tip; can serve as the basis for building complex structures by positional synthesis. A programmable device able to position molecular tools with high precision, for example, to direct a sequence of mechanosynthetic steps; a molecular assembler.

Molecular manufacturing: Manufacturing using molecular machinery, giving molecule-by-molecule control of products via positional chemical synthesis, to produce complex molecular structures manufactured to precise specifications. (2) Manufacturing using molecular machinery, giving molecule-by-molecule control of products and by-products via positional chemical synthesis. Exponential general-purpose molecular manufacturing — that's a mouthful, but what does it mean? Let's take the phrase apart to see why it is so important. The building of complex structures by mechanochemical processes. The production of complex structures via nonbiological mechanosynthesis (and subsequent assembly operations).

Molecular mechanics models: Many of the properties of molecular systems are determined by the molecular potential energy function. Molecular mechanics models approximate this function as a sum of 2-atom, 3-atom, and 4-atom terms, each determined by the geometries and bonds of the component atoms. The 2-atom and 3-atom terms describing bonded interactions roughly correspond to linear springs.

Molecular medicine: A variety of pharmaceutical techniques and gene therapies that address specific molecular diseases or molecular defects in biological systems. A variety of pharmaceutical techniques and therapies in use today. Studying molecules as they relate to health and disease, and manipulating those molecules to improve the diagnosis, prevention, and treatment of disease.

Molecular mill: A mechanochemical processing system characterized by limited motions and repetitive operations without programmable flexibility.

Molecular motors: Protein based machines that are involved in or cause movement such as the rotary devices (flagellar motor and the F1 ATPase) or the devices whose movement is directed along cytoskeletal filaments (myosin, kinesin and dynein motor families). [MeSH, 1999]

Molecular Nanoscience: An emerging interdisciplinary field that combines the study of molecular/ biomolecular systems with the science and technology of nanoscale structures and systems. The potential applications for this research are very broad and include such possibilities as 1) the use of biomolecules and cellular systems to self- assemble nanoelectronic circuitry and other nanoscale structures and 2) the use of lamellar host frameworks containing nano pores that can be tailored to include guest molecules for separation of chemicals for pharmaceutical and other applications. Molecular Nanoscience Alliance for Interdisciplinary Studies and Activities, Univ. of Minnesota, US,

Molecular Nanotechnology (MNT): A term coined by Drexler refering to technology resulting from the ability to thoroughly and inexpensively control matter on the molecular level. The ability to construct shapes, devices, and machines with atomic precision, and to combine them into a wide range of products inexpensively. Thorough, inexpensive control of the structure of matter based on molecule-by-molecule control of products and byproducts; the products and processes of molecular manufacturing, including molecular machinery. Thorough, inexpensive control of the structure of matter based on molecule-by-molecule control of products and byproducts; the products and processes of molecular manufacturing, including molecular machinery; a technology based on the ability to build structures to complex, atomic specifications by mechanosynthesis or other means; most broadly, the engineering of all complex mechanical systems constructed from the molecular level.

Molecular Recognition: A chemical term referring to processes in which molecules adhere in a highly specific way, forming a larger structure; an enabling technology for nanotechnology.

Molecular robotics: Ralph Merkle, A New Family of Six Degree of Freedom Positional Devices, Zyvex.

Molecular self-assembly: The spontaneous formation of molecules into covalently bonded, well-defined, stable structures - is a very important concept in biological systems and has increasingly become a focus of non- biological research.

Molecular sorting rotor: A class of nanomechanical device capable of selectively binding (or releasing) molecules from (or to) solution, and of transporting these bound molecules against significant concentration gradients.

Molecular surgery (molecular repair): In medical nanorobotics, the analysis and physical correction of molecular structures in the body using medical nanomachines. Analysis and physical correction of molecular structures in the body using medical nanomachines.

Molecular Systems Engineering: Design, analysis, and construction of systems of molecular parts working together to carry out a useful purpose.

Molecular Wire: A molecular wire - the simplest electronic component - is a quasi-one-dimensional molecule that can transport charge carriers (electrons or holes) between its ends. [Michael D Ward]

Molecularly imprinted polymers MIPs: A new class of materials that have artificially created receptor structures. Since their discovery in 1972, MIPs have attracted considerable interest from scientists and 3engineers involved with the development of chromatographic absorbents, membranes, sensors and enzyme and receptor mimics.

Molecule: A set of atoms linked by covalent bonds. A macroscopic piece of diamond is technically a single molecule. (Sets of atoms linked by bonds of other kinds are sometimes also termed molecules.) Group of atoms held together by chemical bonds; the typical unit manipulated by nanotechnology. The smallest particle of a chemical substance; typically a group of atoms held together in a particular patter, by chemical bonds. Molecules are groups of atoms which are bonded together. How molecules are formed is governed by the laws of physics. An illustration representing a molecule: above a gold surface. The grey balls are: carbon atoms, the little yellow balls are sulfur,: the big yellow balls are the gold,: the white are hydrogen, and the rods joining: the balls together represent the bonds.

\Moledulcar manufacturing: The automated building of products from the bottom up, molecule by molecule, with atomic precision. This will make products that are extremely lightweight, flexible, durable, and potentially very 'smart'.

Monkeywrenching: In medical nanorobotics, the mechanical or chemical jamming of cellular equilibrium processes, with a cytocidal objective.

Monoclonal antibodies: Produced by injecting animals to elicit a response from lymphocytes to produce antibodies. Lymphocytes which produce antibodies with strong binding capability can be isolated and used to produce only one kind of antibody (monoclonal) on a permanent basis once the lymphocytes are immortalized. This is accomplished by fusing them (combining them genetically) with cancer cells which have the distinction of living indefinitely in a culture. Monoclonal antibodies can be produced repeatedly and collected for use in immunodetection.

Mutation: An inheritable modification in a genetic molecule, such as DNA. Mutations may be good, bad, or neutral in their effects on an organism; competition weeds out the bad, leaving the good and the neutral.

MUX: Device for combining several signals or data streams into a single flow.

Myosin: Myosin is one of the two major proteins responsible for contraction of muscle. In muscle cells, myosin is arranged in long filaments that lie parallel to the microfilaments of actin. In muscle contraction, filaments of actin and myosin alternately unlink and chemically link in a sliding action. The energy for this reaction is supplied by adenosine triphosphate.

NAD+ (nicotine adenine dinucleotide): Coenzyme that participates in an oxidation reaction by accepting a hydride ion (H-) from a donor molecule. The NADH formed is an important carrier of electrons for oxidative phosphorylation. Coenzyme closely related to NAD+ that is used extensively in biosynthetic, rather than catabolic, pathways.

Nanapheresis: In medical nanorobotics, the removal of bloodborne medical nanorobots from the body using aphersis-like processes.

Nanarchist: Someone who circumvents government control to use nanotechnology, or someone who advocates this. [Eli Brandt, October 1991]

Nanite: Machines with atomic-scale components. (Popularized by the Star Trek episode "Evolution") As to their weight, a popular question: "Do you 'feel' heavier after you drink a mouthful of water? A mouthful of water, roughly 5 cm^3, would have the same mass as a ~2 terabot (2 trillion nanites) dose of 1 micron^3 nanorobots. You'll never feel it." Robert A. Freitas Jr. "Nanobot" and "Nanorobot" usually mean the same thing. Slang term for a nanomachine (see nanotechnology), esp. a machine able to replicate itself. [Popularized by the Star Trek episode "Evolution"]

Nano Cubic Technology: An ultra-thin layer coating that results in higher resolution for recording digital data, ultra-low noise and high signal-to-noise ratios that are ideal for magneto-resistive (MR) heads. It is capable of catapulting data cartridge and digital videotape to one-terabyte native (uncompressed) capacities and floppy disk capacities to three gigabytes. To help visualize the potential, 1TB can store up to 200 two-hour movies.

Nanoarray: an ultra-sensitve, ultra-miniaturized array for biomolecular analysis. BioForce Nanosciences' Nanoarrays utilize approximately 1/10,000th of the surface area occupied by a conventional microarray, and over 1,500 nanoarray spots can be placed in the area occupied by a single microarray domain.

Nanoassembler: the Holy Grail of nanotechnology; once a perfected nanoassembler is availble, building anything becomes possible, with physics and the imagination the only limitation (of course each item would have to be designed first, which is another small hurdle). See Bootstrapping a Nanofactory "So how hard is it to build a nanofactory? You need to start with a working fabricator, a nanoscale device that can combine individual molecules into useful shapes. But once you have that, the rest is pretty straightforward." —CRN

Nanocentrifuge: In medical nanorobotics, a proposed nanodevice that can spin materials at very high speed, imparting rotational accelerations of up to one trillion gravities (g's), thus permitting rapid sortation.

Nanocentrifuge: In medical nanorobotics, a proposed nanodevice that can spin materials at very high speed, imparting rotational accelerations of up to one trillion gravities (g's), thus permitting rapid sortation.

Nanochemistry: The goal of the Nanochemistry group is to develop chemistry beyond the realm of individual molecules by exploring how systems of several molecules or other electronically confined entities self-organize into distinct structures on the nanometer scale. The new insight is used to study and manipulate biological nanosystems and to create novel nanoarchitectures with unique structural, electronic, optical and magnetic properties. Specific research areas include bionanotechnology and self- assembling molecular electronics.

Nanochip: In Friday's issue [Aug. 1, 2001] of the journal *Science,* physicists from IBM's Thomas J. Watson Research Center [Philip G. Collins, Michael S. Arnold and Phaedon Avouris]

announce their fabrication of the world's first array of transistors made from carbon nanotube. We are approaching the limits of standard microchip technology; thus, the "nanochip" — a next-smaller microchip. [ed] They are also a next-gen device for mass storage, of significantly higher density, with greater speed, and much lower cost.

Nanochondria: Nanomachines existing inside living cells, participating in their biochemistry (like mitochondria) and/or assembling various structures.

Nanochronometer: In medical nanorobotics, a proposed clock or timing mechanism constructed of nanoscale components.

Nanocircles: Stanford scientists have synthesized a molecule of DNA that is capable of shutting off specific genes in living bacteria. Dubbed the "nanocircle," the new nanometer- size molecule might one day give researchers the ability to target harmful genes that cause cancer and other diseases in humans. A nanocluster or nanocrystal is a fragment of solid comprising somewhere between a few atoms to a few tens of thousands of atoms. Nanoclusters are therefore a novel state of matter with properties that are neither those of a bulk crystal nor those of individual atoms and molecules. Over the past 10 years huge advances have been made both in the synthesis of size- tunable, monodisperse (i.e. size- selected) nanoclusters of various chemical compositions and in the development of techniques for their assembly into nanostructured solids (facilitating the synthesis of what have been termed "designer materials"). Nanoparticles, Nanoclusters, Nanoscience Group, Nottingham Univ., UK, 2002

Nanocomputer: A computer made from components (mechanical, electronic, or otherwise) built at the nanometer scale. These computers could be many orders-of-magnititude faster than today's, which enables software to take proportional leaps. (2) A computer whose fundamental components measure only a few nanometers in size. State of the art current computer components are no smaller than about 350 nm. [Mnemosyne Mnews 21 (3) January 2001 " The Nanotechnology Initiative and Future Electronics" Presentation by Gail J. Brown, Air Force Research Laboratory, Wright- Patterson Air Force Base, November 16, 2000] (3) A computer with parts built on a molecular scale.

Nanocones: Nonplanar graphitic structures. Carbon-based structures with five-fold symmetry that form due to disclination defects in two-dimensional graphene sheets. They have been observed as nanotube caps and as freestanding structures. [North Carolina State University].

Nanocontainers: "Micellar nanocontainers" or "Micelles," these are nanoscale polymeric containers that could be used to selectively deliver hydrophobic drugs to specific sites within individual cells.

Nanocrit (Nct)**:** In medical nanorobotics, volume-fraction or bloodstream concentration of medical nanorobots, expressed as a percentage.

Nanomachining: like traditional machining, where portions of the structure are removed or modified, nanomachining involves changing the structure of nano-scale materials or molecules.

Nanomanipulation: The process of manipulating items at an atomic or molecular scale in order to produce precise structures.

NanoManipulator: uses virtual reality (VR) goggles and a force feedback probe as an interface to a scanning probe microscope, providing researchers with a new way to interact with the atomic world. Researchers can travel over genes, tickle viruses, push bacteria around, and tap on molecules - the nanoManipulator simplifies the process and allows researchers to play with their atoms. University of North Carolina at Chapel Hill (UNC-CH) *The Nanomanipulator* from the Center for Computer Integrated Systems for Microscopy and Manipulation (CISMM) at UNC Chapel Hill. Part of the Nanoscale Science Research Group (NSRG).

Nanopens & Nanopencils: (AKA: Atomic Pencil) *"Analogous to using a quill pen but on a billionth the scale"*, and may transform dip-pen nanolithography. Allows for drawing electronic circuits a thousand times smaller than current ones. The "pen" is an atomic force microscope (AFM).

NanoPGM-nanometer-scale patterned granular motion: The goal of NanoPGM is to generate millions of ìnanofingers, finger-like structures each only a few nanometers long, that might someday perform precise, massively parallel manipulation of molecules and directed assembly of other nanometer-scale objects. This ability answers one of the biggest technical challenges facing builders of nanocomputers: how to arrange as many as a trillion molecular computing components in an area only a few millimeters square. [MITRE / Alex Wissner-Gross]

Nanopharmaceuticals: Nanoscale particles used to modulate drug transport for drug uptake and delivery applications.

Nanophase Carbon Materials (carbon nanotubes, nanodiamond, nanocomposite]: A form of matter in which small clusters of atoms form the building blocks of a larger structure. These structures differ from those of naturally occurring crystals, in which individual atoms arrange themselves into a lattice.

Nanosensor: A chemical or physical sensor constructed using nanoscale components, usually microscopic or submicroscopic in size.

Nanoshells: Nanoscale metal spheres, which can absorb or scatter light at virtually any wavelength. "The nanoshells act as an amazingly versatile optical component on the nanometer scale: they may provide a whole new approach to optical materials and components," Professor Naomi Halas.

Nanosieving: In medical nanorobotics, a nanodevice that can sort molecules or other nanoscale objects by physical sieving.

Nanosieving: In medical nanorobotics, a nanodevice that can sort molecules or other nanoscale objects by physical sieving.

Nanosome: Nanodevices existing symbiotically inside biological cells, doing mechanosynthesis and disassembly for it and replicating with the cell. Similar to nanochondria. [AS January 1998]

Nanosome: Nanodevices existing symbiotically inside biological cells, doing mechanosynthesis and disassembly for it and replicating with the cell. Similar to nanochondria. [Anders Sandberg January 1998]

Nanosources: sources that emit light from nanometre-scale volumes.

Nanospheres: Self-assembling nanospheres that fit inside each other like Russian dolls are one form of a broad range of submicroscopic spheres ... The durable silica spheres, which range in size from 2 to 50 nanometers, form in a few seconds, are small enough to be introduced into the body, and have uniform pores that could enable controlled release of drugs. The spheres can absorb organic and inorganic substances including small particles of iron, which means they can be controlled by magnets and the contents released as needed. [Sandia National Labs news release "Self- assembled spheres may be helpful against disease or terrorism".

Nanosprings: A nanowire wrapped into a helix. Speculation is that they "may someday make highly sensitive magnetic field detectors, perhaps finding application in hard drive read heads. Alternatively, nanosprings could serve as positioners, or even as tiny conventional springs, for nanomachines of the future."

Nanosurgery: A generic term including molecular repair and cell surgery.

Nanoswarm: UFog and Goo

Nanosystem: A eutactic set of nanoscale components working together to serve a set of purposes; complex nanosystems can be of macroscopic size.

Nanotech: Slang for nanotechnology.

Nanotechism: the religion of nanotech, as opposed to the science of nanotech

Nanoterrorism: using MNT derived nanites to do damage to people or places.

Nernst equation: An equation relating the potential of an electrochemical cell to the concentrations of the cell components: $E = Eo + RT/zF \ln C_1/C_2$ with z the charge exchanged at the electrode and C_1 and C_2 concentrations of two electro-active compounds.

Neroprosthesis: Implanted cybernetic brain augmentation.

Neural Simulation: Imitating the functions of a neural system—such as the brain—by simulating the function of each cell.

Neurocomputation: The study of how natural and artificial neural networks process information.

Neuron Star: A neutron star used as a basis for a densely packed mind (similar to a jupiter brain or a omegon) exploiting neutronium or quark matter for computation [Damien Broderick, 1996]

Neuron: A nerve cell, such as those found in the brain.

Neuronaut: A person who explores their own neural functioning and internal mentational processes by various means, including deep introspection and meditation, psychoactive drugs, mind machines, and neuroscientific understanding.

Neurosuspension: Cryonically suspending only the head or brain of a person.

NMOS: An acronym for n-*channel metal-oxide-semiconductor*, as in NMOS transistor and NMOS logic.

Nono: A prefix meaning ten to the minus ninth power, or one billionth.

Nootropic: A cognition-enhancing drug that has no significant side-effects.

NOR: NOT-OR — logic gate whose output is the negation of that of the OR gate.

NOT: Logic gate whose output is binary 1 when its input is 0, and whose output is a 0 when its input is a 1.

Now shock: Being shocked or confused by the rapid changes that has already taken place, a kind of future shock before the present. [Michael Rothschild 1995, "Cornucopia Or Black Hole?", *Upside*]

NRAM™ - Nanotube-based/Nonvolatile RAM, developed by Nantero, using proprietary concepts and methods derived from leading-edge research in nanotechnology.

n-type: Characterizes a semiconductor containing predominantly mobile electrons (also see"p-type").

Pathogen (adjective pathogenic): An organism or other agent that causes diseases.

PCR (polymerase chain reaction): Technique for amplifying specific regions of DNA by multiple cycles of DNA polymerization, each followed by a brief heat treatment to separate complementary strands.

Peptide bond: Chemical bond between the carbonyl group of one amino acid and the amino group of a second amino acid - a special form of amide linkage.

Peptide map: Characteristic two-dimensional pattern (on paper or gel) formed by the separation of the mixture of peptides produced by the partial digestion of a protein.

Phagocytosis: Process by which particulate material is endocytosed ("eaten") by a cell. Prominent in carnivorous cells, such as *Amoeba proteus*, and in vertebrate macrophages and neutrophils. (From Greek *phagein*, to eat.)

Pharmacyte: In medical nanorobotics, a theorized (nanorobotic) device capable of delivering precise doses of biologically active chemicals to individually-addressed human body tissue cells (e.g. cell-by-cell drug delivery).

Phase shift: A time difference between the input and output signals.

Piezoelectric material: A ferroelectric material in which an electrical potential difference is created due to mechanical deformation, or conversely, in which the application of a voltage causes dimensional changes in the material.

Piezoelectric: A piezoelectric material expands or contracts when a voltage is applied across it. Piezoelectric tubes are often used to move the tip in an AFM or STM. If a piezoelectric crystal is compressed, a voltage appears across it. This effect is sometimes used to measure displacements. A commonly used piezoelectric is PZT (Pb,Zr)TiO_3.

Pinhole: The term pinhole embraces a wide variety of oxide defects and is used in a broad sense today. Listed in this category are cracks caused by thermal contraction after oxidation or by handling, and regions of oxide with low dielectric strength caused by dust particles, inadequate masking, contamination, or poor resist adhesion.

Pin-out: Diagram showing for electronic components the relations between connecting pins and internal components.

Pitting: A form of very localized corrosion wherein small pits or holes form, usually in a vertical direction.

Pixel: Picture Element — smallest element of an image, such as a dot on a computer monitor screen.

pK value: A measure of the strength of an acid on a logarithmic scale. The pK value is given by log10 (1/Ka), where Ka is the acid dissociation constant pK values often are used to compare the strengths of different acids.

Plastic deformation: Permanent or nonrecoverable deformation, accompanied by permanent atomic displacement.

Plasticizer: A low molecular weight polymer additive that enhances flexibility and workability and reduces stiffness and brittleness.

Point defect: A crystalline defect associated with one or several atomic sites.

Poisson's ratio (í): For elastic deformation, the negative ratio of lateral and axial strains that result from an applied axial stress.

Poisson's ratio: A bar of an isotropic, elastic material ordinarily shrinks laterally when it is stretched longitudinally. The lateral contracting strain divided by the applied tensile strain is Poisson's ratio, which varies from material to material.

Polarization (P): The total electric dipole moment per unit volume of dielectric material.

Polyclonal antibodies: Antibodies produced by an animal's white blood cells (lymphocytes, specifically) in response to an antigen. This response occurs naturally or can purposely be created by injecting an animal, such as a rabbit or goat, with a specific antigen. More than one kind of anti-body is produced since more than one lymphocyte is producing antibodies. This is referred to as"polyclonal". The polyclonal antibodies are isolated from the animal and can be used for detection purposes. Because the antibodies are actually a mixture with different affinities (binding capability) for the antigen of interest, some variability in performance can occur from one test to another or one batch of antibodies to another.

Polycyclic: A cyclic structure contains rings of bonds; a structure having many such rings is termed polycyclic. In the polycyclic structures of interest in this volume, a large fraction of the atoms are members of multiple small rings, resulting in considerable rigidity.

Polymer: A molecule consisting of many smaller sub-units covalently linked together.

Polymerase Chain Reaction (PCR): A chemical reaction that uses the polymerase enzyme to carry out *in vitro* replication of DNA.

Potentiometric Device: Monitors the voltage between a sensing electrode and a reference electrode. A high input impedance voltmeter is used to minimize current flow. The voltage typically is proportional to the logarithm of the analyte concentration.

Power [W]: Product of voltage and current in a component; also, refers to the field of electric energy supply.

Precision: The degree of reproducibility among several independent measurements of the same true value under specified conditions.

Presentation semaphore: In medical nanorobotics, a mechanical device used to display specific antigens, chemical ligands, or other molecular objects to the external environment, with the purpose of selectively modifying the chemical or other surface characteristics of a nanorobot exterior.

Printed circuit board: PCB — selectively metallized insulating sheet for supporting and interconnecting circuit components.

Probability density function: Consider an uncertain physical property and a corresponding space describing the range of values that the property can have (e.g., the configuration of a thermally excited *N* particle system and the corresponding 3*N* dimensional configuration space). The probability density function associated with a property is defined over the corresponding space; its value at a particular point is the probability per unit volume that the property has a value in an infinitesimal region around that point.

Protein Design, Protein Engineering: The design and construction of new proteins; an enabling technology for nanotechnology.

Protein Folding: "The process by which proteins acquire their functional, preordained, three-dimensional structure after they emerge, as linear polymers of amino acids, from the ribosome." [The Scientist]

Protein: Living cells contain many molecules that consist of amino acid polymers folded to form more-or-less definite three-dimensional structures; these are termed proteins. Short polymers lacking definite three-dimensional structures are termed peptides. Many proteins incorporate structures other than amino acids, either as covalently attached side chains or as bound ligands. Molecular objects made of protein form much of the molecular machinery of living cells.

Proteomics: The study of the full expression of proteins in an organism The term proteome refers to all the proteins expressed by a genome, and thus proteomics involves the identification of proteins in the body and the determination of their role in physiological and pathophysiological functions. ... Ultimately it is believed that through proteomics new disease markers and drug targets can be identified that will help design products to prevent, diagnose and treat disease. [e-proteomics.net]

Quantum Cryptography: A system based on quantum- mechanical principles. Eavesdroppers alter the quantum state of the system and so are detected. Developed by Brassard and Bennett, only small laboratory demonstrations have been made.

Quaternary Structure: Three-dimensional relationship of the different polypeptide chains in a protein complex.

Qubit: The quantum computing analog to a bit. Qubits exhibit superposition. Thus, unlike normal bits, qubits can be both 1 and 0 at the same time.

Resolution: The minimum distance between two objects that can be distinguished in microscopy or the minimum spacing between two features that can be fabricated with lithography. The smallest measurable change in input that will produce a small but noticeable change in the output. In the context of chemical separations, defines the completeness of separation.

Resonant frequency: The frequency at which a moving member or a circuit has a maximum output for a given input.

Reverse bias: The insulating bias for a p-n junction rectifier; electrons flow into the p side of the junction.

RF: Radio Frequency — refers to alternating voltages and currents having frequencies between 9 kHz and 3 MHz.

Ribonuclease: An enzyme that cuts RNA molecules into smaller pieces.

Ribosome: A molecular machine, found in all cells, which builds protein molecules according to instructions read from RNA molecules. Ribosomes are complex structures built of protein and RNA molecules. A naturally occurring molecular machine that manufactures proteins according to instructions derived from the cell's genes.

Rigid structure: As used in this volume, a covalent structure that is reasonably stiff. In a typical rigid structure, all modes of deformation encounter first-order restoring forces resulting from some combination of bond stretching and angle bending; such a structure cannot undergo deformation by bond torsion alone. Meeting this condition usually requires a polycyclic diamondoid structure.

RNA: Ribonucleic acid; a molecule similar to DNA. In cells, the information in DNA is transcribed to RNA, which in turn is "read" to direct protein construction. Some viruses use RNA as their genetic material.

Step response: The response of a system to an instantaneous jump in the input signal.

TATA box: Consensus sequence in the promoter region of many eucaryotic genes that binds a general transcription factor and hence specifies the position where transcription is initiated.

Technocyte: A nanoscale artificial device (especially a nanite) in the human bloodstream used for repairs, cancer protection, as an artificial immune system or for other uses. [AS 1995]

Tight junction: Cell-cell junction that seals adjacent epithelial cells together, preventing the passage of most dissolved molecules from one side of the epithelial sheet to the other.

Tight-receptor structures: A receptor structure in which a bound ligand of a particular kind is confined on all sides by repulsive interactions (note that favorable binding energies are compatible with repulsive forces). A tight-receptor structure discriminates strongly against all molecules larger than the target.

Time constant: The time it takes for the output change to reach 63% of its final value.

Triple bond: A double bond is formed when a pi bond is superimposed on a single bond; adding a second pi bond results in a triple bond. The two pi bonds have perpendicular nodal planes, and their sum has roughly cylindrical symmetry, permitting rotation in much the same manner as a single bond.

Virtual reality system: A combination of computer and interface devices (goggles, gloves, etc.) that presents a user with the illusion of being in a three dimensional world of computer-generated objects.

Viscoelasticity: A type of deformation exhibiting the mechanical characteristics of viscous flow and elastic deformation.

Viscosity: The ratio of the magnitude of an applied shear stress to the velocity gradient that it produces; in other words: a measure of a noncrystalline material's resistance to permanent deformation.

Volitional normative model of disease: In medical nanorobotics, disease is said to be present in a human being upon either (1) the failure of optimal physical (e.g. biological) functioning, or (2) the failure of desired (by the patient) functioning.

Voltage [V]: Potential difference between two points: energy to move a 1-C charge through a 1-V potential difference is 1-J.

Wave function: In quantum mechanics, a complex function extending over the configuration space of a system; its complex conjugate yields the probability density function, and other mathematical operations yield other physical quantities.

Weber: Unit of magnetic flux. One weber is a magnetic flux that, linking a circuit of 1 turn, would produce in it an electromotive force of 1 volt if it were reduced to zero at a uniform rate in 1 second.

Weight percent (wt%): Concentration specification on the basis of weight (or mass) of a particular element relative to the total alloy weight (or mass).

Western blotting: Technique by which proteins are separated and immobilized on a paper sheet and then analyzed, usually by means of a labeled antibody.

Bibliography

Averback, R.S., J. Bernholc, and D.L. Nelson. 1991. *MRS Symposium Proceedings* Vol. 206.

Baird, D. & Shew, A.: 2004, 'Probing the History of Scanning Tunneling Microscopy', in: D. Baird, A. Nordmann & J. Schummer (eds.), *Discovering the Nanoscale*, Amsterdam: IOS Press, pp. 145-156.

Baird, D.; Nordmann, A. & Schummer, J. (eds.): 2004, *Discovering the Nanoscale*, Amsterdam: IOS Press.

Baker, R.T.K. Synthesis, properties and applications of graphite nanofibers. In *R&D status and trends*, ed. Siegel et al.

Bate, R., Frazier, G., Frensley, W., Reed., M., "An Overview of Nanoelectronics," *Texas Instruments Technical Journal*, July-August 1989, pp. 13-20.

Beck, J.S., J.C. Vartuli, W.J. Roth, M.E. Leonowicz, C.T. Kresge, K.D. Schmitt, C.T.-W. Chu, D.H. Olsen, E.W. Shepard, S.B. McCullen, J.B. Higgins, and J.L. Schlenker. 1992. *J. Am. Chem. Soc*. 114:10834.

Bensaude-Vincent, B.: 2001, 'The Construction of a Discipline: Materials Science in the United States', *Historical Studies in the Physical and Biological Sciences*, 31, 223-248.

Bensaude-Vincent, B.: 2004, 'Two Cultures of Nanotechnology?', *Hyle: International Journal for Philosophy of Chemistry*, 10(2), 65-82.

Berube, D.M.: 2004, 'The Rhetoric of Nanotechnology', in: D. Baird, A. Nordmann & J. Schummer (eds.), *Discovering the Nanoscale*, Amsterdam: IOS Press, pp. 173-192.

Bowes, C.L., A. Malek, and G.A. Ozin. 1996. *Chem. Vap. Deposition* 2:97.

Braun, P.V., P. Osenar, and S.I. Stupp. 1996. *Nature*. 368: 2.

Braun, T.; Schubert, A. & Zsindely, S.: 1997, 'Nanoscience and nanotechnology on the balance', *Scientometrics*, 38, 321-325.

Brinker, C.J. 1996. *Curr. Opin. Solid State Mater. Sci.* 1:798.

Brown, N.: 2004, 'Needle on the Real: Technoscience and Poetry at the Limits of Fabrication', in: N.K. Hayles (ed.), *Nanoculture: Implications of the New Technoscience*, Bristol, UK: Intellect Books, pp. 173-190.

Bueno, O.: 2004, 'The Drexler-Smalley Debated on Nanotechnology: Incommensurability at Work?', *Hyle: International Journal for Philosophy of Chemistry*, 10(2), 83-98.

Bueno, O.: 2004, 'Von Neumann, Self-Reproduction and the Constitution of Nanophenomena', in: D. Baird, A. Nordmann & J. Schummer (eds.), *Discovering the Nanoscale*, Amsterdam: IOS Press, pp. 101-115.

Carroll, J.S.: 2001, 'Social Science Research Methods for Assessing Societal Implications of Nanotechnology', in: M.C. Roco and W.S. Bainbridge (eds.), *Societal Implications of Nanoscience and Nanotechnology*, Dordrecht: Kluwer, pp. 188-192.

Chen, C-Y., S.L. Burkett, H.-X. Li, and M.E. Davis. 1993. *Microporous Mater*. 2:27.

Chianelli, R.R. 1998. Synthesis, fundamental properties and applications of nanocrystals, sheets, and fullerenes based on layered transition metal chalcogenides. In *R&D status and trends,* ed. Siegel et al.

Chianelli, R.R., M. Daage, and M.J. Ledoux. 1994. *Advances in Catalysis* 40:177.

Claeson, T., and Likharev, K.," Single Electronics," *Scientific American,* June 1992, pp. 80-85.

Coenen, C.: 2004, 'Nanofuturismus: Anmerkungen zu seiner Relevanz, Analyse und Bewertung', *Technikfolgenabschätzung ' Theorie und Praxis*, 13 (2), 78-85.

Crandall, B.C. and Lewis, J., *Nanotechnology: Research and Perspectives*, MIT Press, Cambridge, 1992.

Dupuy, J.P.: 2004, 'Complexity and Uncertainty: A Prudential Approach To Nanotechnology', in: European Commission (Community Health and Consumer Protection): *Nanotechnologies: A Preliminary Risk Analysis on the Basis of a Workshop, Brussels, 1-2 March 2004*, pp. 71-93.

Dupuy, J.P.: 2004, 'Pour une évaluation normative du programme nanotechnologique', *Annales des Mines*, (February 2004), 27-32.

Einsiedel, E.F.; Goldenberg, L.: 2004, 'Dwarfing the Social? Nanotechnology Lessons from the Biotechnology Front', *Bulletin of Science, Technology & Society*, 24, 28-33.

European Commission: 2004, *Converging Technologies: Shaping the Future of European Societies (Report of the High Level Expert Group "Foresighting the New Technology Wave")*, Brussels: European Commission Research.

F. Buot, "Mesoscopic Physics and Nanoelectronics: Nanoscience and Nanotechnology," *Physics Reports*, pp.73-174, 1993.

F. Capasso and S. Datta, "Quantum Electron Devices," *Physics Today*, pp.74-82, February 1990.

Feynman, R., "There's Plenty of Room at the Bottom: An invitation to Enter a New Field of Physics," Talk at the Annual Meeting of the American Physical Society, 29 December 1959. Reprinted in Appendix B of Crandall and Lewis. See citation above.

Fielder, F.A.; Reynolds, G.H.: 1994, 'Legal Problems of Nanotechnology: An Overview', *Southern California Interdisciplinary Law Journal*, 3, 593-629.

Fogelberg, H. & Glimell, H.: 2003, 'Molecular Matters: In Search of the Real Stuff', in: H. Fogelberg & H. Glimell, *Bringing Visibility to the Invisible: Towards A Social Understanding of Nanotechnology*, Göteborg: Göteborg University, pp. 5-32 [www.sts.gu.se/publications/STS_report_6.pdf].

Fogelberg, H. & Glimell, H.: 2003, *Bringing Visibility to the Invisible: Towards A Social Understanding of Nanotechnology*, Göteborg: Göteborg University [www.sts.gu.se/publications/STS_report_6.pdf].

Fogelberg, H.: 2003, 'The Grand Politics of Technoscience: Contextualizing Nanotechnology', in: H. Fogelberg & H. Glimell, *Bringing Visibility to the Invisible: Towards A Social Understanding of Nanotechnology*, Göteborg: Göteborg University, pp. 33-54 [www.sts.gu.se/publications/STS_report_6.pdf].

Fogelberg, H.: 2003, 'The Material Culture of Nanotechnology', in: H. Fogelberg & H. Glimell, *Bringing Visibility to the Invisible: Towards A Social Understanding of Nanotechnology*, Göteborg: Göteborg University, pp. 99-114.

Frazier, G., "An Ideology For Nanoelectronics," in *Concurrent Computations: Algorithms, Architecture, and Technology*, Plenum Press, New York, 1988.

Glimell, H.: 2001, 'Challenging Limits ' Excerpts from an Emerging Ethnography of Nano Physicists', in: H. Glimell & O. Johlin (eds.), *The Social Production of Technology: On the Everyday Life with Things*, Göteburg, SE: BAS Publisher, chapter 7, pp. 111-131 (reprinted in: H. Fogelberg & H. Glimell, *Bringing Visibility to the Invisible: Towards A Social Understanding of Nanotechnology*, Göteborg: Göteborg University, pp. 115-139.

Glimell, H.: 2001, 'Dynamics of the Emerging Field of Nanoscience', in: M.C. Roco and W.S. Bainbridge (eds.), *Societal Implications of Nanoscience and Nanotechnology*, Dordrecht: Kluwer, pp. 156-160.

Glimell, H.: 2003, 'A Nano Narrative: The Mircopolitics of a New Generic Enabling Technology', in: H. Fogelberg & H. Glimell, *Bringing Visibility to the Invisible: Towards A Social Understanding of Nanotechnology*, Göteborg: Göteborg University, pp. 55-77 [www.sts.gu.se/publications/STS_report_6.pdf].

Glimell, H.: 2003, 'Dynamics of the Emerging Field of Nanoscience', in: H. Fogelberg & H. Glimell, *Bringing Visibility to the Invisible: Towards A Social*

Understanding of Nanotechnology, Göteborg: Göteborg University, pp. 79-85 [www.sts.gu.se/publications/STS_report_6.pdf].

Glimell, H.: 2004, 'Grand Visions and Lilliput Politics: Staging the Exploration of the 'Endless Frontier", in: D. Baird, A. Nordmann & J. Schummer (eds.), *Discovering the Nanoscale*, Amsterdam: IOS Press, pp. 231-246.

Goldhaber-Gordon, D., Montemerlo, M.S., Love, J.C., Opiteck, G.J., and Ellenbogen, J.C., "Overview of Nanoelectronic Devices," submitted to the *Proceedings of the IEEE*, February 1997. For more information, please send e-mail to nanotech@mitre.org.

Gorman, M.E.; Groves, J.F. & Shrager, J.: 2004, 'Societal Dimensions of Nanotechnology as a Trading Zone: Results from a Pilot Project', in: D. Baird, A. Nordmann & J. Schummer (eds.), *Discovering the Nanoscale*, Amsterdam: IOS Press, pp. 63-73.

Grinbaum, A. & Dupuy, J.-P.: 2004, 'Living with Uncertainty: Toward the Ongoing Normative Assessment of Nanotechnology', *Techne: Research in Philosophy and Technology*, 8(3) (forthcoming).

Grinbaum, A.: 2004, 'La condition de l'homme moderne et les nanotechnologies', in: G. Nivat (ed.), *Les limites de l´humain. 39èmes Rencontres Internationales de Genève*, L´Age d´Homme, Genève, p. 141.

Grunwald, A.: 2004, 'Ethische Aspekte der Nanotechnologie. Eine Felderkundung', *Technikfolgenabschätzung ' Theorie und Praxis*, 13 (2), 71-78.

Hullmann, A. & Meyer, M.: 2003, 'Publications and Patents in Nanotechnology. An overview of previous studies and the state of the art', *Scientometrics*, 58 (3) 507-527.

Jones, C.W., and W.J. Koros. 1994. *Carbon* 32:1419.

Jounet, C., W.K. Maser, P. Bernier, A. Loiseau, M. Lamy de la Chapelle, S. Lefrant, P. Deniard, R. Lee, and J.E. Fischer. 1997. *Nature* 388:756.

K.E. Drexler, *Nanosystems: Molecular Machinery, Manufacturing, and Computation*, Wiley, New York, 1992.

Karger, J., and D.M. Ruthven. 1992. *Diffusion in zeolites*. New York: J. Wiley.

Keiper, A.: 2003, 'The Nanotechnology Revolution', *The New Atlantis*, 2, 17-34.

Khushf, G.: 2004, 'A Hierarchical Architecture for Nano-scale Science and Technology: Taking Stock of the Claims About Science Made By Advocates of NBIC Convergence', in: D. Baird, A. Nordmann & J. Schummer (eds.), *Discovering the Nanoscale*, Amsterdam: IOS Press, pp. 21-33.

Khushf, G.: 2004, 'Systems Theory and the Ethics of Human Enhancement: A Framework for NBIC Convergence', *Annals of the New York Academy of Sciences*, 1013, 124-149.

Krätschmer, W., L.D. Lamb, K. Fostiropoulos, and D.R. Huffman. 1990. *Nature* 347:354.

Kupperman, A., S. Nadimi, S. Oliver, G. Ozin, J. Garcés, and M. Olken. 1993. *Nature* 365:239.

Kuusi, O.; Meyer, M.: 2002, 'Technological generalizations and leitbilder ' the anticipation of technological opportunities', *Technological Forecasting & Social Change*, 69, 625-639.

Landon, B.: 2004, 'Less is More: Much Less is Much More: The Insistent Allure of Nanotechnology Narratives in Science Fiction', in: N.K. Hayles (ed.), *Nanoculture: Implications of the New Technoscience*, Bristol, UK: Intellect Books, pp. 131-146.

Laszlo, P.: 2004, 'Is There Life After Partington?', *Hyle: International Journal for Philosophy of Chemistry*, 10(2), 169-178.

Lenhard, J.: 2004, 'Nanoscience and the Janus-Faced Character of Simulations', in: D. Baird, A. Nordmann & J. Schummer (eds.), *Discovering the Nanoscale*, Amsterdam: IOS Press, pp. 93-100.

Lent, C.S., Tougaw, P.D., Porod, W., Bernstein, G.H., "Quantum Cellular Automata," *Nanotechnology*, Vol. 4, p. 49, 1993.

Lewak, S.E.: 2004, 'What's the Buzz? Tell Me What's A-Happening: Wonder, Nanotechnology, and Alice's Adventures in Wonderland', in: N.K. Hayles (ed.), *Nanoculture: Implications of the New Technoscience*, Bristol, UK: Intellect Books, pp. 201-.

Lide, D.R., ed. 1993-1994. *CRC Handbook of Chemistry and Physics*, 74th ed.

Lin-Easton, P.C.: 2001, 'It's Time for Environmentalists to Think Small ' Real Small: A Call for the Involvement of Environmental Lawyers in Developing Precautionary Policies Molecular Nanotechnology', *Georgetown International Law Review*, 14, 106-134.

López, J.: 2004, 'Bridging the Gaps: Science Fiction in Nanotechnology', *Hyle: International Journal for Philosophy of Chemistry*, 10(2), 129-152.

Lösch, A.: 2004, 'Nanomedicine and Space: Discursive Orders of Mediating Innovations', in: D. Baird, A. Nordmann & J. Schummer (eds.), *Discovering the Nanoscale*, Amsterdam: IOS Press, pp. 193-202.

M.C. Hersam, G.C. Abeln, J.W. Lyding, "An approach for efficiently locating and electrically contacting nanostructures fabricated via UHV-STM lithography on Si(100)," *Microelectronic Engineering* 47(1999):235-.

M.C. Hersam, G.C. Abeln, J.W. Lyding, "An approach for efficiently locating and electrically contacting nanostructures fabricated via UHV-STM lithography on Si(100)," *Microelectronic Engineering* 47(1999):235-.

M.C. Roco, "National Nanotechnology Initiative: Overview," September 20, 2004.

Mackenzie, D.: 1996a, 'Economic and Sociological Explanations of Technological Change', in: *Knowing Machines: Essays on Technical Change*, MIT Press, Cambridge, MA, pp. 49-65.

Mahmood U., C.H. Tung, A. Bogdanov Jr., R. Weissleder. 1999. "Near-Infrared Optical Imaging of Protease Activity for Tumor Detection." *Radiology*, Dec; 213 (3): pp. 866-870.

Majetich, S.A. and A.C. Canter. 1993. *J. Phys. Chem*. 97:8727.

Malinow, R., Z.F. Mainen, and Y. Hayashi. 2000. LTP mechanisms: from silence to four-lane traffic. *Curr. Opin. Neurobiol*. 10(3): p. 352-7.

Mark, H., Statement of The Honorable Hans Mark, Director, Defense Research and Engineering, before the House Armed Service Committee, October 20, 1999. Ref: http://www.house.gov/hasc/testimony/ 106thcongress/99-10-20mark.htm.

Marty, A.: 2003, *Testimony before the Committee on Science on HR 766, Nanotechnology Research and Development Act*, US Government Printing Office, Washington, DC.

Marx, K.: 1963 [1847], *The Poverty of Philosophy*, International Publishers, New York.

Matsuki, H., K. Ofuji, N. Chubachi, and S. Nitta. 1996. Signal transmission for implantable medical devices using figure-of-eight coils. *IEEE Trans. on Magnetics.*, vol. 32. No.5, 5121. September.

Matsuki, H., Y. Yamakata, N. Chubachi, S. Nitta, and H. Hashimoto. 1996. Transcutaneous DC-DC converter for totally implantable artificial heart using synchronous rectifier. *IEEE Trans. On Magnetics*. 32, No.5, 5118 (Sept.).

McCulloch, W.S.: 1962, 'The imitation of one form of life by another – Biomimesis', in: E.E. Bernard & M.R. Kare (eds.), *Biological prototypes and synthetic systems*, Plenum Press, New York, vol. 1., p. 393-97.

McHale, B.: 1987, *Postmodern Fiction*, Methuen, New York & London.

Meadows, D.L., et al. 1972. *The Limits to Growth*, New York: University.

Mehl, R.F., and R.W. Cahn. 1983. Historical development. In *Physical metallurgy*. North Holland. Milligan, W.W., S.A. Hackney, M. Ke, and E.C. Aifantis. 1993. *Nanostructured Materials* 2:267.

Mehta, M.D.: 2002, 'Nanoscience and Nanotechnology: Assessing the Nature of Innovation in These Fields', *Bulletin of Science, Technology & Society*, 22(4), 269-273.

Mehta, M.D.: 2004, 'From Biotechnology to Nanotechnology: What Can We Learn from Earlier Technologies?, *Bulletin of Science, Technology & Society*, 24, 34-39.

Merz, J.L. 1986. "The Optoelectronics Joint Research Laboratory: Light Shed on Cooperative Research in Japan." *Scientific Bulletin* (ONR Far East Office). 11 (4), 1-30 (Oct.-Dec.).

Methods of Molecular Quantum Mechanics, by R. McWeeny, Second Edition, Academic Press 1989.

Meyer, G. and Amer, N.M. (1988) Novel optical approach to atomic force microscopy. Appl. Phys. Lett. 53(12), 1045-1047

Michael E.J. Holwill, Peter Satir, "II.4 Generation of Propulsive Forces by Cilia and Flagella," in J. Bereiter-Hahn, O.R. Anderson, W.-E. Reif, eds., *Cytomechanics: The Mechanical Basis of Cell Form and Structure*, Springer-Verlag, Berlin, 1987, pp. 120-130.

Michael Page, Donald W. Brenner, "Hydrogen abstraction from a diamond surface: *Ab initio* quantum chemical study using constrained isobutane as a model," *J. Am. Chem. Soc.* 113(1991):3270-3274.

Milburn, C.: 2004, 'Nano/Splatter: Disintegrating the Postbiological Body', *New Literary History*, 35 (in print).

Minsky, M.: 1995, 'Virtual Molecular Reality', in: Krummenacker, M. & Lewis, J. (eds.), *Prospects in Nanotechology. Proceedings of the 1st general conference on nanotechnology: developments, applications, and opportunities, November 11-14, 1992, Palo-Alto*, John Wiley & Sons, New York, pp. 187-205.

Mody, C.C.M.: 2000, "A New Way of Flying': Differance, Rhetoric, and the Autogiro in Interwar Aviation', *Social Studies of Science*, 30, 513-543.

Mody, C.C.M.: 2001, 'A Little Dirt Never Hurt Anyone: Knowledge-Making and Contamination in Materials Science', *Social Studies of Science*, 31, 7-36.

Mody, C.C.M.: 2004, 'How Probe Microscopists Became Nanotechnologists', in: D. Baird, A. Nordmann & J. Schummer (eds.), *Discovering the Nanoscale*, Amsterdam: IOS Press, pp. 119-133.

Mody, C.C.M.: 2004, 'How Probe Microscopists Became Nanotechnologists', in: D. Baird, A. Nordmann, J. Schummer (eds.), *Discovering the Nanoscale*, IOS Press, Amsterdam, pp. 119-133.

Mody, C.C.M.: 2004, 'Instruments in Training: The Growth of American Probe Microscopy in the 1980s', in: D. Kaiser (ed.), *Pedagogy and the Practice of Science: Producing Physical Scientists, 1800-2000*, Cambridge, MA: MIT Press (forthcoming).

Mody, C.C.M.: 2004, 'Small, but Determined: Technological Determinism in Nanoscience', *Hyle: International Journal for Philosophy of Chemistry*, 10(2), 99-128.

Mody, C.C.M.: 2004, *Crafting the Tools of Knowledge: The Invention, Spread, and Commercialization of Probe Microscopy, 1960-2000*, Ph.D. dissertation, Cornell University.

Molecular engineering: An approach to the development of general capabilities for molecular manipulation, by K. Eric Drexler, Proceedings of the National Academy of Sciences U.S.A. 78(9): 5275-5278, 1981.

Molecular Mechanics, by Ulrich Burkert and Norman L. Allinger, ACS Monograph 177, American Chemical Society, 1982.

Molecular Mechanics: The Art and Science of Parameterization, by J. Phillip Bowen and Norman L. Allinger, page 81-97 of reference [15].

Montemerlo, M.S., Love, J.C., Opiteck, G.J., Goldhaber, D. J., and Ellenbogen, J.C., "Technologies and Designs for Electronic Nanocomputers," MITRE Technical Report 96W0000044, The MITRE Corporation, McLean, VA, July 1996. For more information, send e-mail to nanotech@mitre.org.

Moor, J.H. & Weckert, J.: 2004, 'Nanoethics: Assessing the Nanoscale From an Ethical Point of View', in: D. Baird, A. Nordmann & J. Schummer (eds.), *Discovering the Nanoscale*, Amsterdam: IOS Press, pp. 301-310.

Moore, J.C., H.M. Jin, O. Kuchner, and F.H. Arnold. 1997. Strategies for the in vitro evolution of protein function: Enzyme evolution by random recombination of improved sequences. *J. Mol. Biol*. 272:336-347.

Moronne, M.M.; Ben Dahan, M.; Bruchez, M.; Hamamoto, D.J.; Weiss, S.; Alivisatos, A.P. 1999. "Nanocrystal phosphors as biological probes for fluorescence microscopy", *Biophysical Journal*, 76, A450-A450.

Morris, D.G., and M.A. Morris. 1997. *Materials Science Forum* 235-238:861. Nagpal, P., and I. Baker. 1990. *Scripta Metall. Mater*. 24:2381.

Moshe Sipper, "Fifty Years of Research on Self-Replication: An Overview," *Artificial Life 4*(Summer 1998):237-257, in "Special Issue on Self-Replication." See also: Moshe Sipper, "The Artificial Self-Replication Page," http://lslwww.epfl.ch/~moshes/selfrep/

Mui, C. and C. Musgrave, "The Hydroxylation and Oxidation of the Ge(100)-2x1 Surface by Hydrogen Peroxide and Water," *Langmuir* , 20, 7604-7609 (2004).

Mui, C., S. Bent and C. Musgrave, "A Quantum Chemistry Based Statistical Mechanical Model of Hydrogen Deposition from Si(100)-2x1, Ge(100)-2x1, and SiGe Alloy Surfaces," *Journal of Physical Chemistry B* , 108, 18243-18253, (2004).

Mui, C., Y. Widjaja, J. Kang, and C. Musgrave, "Surface Reaction Mechanisms for Atomic Layer Deposition of Silicon Nitride," *Surface Science,* 557, 159-170, (2004).

Mumford, Lewis. 1934. *Technics and Civilization* (New York: Harcourt Brace Jovanovich, Inc.). National Science and Technology Council (NSTC). 1999. *Nanotechnology: Shaping the World Atom by Atom* (Washington, DC: September; available at http://www.nano.gov).

Munn Sanchez, E.: 2004, 'The Expert's Role in Nanoscience and Technology', in: D. Baird, A. Nordmann & J. Schummer (eds.), *Discovering the Nanoscale*, Amsterdam: IOS Press, pp. 257-266.

Murray, C.B.; Kagan, C.R.; Bawendi, M.G. 2000. "Synthesis and characterization of monodisperse nanocrystals and close-packed nanocrystal assemblies", *Annual Review of Materials Science*, 30, 545- 610.

N. Hanai, K. Nakamura, K. Shitara, "Recombinant antibodies against gangliosides expressed on tumor cells," *Cancer Chemother. Pharmacol.* 46(2000):S13-S17.

N. Jonoska, P. Sa-Ardyen and N.C. Seeman, Computation by Self-Assembly of DNA Graphs, Journal of Genetic Programming and Evolvable Machines 4, 123-137 (2003).

N. Jonoska, S. Liao, & N.C. Seeman, Transducers with programmable input by DNA Self-assembly, Aspects of Molecular Computing, Lecture Notes in Computer Science 2340, Springer-Verlag, Berlin, 219-240 (2004).

N. Koumura, R.W. Zijlstra, R.A. van Delden, N. Harada, B.L. Feringa, "Light-driven monodirectional molecular rotor," *Nature* 401(9 September 1999):152-155.

N.C. Seeman & A.M. Belcher, Emulating Biology: Nanotechnology from the Bottom Up, Proceedings of the National Academy of Sciences (USA) 99 (supp. 2), 6451-6455 (2002).

N.C. Seeman and N.R. Kallenbach, Design of Immobile Nucleic Acid Junctions. Biophysical Journal 44, 201-209 (1983).

N.C. Seeman and N.R. Kallenbach, Nucleic Acid Junctions, The Tensors of Life? In: Nucleic Acids: The Vectors of Life, ed. by B. Pullman and J. Jortner, D. Reidel, Dordrecht, 183-200 (1983).

N.C. Seeman and N.R. Kallenbach, Nucleic Acid Junctions: A Successful Experiment in Macromolecular Design. In: Molecular Structure: Chemical Reactivity and Biological Activity, ed. by J.J. Stezowski J.-L. Huang and M.-C. Shao, Oxford University Press, Oxford, (1988) pp. 189-194.

N.C. Seeman and P.S. Lukeman, Nucleic Acid Nanostructures, Reports on Progress in Physics 68, 237-270 (2005).

N.C. Seeman, At the Crossroads of Chemistry, Biology and Materials: Structural DNA Nanotechnology, Chemistry & Biology 10, 1151-1159 (2003).

N.C. Seeman, B. Ding, S. Liao, T. Wang, W.B. Sherman, P.E. Constantinou, J. Kopatsch, C. Mao, R. Sha, F. Liu, H. Yan & P.S. Lukeman, Experiments in Structural DNA Nanotechnology: Arrays and Devices , Proc. SPIE; Nanofabrication: Technologies, Devices and Applications 5592, 71-81 (2005).

N.C. Seeman, Biochemistry and Structural DNA Nanotechnology: An Evolving Symbiotic Relationship, Biochemistry 42, 7259-7269 (2003).

N.C. Seeman, Branched DNA: A 3-D Structural Design System, Clinical Chemistry 39, 722-724 (1993).

N.C. Seeman, De Novo Design of Sequences for Nucleic Acid Structure Engineering, Journal of Biomolecular Structure and Dynamics 8, 573-581 (1990).

N.C. Seeman, Design and Engineering of Nucleic Acid Nanoscale Assemblies, Current Opinion in Structural Biology, 6, 519-526 (1996).

N.C. Seeman, Directing the Structure of Matter through Nanotechnology, Proceedings of IEEE International Joint Symposia on Intelligence and Systems IEEE Computer Society, Los Alamitos, CA, 146-150 (1998).

N.C. Seeman, DNA Components for Molecular Architecture. Accounts of Chemical Research 30, 357-363 (1997).

N.C. Seeman, DNA Engineering and its Application to Nanotechnology, Trends in Biotechnology 17, 437-443 (1999).

N.C. Seeman, DNA in a Material World, Nature 421, 33-37 (2003).

N.C. Seeman, DNA Nanostructures for Mechanics and Computing: Nonlinear Thinking With Life's Central Molecule. NanoBiotechnology. Editors, Chad Mirkin and Christof Niemeyer, Wiley-VCH Verlag GmbH & Co., KgaA, Weinheim, Chapter 20, pp 308-318 (2004).

N.C. Seeman, DNA Nanostructures for Mechanics and Computing: Nonlinear Thinking With Life's Central Molecule. NanoBiotechnology. Editors, Chad Mirkin and Christof Niemeyer, Wiley-VCH Verlag GmbH & Co., Weinheim, Chapter 20, pp 308-318 (2004).

N.C. Seeman, DNA Nanotechnology, Encyclopedia of Supramolecular Chemistry, Marcel Dekker, New York, 475-483 (2004).

N.C. Seeman, DNA Nanotechnology, In: WTEC Workshop Report on R & D Status and Trends in Nanoparticles, Nanostructured Materials, and Nanodevices in the United States, ed. by R.W. Siegel, E. Hu and M.C. Roco, International Technology Research Institute,Baltimore (1998).

N.C. Seeman, DNA Nanotechnology: From Topological Control to Structural Control, in Pattern Formation in Biology, Vision and Dynamics, ed. by A.Carbone, M.Gromov, P.Pruzinkiewicz, World Scientific Publishing Company, Singapore, 271-309 (2000).

N.C. Seeman, DNA Nanotechnology: Life's Central Performer in a New Role, Biological Physics Newsletter 2 (1) 2-6 (2002).

N.C. Seeman, DNA Nanotechnology: Novel DNA Constructions. Annual Review of Biophysics and Biomolecular Structure 27, 225-248 (1998).

N.C. Seeman, DNA Nicks and Nodes and Nanotechnology, NanoLetters 1, 22-26 (2001).

N.C. Seeman, DNA Structural Engineering Utilizing Immobile Junctions, Current Opinion in Structural Biology 1, 653-661 (1991).

N.C. Seeman, DNA: Beyond the Double Helix, Macromolecular Symposia 201, 237-244 (2003).

N.C. Seeman, F. Liu, C. Mao, X. Yang, L. A. Wenzler, R. Sha, W. Sun, Z. Shen, X. Li, J. Qi, Y. Zhang, T.-J. Fu, J. Chen and E. Winfree, Two Dimensions and Two States in DNA Nanotechnology, Proceedings of the 11th Conversation in Biomolecular Stereodynamics, ed. by R.H. Sarma and M.H. Sarma, Adenine Press, New York, 253-262 (2000).

N.C. Seeman, From Genes to Machines: DNA Nanomechanical Devices, Trends in Biochemical Sciences 30, 119-235 (2005).

N.C. Seeman, H. Wang, B. Liu, J. Qi, X. Li, X. Yang, F. Liu, W. Sun, Z. Shen, Y. Wang, R. Sha, C. Mao, S. Zhang, T.-J. Fu, S. M. Du, J.E. Mueller, Y. Zhang and J. Chen, The Perils of Polynucleotides: The Experimental Gap Between the Design and Assembly of Unusual DNA Structures, In: DNA Based Computers II, ed.by L.F. Landweber and E.B. Baum, Am. Math. Soc., Providence, pp. 215-233 (1998).

N.C. Seeman, H. Wang, J. Qi., X.J. Li, X.P. Yang, Y. Wang, H. Qiu, B. Liu, Z. Shen, W. Sun, F. Liu, J.J. Molenda, S.M. Du, J. Chen, J.E. Mueller., Y. Zhang, T.-J. Fu, and S. Zhang, DNA Nanotechnology and Topology. In: Biological Structure and Dyinamics, ed. by R.H. Sarma and M.H. Sarma, Adenine Press, New York, vol. 2, pp. 319-341 (1996).

N.C. Seeman, H. Wang, X. Yang, F. Liu, C. Mao, W. Sun, L. Wenzler, Z. Shen, R. Sha, H. Yan, M.H. Wong, P. Sa-Ardyen, B. Liu, H. Qiu, X. Li, J. Qi, S.M. Du, Y. Zhang, J.E. Mueller, T.-J. Fu, Y. Wang, and J. Chen, New Motifs in DNA Nanotechnology, Nanotechnology 9, 257-273 (1998).

N.C. Seeman, In the Nick of Space: Generalized Nucleic Acid Complementarity and the Development of DNA Nanotechnology, Synlett 2000, 1536-1548 (2000).

N.C. Seeman, It Started with Watson and Crick, But it Sure Didn't End There: Pitfalls and Possibilities beyond the Classic Double Helix, Natural Computing 1, 53-84 (2002).

N.C. Seeman, J. Chen, S.M. Du, J.E. Mueller, Y. Zhang, T.-J. Fu, H. Wang, Y. Wang, and S. Zhang, Synthetic DNA Knots and Catenanes, New Journal of Chemistry 17, 739-755 (1993).

N.C. Seeman, J. Chen, Y. Zhang, B. Liu, H. Qiu, T.-J. Fu, Y. Wang, X. Li, X. Yang, J. Qi, F. Liu, L.A. Wenzler, S, Du, J.E. Mueller, H. Wang, C. Mao, W. Sun, Z. Shen, M.H. Wong, H, Yan and R. Sha, A Bottom-Up Approach to Nanotechnology Using DNA, In: Biological Molecules in Nanotechnology, ed. by S.C. Lee, IBC Libraries, Southborough, MA, Chapter 8 (1998).

N.C. Seeman, J. Qi, X. Li, X. Yang, N.B. Leontis, B. Liu, Y.Zhang, S.M. Du and J. Chen, The Control of DNA Structure: From Topological Modules to Geometrical Modules. In: Modular Chemistry, ed. by J. Michl, Kluwer, 95-104 (1997).

N.C. Seeman, Key Experimental Approaches in DNA Nanotechnology, Current Protocols in Nucleic Acid Chemistry, Unit 12.1, John Wiley & Sons, New York (2002).

N.C. Seeman, Macromolecular Design, Nucleic Acid Junctions and Crystal Formation. Journal of Biomolecular Structure and Dynamics 3, 11-34 (1985).

N.C. Seeman, Nanoengineering with DNA, Biomolecular Materials: Materials Research Society Symposium Proceedings 292, 123-134 (1993).

N.C. Seeman, Nanoscale Assembly and Manipulation of Branched DNA: A Biological

Starting Point for Nanotechnology, NANOCON Proceedings, ed. by J. Lewis and J.L. Quel, NANOCON, P.O. Box 40176, Bellevue, WA 98004, pp. 101-123; transcript of oral presentation, pp. 30-36 (1989).

N.C. Seeman, Nucleic Acid Junctions and Lattices. Journal of Theoretical Biology 99, 237-247 (1982).

N.C. Seeman, Nucleic Acid Junctions: Building Blocks for Genetic Engineering in Three Dimensions. In: Biomolecular Stereodynamics, ed. by R.H. Sarma, Adenine Press, New York (1981), pp. 269-277.

N.C. Seeman, Nucleic Acid Nanostructures and Topology. Angewandte Chemie. 110, 3408-3428 (1998); Angewandte Chemie International Edition 37, 3220-3238 (1998).

N.C. Seeman, Physical Models for Exploring DNA Topology. Journal of Biomolecular Structure and Dynamics 5, 997-1004 (1988).

N.C. Seeman, Structural Control and Engineering of Nucleic Acids, In: Concepts in Protein Engineering and Design, P. Wrede and G. Schneider, eds., Walter-de-Gruyter, Berlin, 319-343 (1994).

N.C. Seeman, Structural DNA Nanotechnology: A New Organizing Principle for Advanced Nanomaterials, Materials Today 6 (7), 24-29 (2003).

N.C. Seeman, Synthetic DNA Topology, Molecular Catenanes, Rotaxanes and Knots, ed. by J.-P. Sauvage and C. Dietrich-Buchecker, Wiley-VCH, Weinheim pp. 323-356 (1999).

N.C. Seeman, The Construction of 3-D Stick Figures from Branched DNA, DNA and Cell Biology 10, 475-486 (1991).

N.C. Seeman, The Interactive Manipulation and Design of Macromolecular Architecture Utilizing Nucleic Acid Junctions. Journal of Molecular Graphics 3, 34-39 (1985).

N.R. Kallenbach, R.-I. Ma and N.C. Seeman, An Immobile Nucleic Acid Junction Constructed from Oligonucleotides. Nature 305, 829-831 (1983).

Nardi, B., O'Day, V. 1999. *Information Ecologies: Using Technology with Heart*. MIT Press.

National Coordination Office for Information Technology, *Ethical, Legal, and Societal Implications of Information Technology Research and Development*. Available at: http://www.itrd.gov, October 1999.

National Human Genome Research Institute, *Ethical, Legal, and Social Implications of Human Genetics Research of the National Institutes of Health*. Available at:

National Science and Technology Council (NSTC). 2000. *National Nanotechnology Initiative: Leading to the Next Industrial Revolution* (Supplement to the President's FY 2001 Budget) (Washington, DC: February; available at http://www.nano.gov).

National Science and Technology Council (NSTC). 2000. *National Nanotechnology Initiative: Leading to the Next Industrial Revolution*. February. (http://www.nano.gov/nni.pdf).

Naughton, G. 1999. The Advanced Tissue Sciences story. *Sci. Am.* 280(4): p. 84-5.

Nauta, K. and R.E. Miller. 1999. Non-equilibrium self-assembly of long chains of polar molecules in superfluid helium. *Science*. Vol. 283. 1895-1897.

New Approaches to Empirical Force Fields, by Uri Dinur and Arnold T. Hagler, pages 99-164, reference [15].

Nicholas Wade, "A DNA Success Raises Bioterror Concern," New York Times, January 12, 2005.

Nieman, G.W., J.R. Weertman, and R.W. Siegel. 1991a. *Mater. Res. Soc. Symp. Proc.* 206:581-586.

Nomura, M. & Held, W. (1974) in *Ribosomes,* eds. Nomura, M., Tissiers, A. & Lengyel, P. (Cold Spring Harbor Laboratory, Cold Spring Harbor, NY), pp. 193-203.

Nordmann, A. (rapp.) Converging Technologies—Shaping the Future of European Societies, European Commission report, 2004.

Nordmann, A.: 2003, 'Shaping the World Atom by Atom: Eine nanowissenschaftliche WeltBildanalyse', in: A. Grunwald (ed.), *Technikgestaltung zwischen Wunsch und Wirklichkeit*, Berlin: Springer, pp. 191-199.

Nowotny, H.; Scott, P.; Gibbons, M.: 2001, *Re-thinking Science*, Polity, Cambridge.

NSET. 2000. *National Nanotechnology Initiative: Leading to the Next Industrial Revolution*. 2000. A report by the Interagency Working Group on Nanoscience, Engineering and Technology (NSET), of the National Science and Technology Council's Committee on Technology. Washington, D.C., February.

NSTC (National Science and Technology Council): 1999, *Nanotechnology: Shaping the World Atom by Atom* [www.wtec.org/loyola/nano/IWGN.Public.Brochure/IWGN.Nanotechnology.Brochure.pdf, accessed: 15 Nov. 2003].

Nye, D.E.: 1994, *American Technological Sublime*, MIT Press, Cambridge, MA.

Nye, D.E.: 2003, *America as Second Creation: Technology and Narratives of New Beginnings*, MIT Press, Cambridge, MA.

P. Livingston, "Ganglioside vaccines with emphasis on GM2," *Semin. Oncol.* 25(December 1998):636-645.

P. Sa-Ardyen, A.V. Vologodskii and N.C. Seeman, The Flexibility of DNA Double Crossover Molecules. Biophysical Journal 84, 3829-3837 (2003).

P. Sa-Ardyen, N. Jonoska and N.C. Seeman, Self-Assembling DNA Graphs, DNA-Based Computers VIII, LNCS 2568, Springer-Verlag, Berlin, 1-9 (2003).

P. Sa-Ardyen, N. Jonoska and N.C. Seeman, Self-Assembling DNA Graphs, Natural Computing, 2, 427-438 (2003).

P. Sa-Ardyen, N. Jonoska and N.C. Seeman, The Construction of Graphs Whose Edges are DNA Helix Axes, J. Am. Chem. Soc., 126, 6648-6657 (2004).

P.S. Lukeman, A. Mittal, & N.C. Seeman, Two Dimensional PNA/DNA Arrays: Estimating the Helicity of Unusual Nucleic Acid Polymers, Chemical Communications 2004, 1694-1695 (2004).

Parikh, A.N.; Schivley, M.A.; Koo, E.; Seshadri, K.; Aurentz, D.; Mueller, K.; Allara, D.L. 1997. "nalkylsiloxanes: From single monolayers to layered crystals. The formation of crystalline polymers from the hydrolysis of n- octadecyltrichlorosilane", *Journal of the American Chemical Society*, 119, 3135-3143.

Parker, J.C., et al. 1995. U.S. Patent 5,460,701. POST (Parliamentary Office of Science and Technology). 1995.

Parr, R.G. and W. Yang. 1989. *Density-Functional Theory of Atoms and Molecules.* Oxford University Press. Oxford.

Paschen, H.; Coenen, C.; Fleischer, T.; Grünwald, R.; Oertel, D. & Revermann, C.: 2004, *Nanotechnologie: Forschung, Entwicklung, Anwendung*, Berlin: Springer (*Nanotechnologie, TAB-Arbeitsbericht 92*, Berlin: Büro für Technikfolgen-Abschätzung beim Deutschen Bundestag).

Paul A. Schulte, Frederica P. Perera, eds., *Molecular Epidemiology: Principles and Practices*, Academic Press, San Diego, CA, 1993.

Peng, X.G.; Manna, L.; Yang, W.D.; Wickham, J.; Scher, E.; Kadavanich, A.; Alivisatos, A.P. 2000. "Shape control of CdSe nanocrystals", *Nature*, 404, 59-61.

Perkal M., C. Marks, M.I. Lorber, W.H. Marks. 1992. "A three-year experience with serum anodal trypsinogen as a biochemical marker for rejection in pancreatic allografts. False positives, tissue biopsy, comparison with other markers, and diagnostic strategies." *Transplantation*, Feb; 53 (2): pp. 415-419.

Perspectives on the Computer Revolution, Zenon W. Pylyshyn, Prentice Hall 1970.

Philip Ball, "Synthetic Biology: Starting from Scratch," Nature, 431, pp. 624-626, 7 October 2004.

Philip G. Collins, Michael S. Arnold, Phaedon Avouris, "Engineering carbon nanotubes and nanotube circuits using electrical breakdown," *Science* 292(27 April 2001):706-709.

Philip G. Collins, Phaedon Avouris, "Nanotubes for electronics," *Scientific American* 283(December 2000):62-69.

Philipse, A.P. 1998. "Colloidal Dispersions," section 3.4, A. ten Wolde (ed.), *Nanotechnology: Towards a molecular construction kit*. STT Report #60, The Hague, 171-178.

Phoenix, C.; Drexler, E.: 2004, 'Safe Exponential Manufacturing', *Nanotechnology*, 15, 869-72.

Phoenix, C.; Drexler, E.: 2004, 'Safe Exponential Manufacturing', *Nanotechnology*, 15, 869-872.

Pierce, N.A. and M.B Giles. 2000. Adjoint reconstruction of superconvergent functionals from PDE approximations. *SIAM Rev.* Vol. 42(2). 247-264.

Pinch, T.: 1986, *Confronting Nature*, Reidel, Dordrecht. -

Pinch, T.: 1993, "Testing – One, Two, Three ... Testing!': Toward a Sociology of Testing', *Science, Technology, and Human Values*, 18, 25-41.

Pitt, J.C.: 2004, 'The Epistemology of the Very Small', in: D. Baird, A. Nordmann & J. Schummer (eds.), *Discovering the Nanoscale*, Amsterdam: IOS Press, pp. 157-163.

Piveteau, Laurent-Dominique, Anthe S. Zandvliet and Robert Langer. 2001. "Nitroxide and Galactose Grafted Dendrimer: A Specifically Targeted Contrast Agent for MRI and EPR Imaging of the Liver," *in preparation.*

Polanyi, Karl. 1944. *The Great Transformation: The Political and Economic Origins of Our Time* (Boston: Beacon Press, 1957; orig. ed. 1944).

Polanyi, M.: 1962, *Personal Knowledge: Towards a Post-Critical Philosophy*, Harper Torchbooks, New York.

Poldrack, R.A. 2000. Imaging brain plasticity: conceptual and methodological issues — a theoretical review. *Neuroimage*. 12(1): p. 1-13.

Pollock, C.R. 1995. *Fundamentals of Optoelectronics*. Irwin, Inc. pp. 3-5.

Porod, W., C.S. Lent, G.H. Bernstein, A.O. Orlov, I. Amlani, G.L. Snider, and J.L. Merz. 1999. Quantumdot cellular automata: computing with coupled quantum dots. *Int. Journal of Electronics* 86, No.5, 549-590.

pp.140-145.

Pressman, J.: 2004, 'Nano Narrative: A Parable from Electronic Literature', in: N.K. Hayles (ed.), *Nanoculture: Implications of the New Technoscience*, Bristol, UK: Intellect Books, pp. 191-200.

Prigogine, I., and S. Rice. 1988. *Advances in chemical physics*, Vol. 70, Parts 1 & 2. New York: J. Wiley.

Pritchard, M.S. 1996. "Conflicts of Interest: Conceptual and Normative Issues," *Academic Medicine*, 71:12, December, pp. 1305-1313.

Putman, C.A.J., De Grooth, B.G., Van Hulst, N.F., and Greve, J. (1992) A detailed analysis of the optical beam deflection technique for use in atomic force microscopy. J. App. Phys. 72(1), 6-12

Putnam, D., Gentry, C.A., Pack, D.W., Langer, R. 2001. "Polymer-based Gene Delivery with Low Cytotoxicity by a Unique Balance of Side-chain Termini." *Proceedings of the National Academy of Sciences*, U.S.A., Volume 98, Number 3, pages 1200-1205.

R. Sha, F. Liu and N.C. Seeman, Atomic Force Measurement of the Interdomain Angle in Symmetric Holliday Junctions, Biochemistry 41, 5950-5955 (2002).

R. Sha, F. Liu, D.P. Millar N.C. Seeman, Atomic Force Microscopy of Parallel DNA Branched Junction Arrays, Chemistry & Biology 7, 743-751 (2000).

R. Sha, F. Liu, M. Bruist and N.C. Seeman, Parallel Helical Domains in DNA Branched Junctions Containing 5', 5' and 3', 3' Linkages, Biochemistry 38, 2832-2841 (1999).

R. T. Bate, "Nanoelectronics," *Nanotechnology*, Vol. 1, pp. 1-7, 1990.

R.G. Postier, "Antibiotic-resistant organism infection," *Am. Surg.* 66(February 2000):112-116.

R.-I. Ma, N.R. Kallenbach, R.D. Sheardy, M.L. Petrillo and N.C. Seeman, 3-Arm Nucleic Acid Junctions Are Flexible. Nucleic Acids Research 14, 9745-9753 (1986).

R.J. Jackson, A.J. Ramsay, C.D. Christensen, S. Beaton, D.F. Hall, I.A. Ramshaw, "Expression of mouse interleukin-4 by a recombinant ectromelia virus suppresses cytolytic lymphocyte responses and overcomes genetic resistance to mousepox," *J. Virol.* 75(February 2001):1205-1210.

R.N. Jones, "Resistance patterns among nosocomial pathogens: trends over the past few years," *Chest* 119(February 2001):397S-404S.

Rademann, K., B. Kaiser, U. Even, F. Hensel. 1987. *Phys. Rev. Lett.* 59: 2319.

Rademann, K., B. Kaiser, U. Even, F. Hensel. 1987. *Phys. Rev. Lett.* 59:2319. Rao, C.N.R., B.C. Satishkumar, and A. Govindaraj. 1997. *Chem. Commun.* 1581.

Ralph C. Merkle, "A new family of six degree of freedom positional devices," *Nanotechnology* 8(1997):47-52; http://www.zyvex.com/nanotech/6dof.html

Ralph C. Merkle, "A Proof About Molecular Bearings," *Nanotechnology* 4(1993):86-90; http://www.zyvex.com/nanotech/bearingProof.html

Ralph C. Merkle, "A proposed 'metabolism' for a hydrocarbon assembler," *Nanotechnology* 8(1997):149-162; http://www.zyvex.com/nanotech/hydroCarbonMetabolism.html

Ralph C. Merkle, "A proposed 'metabolism' for a hydrocarbon assembler," *Nanotechnology* 8(1997):149-162; http://www.zyvex.com/nanotech/hydroCarbonMetabolism.html

Ralph C. Merkle, "Binding sites for use in a simple assembler," *Nanotechnology* 8(1997):23-28; http://www.zyvex.com/nanotech/bindingSites.html

Ralph C. Merkle, "Casing an assembler," Nanotechnology 10(1999):315-322; http://www.zyvex.com/nanotech/casing

Ralph C. Merkle, "Casing an assembler," *Nanotechnology* 10(1999):315-322; http://www.zyvex.com/nanotech/casing

Ralph C. Merkle, "Convergent assembly," *Nanotechnology* 8(1997):18-22; http://www.zyvex.com/nanotech/convergent.html

Ralph C. Merkle, "Design considerations for an assembler," *Nanotechnology* 7(1996):210-215; http://www.zyvex.com/nanotech/nano4/merklePaper.html

Ralph C. Merkle, "Design-ahead for nanotechnology," in Markus Krummenacker, James Lewis, eds., *Prospects in Nanotechnology: Toward Molecular Manufacturing*, John Wiley & Sons, New York, 1995, pp. 23-52.

Ralph C. Merkle, "Molecular building blocks and development strategies for molecular nanotechnology," *Nanotechnology* 11(2000):89-99; http://www.zyvex.com/nanotech/mbb/mbb.html

Ralph C. Merkle, "Self replicating systems and molecular manufacturing," *J. Brit. Interplanet. Soc.* 45(1992):407-413; http://www.zyvex.com/nanotech/selfRepJBIS.html

Ralph C. Merkle, "Self-replicating systems and low cost manufacturing," in M.E. Welland, J.K. Gimzewski, eds., The Ultimate Limits of Fabrication and Measurement, Kluwer, Dordrecht, 1994, pp. 25-32; http://www.zyvex.com/nanotech/selfRepNATO.html

Ramanan, V.R. 1998. Nanocrystalline soft magnetic alloys for applications in electrical and electronic devices. In *R&D Status and Trends*, ed. Siegel et al.

Ramos, L.; Lubensky, T.C.; Dan, N.; Nelson, P.; Weitz, D.A. 1999. "Surfactant-mediated two-dimensional crystallization of colloidal crystals", *Science*, 286, 2325-2328.

RAND Corporation. *The Potential of Nanotechnology for Molecular Manufacturing*. 1995.

Rao, C.N.R., B.C. Satishkumar, and A. Govindaraj. 1997. *Chem. Commun*. 1581.

Rao, M.B., and S. Sircar. 1993. *Gas Separation and Purification* 7: 279.

Rao, M.B., and S. Sircar. 1993. *Gas Separation and Purification* 7:279. Reetz, M.T., et al. 1995. *Science* 267:367.

Ratner, M.; Ratner, D.: 2003, *Nanotechnology*, Prentice Hall, New Jersey.

Ratner, M.; Ratner, D.: 2003, *Nanotechnology: A Gentle Introduction to the Next Big Idea*, Prentice Hall, Upper Saddle River, NJ.

Ravetz, J.: 1999, 'What is Post-normal Science?', *Futures*, 31, 647-653.

Rechard, R.P. "Historical Relationship Between Performance Assessment for Radioactive Waste Disposal and Other Types of Risk Assessment." Risk Analysis 19, 5 (1999): 763-807.

Reed, M.A., "Quantum Dots," *Scientific American,* January 1993, pp. 118-123.

Reed, Mark A. and James M. Tour. 2000. "Computing with Molecules." *Scientific American*. June.. SIA (Semiconductor Industry Association). 1999. "1999 International Technology Roadmap for Semiconductors."

Regis, E.: 1995, *Nano: The Emerging Science of Nanotechnology*, Little, Brown, Boston.

Reinhardt, C.: 2004, 'Chemistry in a Physical Mode: Molecular Spectroscopy and the Emergence of NMR', *Annals of Science*, 61, 1-32.

Reviews in Computational Chemistry, Vol 2, Kenny B. Lipkowitz and Donald B. Boyd, VCH Publishers, 1991.

Rheinberger, H.-J.: 1997, *Toward a History of Epistemic Things: Synthesizing Proteins in the Test Tube*, Stanford University Press, Stanford.

Richard A. Laing, "Automation introspection," *J. Comp. System Sci.* 13(1976):172-183.

Richard A. Laing, "Automaton models of reproduction by self-inspection," *J. Theoret. Biol.* 66(1977):437-456.

Richard A. Laing, "Some alternative reproductive strategies in artificial molecular machines," *J. Theoret. Biol.* 54(1975):63-84.

Richard E. Smalley, "Of chemistry, love, and nanobots," *Scientific American* 285(September 2001):76-77.

Richard P. Feynman, "There's Plenty of Room at the Bottom," *Engineering and Science* (California Institute of Technology), February 1960, pp. 22-36. Reprinted in B.C. Crandall, James Lewis, eds., *Nanotechnology: Research and Perspectives*, MIT Press, 1992. pp. 347-363, and in D.H. Gilbert, ed., *Miniaturization*, Reinhold, New York, 1961, pp. 282-296; http://www.zyvex.com/nanotech/feynman.html

Rietman, E.A.: 2001, 'Drexler hypothesis of a universal assembler is supported not by theoretical arguments alone but by existence proof in the form of biological life', in: *Molecular Engineering of Nanosystems*, Springer, New York & Berlin.

Rinzler, A.G.; Liu, J.; Dai, H.; Nikolaev, P.; Huffman, C.B.; Rodriguez-Macias, F.J.; Boul, P.J.; Lu, A.H.; Heymann, D.; Colbert, D.T.; Lee, R.S.; Fischer, J.E.; Rao, A.M.; Eklund, P.C.; Smalley, R.E. 1998.

Rip, A., Misa, Th. J.,&Schot, J. W. (eds.). Managing Technology in Society. The Approach of Constructive Technology Assessment. London: Pinter Publishers, 1995.

Robert A. Freitas Jr., "Clottocytes: Artificial Mechanical Platelets," Foresight Update No. 41, 30 June 2000, pp. 9-11; http://www.imm.org/Reports/Rep018.html

Robert A. Freitas Jr., "Exploratory design in medical nanotechnology: A mechanical artificial red cell," *Artificial Cells, Blood Substitutes, and Immobil. Biotech.* 26(1998):411-430; http://www.foresight.org/Nanomedicine/Respirocytes.html

Robert A. Freitas Jr., "Microbivores: Artificial Mechanical Phagocytes using Digest and Discharge Protocol," Zyvex preprint, March 2001; http://www.zyvex.com/Publications/articles/Microbivores.html. See also: Robert A. Freitas Jr., "Microbivores: Artificial Mechanical Phagocytes," Foresight Update No. 44, 31 March 2001, pp. 11-13; http://www.imm.org/Reports/Rep025.html.

Robert A. Freitas Jr., "Nanodentistry," *J. Amer. Dent. Assoc.* 131(November 2000):1559-1566.

Robert A. Freitas Jr., "Some Limits to Global Ecophagy by Biovorous Nanoreplicators, with Public Policy Recommendations," Zyvex preprint, April 2000; http://www.foresight.org/NanoRev/Ecophagy.html

Robert A. Freitas Jr., *Nanomedicine, Volume I: Basic Capabilities*, Landes Bioscience, Georgetown, TX, 1999; http://www.nanomedicine.com/

Robert A. Freitas Jr., *Nanomedicine, Volume I: Basic Capabilities*, Landes Bioscience, Georgetown, TX, 1999, Section 2.4.2.

Robert A. Freitas Jr., *Nanomedicine, Volume I: Basic Capabilities*, Landes Bioscience, Georgetown, TX, 1999, Sections 6.5.3 and 9.3.5.3.4.

Robert A. Freitas Jr., William P. Gilbreath, eds., *Advanced Automation for Space Missions, NASA Conference Publication CP-2255*, 1982; http://www.islandone.org/MMSG/aasm/

Robert F. Service, "Is nanotechnology dangerous?" *Science* 290(24 November 2000):1526-1527.

Roberts, J.A.: 2004, 'Deciding the Future of Nanotechnologies: Legal Perspectives on Issues of Democracy and Technology', in: D. Baird, A. Nordmann & J. Schummer (eds.), *Discovering the Nanoscale*, Amsterdam: IOS Press, pp. 247-255.

Robinson, C.: 2004, 'Images in NanoScience/Technology', in: D. Baird, A. Nordmann & J. Schummer (eds.), *Discovering the Nanoscale*, Amsterdam: IOS Press, pp. 165-169.

Robison, W.L.: 2004, 'Nano-Ethics', in: D. Baird, A. Nordmann & J. Schummer (eds.), *Discovering the Nanoscale*, Amsterdam: IOS Press, pp. 285-300.

Robot Factory, http://www.fanuc.co.jp/eannai/servo.htm; citation courtesy of Tihamer Toth-Fejel, 2000.

Roco, M. C., R. S. Williams, and P. Alivisatos, *Nanotechnology Research Directions: Vision for Nanotechnology in the Next Decade*, Kluwer Academic Publishers (2000).

Roco, M.&Bainbridge, W. (eds.). Converging Technologies for Improving Human Performances, National Science Foundation report, 2002.

Roco, M.C. & Bainbridge, W.S. (eds.): 2001, *Societal implications of nanoscience and nanotechnology*, (Proceedings of a workshop organized by the National Science Foundation, September 28-29, 2000), Kluwer: Dordrecht [available online at http://itri.loyola.edu/nano/societalimpact/nanosi.pdf]

Roco, M.C. & Bainbridge, W.S., (eds.): 2001, *Societal Implications of Nanoscience and Nanotechnology*, National Science Foundation [online: www.wtec.org/loyola/nano/societalimpact/nanosi.pdf , accessed: 15 November 2003].

Roco, M.C. & Bainbridge, W.S., (eds.): 2003, *Converging Technologies for Improving Human Performance: Nanotechnology, Biotechnology, Information Technology and Cognitive Science*, Kluwer, Dordrecht.

Roco, M.C. & Tomellini, R. (eds.): 2002, *Nanotechnology: Revolutionary Opportunities and Societal Implications, Workshop, Lecce (Italy), 31 January - 1 February 2002*, Luxemburg: Office for Official Publications of the European Communities, [200 pp.]

Roco, M.C., R.S. Williams, and P. Alivisatos (eds). 1999. *Nanotechnology Research Directions: IWGN Workshop Report, Vision for Nanotechnology R&D in the Next Decade*. (Kluwer Academic Publishers.

Roco, M.C.: 2003, 'Broader Societal Issues of Nanotechnology', *Journal of Nanoparticle Research*, 5, 181-189.

Roco, M.C.: 2003a, 'Broader Societal Issues of Nanotechnology', *Journal of Nanoparticle Research*, 5, 181-189.

Roco, M.C.: 2003b, 'Coherence and Divergence in Megatrends in Science and Engineering', in: M.C. Roco, W.S. Bainbridge (eds.), *Converging Technologies for Improving Human Performance*, Kluwer, Dordrecht, pp. 79-96.

Roco, M.C.: 2003c, 'Fundamentally New Manufacturing Processes and Products', in: Roco, M.C. & Bainbridge, W.S., (eds.), *Converging Technologies for Improving Human Performance*, Kluwer, Dordrecht, pp. 300-302.

Roco, M.C.: 2004, 'The Emergence and Policy Implications of Converging Technologies' [www.nsf.gov/home/crssprgm/nano/nbic_roco_04_0422_@aaas_57sl.pdf, accessed May 2004].

Roco, M.C.; Bainbridge, W.S. (eds.): 2002, *Converging Technologies for Improving Human Performance: Nanotechnology, Biotechnology, Information Technology and the Cognitive Science*, Arlington, VA: National Science Foundation.

Rod Logic and Thermal Noise in the Mechanical Nanocomputer, by K. Eric Drexler, in *Proceedings of the 3rd International Symposium on Molecular Electronic Devices*, F.L. Carter, R.E. Siatkowski, H. Wohltjen, eds., North-Holland 1988.

Rodney Brooks, "The Merger of Flesh and Machines," The Next Fifty Years: Science in the First Half of the Twenty-First Century, ed. by John Brockman, 2002, p. 191.

Rofagha, R., R. Langer, A.M. El-Sherik, U. Erb, G. Palumbo, and K.T. Aust. 1991. *Scr. Metall. Mater*. 25:2867.

Roher, H. 1993. *Jpn. J. Appl. Phys*. 32:1335. Rohlfing, E.A., D.M. Cox, and A. Kaldor. 1984. *J. Chem. Phys*. 81:3846.

Rohlfing, E.A., D.M. Cox, and A. Kaldor. 1984. *J. Chem. Phys*. 81:3846.

Rohrer, H.: 1995, 'The Nanometer Age – Challenge and Chance', *Microelectronic Engineering*, 27, 3-15.

Roland Pease, "'Living' robots powered by muscle," BBC News, January 17, 2005.

Romanov, A.E., V.I. Vladimirov. 1992. In *Dislocations in solids*, ed. F.R.N. Nabarro, Vol. 9. Amsterdam: North-Holland. Salishekev, G.A., O.R. Valiakhmetov, V.A. Valitov, and S.K. Mukhtarov. 1994. *Materials Science Forum*. 170-172:121.

Rosen, A. 1998. A periodic table in three dimensions: A sightseeing tour in the nanometer world. In *Advances in quantum chemistry*. In press.

Rosen, P.: 1993, 'The Social Construction of Mountain Bikes: Technology and Postmodernity in the Cycle Industry', *Social Studies of Science*, 23, 479-513.

Rosenberg, N. "Why Technology Forecasts Often Fail." Futurist July/August 1995, 16-21.

Rosenberg, Nathan. 1994. *Exploring the Black Box: Technology, Economics, and History*. New York: Cambridge University Press.

Rothman, B.K.: 1998, *Genetic Maps and Human Imagination*, W.W. Norton, New York & London,

Ruthven, D.M., S. Farooq, K.S. Knaebel. 1994. *Pressure swing adsorption*. New York: VCH Publishers.

Ruthven, D.M., S. Farooq, K.S. Knaebel. 1994. *Pressure swing adsorption*. New York: VCH Publishers. Sayari, A. 1996. *Chem. Mater*. 8:1840.

S. Liao and N.C. Seeman, Translation of DNA Signals into Polymer Assembly Instructions, Science 306, 2072-2074 (2004).

S. P. Ho and W. F. DeGrado, "Design of a 4-Helix bundle protein: Synthesis of peptides which self-associate into a helical protein." *J. Am. Chem. Soc.,* 109: 6751-6758 (1987).

S. Sasaki, I. Karube, "The development of microfabricated biocatalytic fuel cells," *Trends Biotechnol.* 17(February 1999):50-52.

S. Xiao, F. Liu, A. Rosen, J.F. Hainfeld, N.C. Seeman, K.M. Musier-Forsyth & R.A. Kiehl, Self-Assembly of Nanoparticle Arrays by DNA Scaffolding, J. Nanoparticle Research 4, 313-317 (2002).

S. Zhang, C. Cordon-Cardo, H.S. Zhang, V.E. Reuter, S. Adluri, W.B. Hamilton, K.O. Lloyd, P.O. Livingston, "Selection of tumor antigens as targets for immune attack using immunohistochemistry: I. Focus on gangliosides," *Int. J. Cancer* 73(26 September 1997):42-49.

S.B. Levy, "The future of antibiotics: facing antibiotic resistance," *Clin. Microbiol. Infect.* 6(2000):101-106 (Suppl. 3).

S.P. Walch, W.A. Goddard III, R.C. Merkle, "Theoretical studies of reactions on diamond surfaces," Fifth Foresight Conference on Molecular Nanotechnology, 1997; http://www.foresight.org/Conferences/MNT05/Abstracts/Walcabst.html

S.P. Walch, W.A. Goddard III, R.C. Merkle, "Theoretical studies of reactions on diamond surfaces," Fifth Foresight Conference on Molecular Nanotechnology, 1997; http://www.foresight.org/Conferences/MNT05/Abstracts/Walcabst.html

San Francisco Chronicle, *Scientist Is Fearful of Computer Mutiny; Sun Micro co-founder says replicating robots could replace humans*, March 13, 2000.

Sanders, P.G., J.A. Eastman, and J.R. Weertman. 1996. In *Processing and properties of nanocrystalline materials*, ed. Suryanarayana et al.

Sanders, P.G., M. Rittner, E. Kiedaisch, J.R. Weertman, H. Kung, and Y.C. Lu. 1997. *Nanostructured Mater*. 9:433.

Sankey, H.: 1994, *The Incommensurability Thesis*, Avebury, Aldershot.

Santini, John T., Jr., Michael J. Cima and Robert Langer. 1999. "A Controlled-release Microchip" *Nature*, Vol 397, 28 Jan, p 335-338.

Sarewitz, D.; Woodhouse, E.: 2003, 'Small is Powerful', in: A. Lightman, D. Sarewitz & Chr. Desser, (eds.), *Living with the Genie: Essays on Technology and the Quest for Human Mastery*, Washington, DC: Island Press, pp. 63-83.

Sarikaya, M.; Aksay, I. (eds.), 1995, *Biomimetics. Design and Processing of Materials*, AIP Press, Woodbury, New York.

Scanlan, R.M., W.A. Fietz, and E.F. Koch. 1975. *J. Appl. Phys*. 46:2244.

Scheraga, H. A. (1978) in *Versatilty of Proteins*, ed. Li, C. H. (Academic, New York), pp. 119-132.

Schiefsky, M.J.: forthcoming, 'Art and Nature in Ancient Mechanics', in: W.R. Newman & B. Bensaude-Vincent (eds.), *The Artificial and the Natural: An Ancient Debate and its Modern Descendants*, MIT Press, Cambridge, MA.

Schiemann, G.: 2004, 'Dissolution of the Nature-Technology Dichotomy? Perspectives on Nanotechnology from an Everyday Understanding of Nature', in: D. Baird, A. Nordmann & J. Schummer (eds.), *Discovering the Nanoscale*, Amsterdam: IOS Press, pp. 209-213.

Schmidt, J.C.: 2004, 'Unbounded Technologies: Working Through Technological Reductionism of Nanotechnology', in: D. Baird, A. Nordmann & J. Schummer (eds.), *Discovering the Nanoscale*, Amsterdam: IOS Press, pp. 35-50.

Schuler, E.: 2004, 'Perception of Risks and Nanotechnology', in: D. Baird, A. Nordmann & J. Schummer (eds.), *Discovering the Nanoscale*, Amsterdam: IOS Press, pp. 279-284.

Schultz, L., J. Wecker, and E. Hellstern. 1987. *J. Appl. Phys*. 61:3583.

Schultz, L., K. Schnitzke, and J. Wecker. 1989. *J. Magn. Mater*. 80:115.

Schummer, J.: 2001, 'Ethics of Chemical Synthesis', *Hyle: International Journal for Philosophy of Chemistry*, 7, 103-124 [online: www.hyle.org/journal/issues/7/schummer.htm].

Schummer, J.: 2003, 'The Notion of Nature in Chemistry', *Studies in History and Philosophy of Science*, 34 (2003), 705-736.

Schummer, J.: 2004, 'Interdisciplinary Issues in Nanoscale Research', in: D. Baird, A. Nordmann & J. Schummer (eds.), *Discovering the Nanoscale*, Amsterdam: IOS Press, pp. 9-20.

Schummer, J.: 2004, 'Interdisciplinary Issues in Nanoscale Research', in: D. Baird, A. Nordmann, J. Schummer (eds.), *Discovering the Nanoscale*, IOS Press, Amsterdam, pp. 9-20.

Schummer, J.: 2004, 'Interdisciplinary Issues in Nanoscale Research', in: D. Baird, A. Nordmann, J. Schummer (eds.), *Discovering the Nanoscale*, IOS Press, Amsterdam, pp. 9-20.

Smith, A., *The Wealth of Nations: An Inquiry into the Nature and Causes*, Random House, New York, 1994.

Smith, C.S. 1981. *A search for structure*. Cambridge, Mass.: MIT Press.

Smith, D., H. Babcock, and S. Chu. 1999. Single-polymer dynamics in steady shear flow. *Science*. Vol. 283. 1724-1727. March.

Smith, M.R.; Marx, L. (eds.): 1994, *Does Technology Drive History? The Dilemma of Technological Determinism*, MIT Press, Cambridge, MA.

Snider, G.L.; Orlov, A.O.; Amlani, I.; Zuo, X.; Bernstein, G.H.; Lent, C.S.; Merz, J.L.; Porod, W. 1999. "Quantum-dot cellular automata: Review and recent experiments (invited)", *Journal of Applied Physics*, 85, 4283-4285.

Snow, E.S., P.M. Campbell, and F.K. Perkins. 1997. Nanofabrication with proximal probes. *Proceedings of the IEEE* 85:601-11.

Soffer, A., J.E. Koresh, S. Saggy. 1987. United States Patent 4,685,940.

Soong, R.; Bachand, G.; Neves, H.; Olkhovets, A.; Craighead, H.; Montemagno, C.: 2000, 'Powering an Inorganic Nanodevice with a Biomolecular Motor', *Science*, 290, 1555-1558.

Soong, R.K.; Bachand, G.D.; Neves, H.P.; Olkhovets, A.G.; Craighead, H.G.; Montemagno, C.D. 2000. "Powering an inorganic nanodevice with a biomolecular motor", *Science*, 290, 1555-1558.

Staden, H.v.: forthcoming, '*Physis* and *Technê* in Greek Medicine', in: W.R. Newman & B. Bensaude-Vincent (eds.), *The Artificial and the Natural: An Ancient Debate and its Modern Descendants*, MIT Press, Cambridge, MA.

Star, S.L.; Griesemer, J.R.: 1989, 'Institutional Ecology, 'Translations' and Boundary Objects: Amateurs and Professionals in Berkeley's Museum of Vertebrate Zoology, 1907-1939', *Social Studies of Science*, 19, 387-420.

Steelman, John R. 1947. *Science and Public Policy* (Washington, DC: US Government Printing Office). Stephenson, Neal. 1995. *The Diamond Age, or, Young Lady's Illustrated Primer* (New York: Bantam Books).

Stephen P. Walch, Ralph C. Merkle, "Theoretical studies of diamond mechanosynthesis reactions," *Nanotechnology* 9(1998):285-296.

Stephen P. Walch, Ralph C. Merkle, "Theoretical studies of diamond mechanosynthesis reactions," *Nanotechnology* 9(1998):285-296.

Sterling, B., *When Robots Act Like Rabbits*, The Wall Street Journal, September 6, 2000.

Sterman, J. D. (1989b). Modeling Managerial Behavior: Misperceptions of Feedback in a Dynamic Decision Making Experiment. *Management Science*, 35(3), 321-339.

T. LaBean, H. Yan, J. Kopatsch, F. Liu, E. Winfree, J.H. Reif and N.C. Seeman, The Construction of DNA Triple Crossover Molecules, Journal of the American Chemical Society 122, 1848-1860 (2000).

T.R. Kelly, H. De Silva, R.A. Silva, "Unidirectional rotary motion in a molecular system," *Nature* 401(9 September 1999):150-152.

Theoretical studies of a hydrogen abstraction tool for nanotechnology, by Charles B. Musgrave, Jason K. Perry, Ralph C. Merkle, and William A. Goddard III, *Nanotechnology*.

Tour, J.M.: 2003, *Molecular Electronics: Commercial Insights, Chemistry, Devices, Architecture and Programming*, World Scientific, Singapore.

Tramontin, A.D. and E.A. Brenowitz. 2000. Seasonal plasticity in the adult brain. *Trends Neurosci*. 23(6): p. 251-8.

Trivelli, A., and W.F. Smith. 1939. *Photog. J.* 79:330,463,609. As referred to in T.H. James. 1977. *The theory of the Photographic process*. New York: MacMillan (p. 100).

Two Types of Mechanical Reversible Logic, by Ralph C. Merkle.

Uyeda, R. 1991. *Prog. in Mater. Sci.* 35:1. Van de Zande, B.M.I., M.R. Bohmer, L.G.J. Fokkink, and C. Shonenberger. 1997. *J. Phys. Chem.* 101:852.

Verbeek, A. & Callaert, J.: 2002, *Detailed analysis of the science-technology interaction in the field of Nanotechnology*, Luxemburg: Office for Official Publications of the European Communities (Linking Science to Technology. Bibliographic References in Patents, Vol. 9) [79 pp.].

Vermaas, P.E.: 2004, 'Nanoscale Technology: A Two-Sided Challenge for Interpretations of Quantum Mechanics', in: D. Baird, A. Nordmann & J. Schummer (eds.), *Discovering the Nanoscale*, Amsterdam: IOS Press, pp. 77-91.

Vettiger, P.; Brugger, J.; Despont, M.; Drechsler, U.; Durig, U.; Haberle, W.; Lutwyche, M.; Rothuizen, H.; Stutz, R.; Widmer, R.; Binnig, G. 1999. "Ultrahigh density, high-data-rate NEMS-based AFM data storage system", *Microelectronic Engineering*, 46, 11-17.

Vettiger, P.; Despont, M.; Drechsler, U.; Durig, U.; Haberle, W.; Lutwyche, M.I.; Rothuizen, H.E.; Stutz, R.; Widmer, R.; Binnig, G.K. 2000. "The "Millipede" - More than one thousand tips for future AFM data storage", *IBM Journal of Research and Development*, 44, 323-340.

Viani, M.B.; Schaffer, T.E.; Chand, A.; Rief, M.; Gaub, H.E.; Hansma, P.K.: 1999, 'Small Cantilevers for Force Spectroscopy of Single Molecules', *Journal of Applied Physics*, 86, 2258-2262.

Vicki Glaser, "Carbohydrate-Based Drugs Move Closer to Market," *Genetic Engineering News*, 15 April 1998, pp. 1, 12, 32, 34.

Vidyasankar, S., M. Ru, and F.H. Arnold. 1997. Molecularly imprinted ligand-exchange adsorbents for the chiral separation of underivatized amino acids. *J. Chromatography A* 775:51-63.

Vincenti, W.G.: 1990, *What Engineers Know and How They Know It: Analytical Studies from Aeronautical History*, Johns Hopkins University Press, Baltimore.

Vogel, S.: 1998, *Cats' paws and catapults*, Norton & Cy, New York, London.

von Neumann, J. "The General and Logical Theory of Automata." Talk given at the Hixon Symposium at the California Institute of Technology, 1948.

Voter, A.F. 1997. Hyperdynamics: accelerated molecular dynamics of infrequent events. *Phys. Rev. Lett.* Vol. 78. 3908.

W. Shen, M. Bruist, S. Goodman & N.C. Seeman, A Nanomechanical Device for Measuring the Excess Binding Energy of Proteins that Distort DNA, Angew. Chem. Int. Ed. 43, 4750-4752 (2004); Angew.Chem. 116, 4854-4856 (2004).

W.B. Sherman and N.C. Seeman, A Precisely Controlled DNA Bipedal Walking Device, NanoLetters 4, 1203-1207 (2004); Erratum, 4, 1801-1801.

W.C. Albrich, M. Angstwurm, L. Bader, R. Gartner, "Drug resistance in intensive care units," *Infection* 27(1999):S19-S23.

Wald, C.A.: 2004, 'Working Boundaries on the NANO Exhibition', in: N.K. Hayles (ed.), *Nanoculture: Implications of the New Technoscience*, Bristol, UK: Intellect Books, pp. 83-108.

Walsh, C. (1979) *Enzymatic Reaction Mechanisms* (Freeman, San Francisco), pp. 33, 38.

Wang, C.C., Z. Zhang, and J.Y. Ying. 1997. *Nanostructured Mater*. 9:583.

Warner, M. and E. M. Terentjev. 1996. Nematic elastomers: a new state of matter? *Progress in Polymer Science*. Vol. 21(5). 853.

Warner, M., E. M. Terentjev, R. B. Meyer, and Y. Mao. 2000. Untwisting of a cholesteric elastomer by a mechanical field. *Phys. Rev. Lett.* Vol. 85(11). 2320.

Washington, London, and Moscow Convention, Convention on the prohibition of the development, production and stockpiling of bacteriological (biological) and toxin weapons and on their destruction, signed Washington, London, and Moscow April 10,1972, entered into force March 26, 1975.

Webb, E. J., Campbell, D. T., Schwartz, R. D., Sechrest, L., and Grove, J. B. *Unobtrusive Measures: Nonreactive Research in the social Sciences* (2nd ed.). Boston: Houghton Mifflin, 1981.

Webster, A. and H. Etzkowitz. 1991. "Academic-industry Relations: The Second Academic Revolution?" Science Policy Support Group, London, 1991.

Weckert, J.: 2001, 'The Control of Scientific Research: The Case of Nanotechnology', *Australian Journal of Professional and Applied Ethics*, 3, 29-44.

Weil, V. 1988. "Afterword," *Biotechnology: Professional Issues and Social Concerns*. (AAAS Publication) co-edited with Paul DeForest, Mark Frankel and Jeanne Poindexter.

Weil, V. 1996. "Biotechnology: Societal Impact and Quandaries," in *Biotechnology and Ethics: A Blueprint for the Future, Report of NSF Workshop*, Northwestern University, Center for Biotechnology.

Weil, V. 2000. "Como Pueden Convivir Los Valores Universitarios Y Las Normes Empresariales?" in *Dos Ejes en La Vinculacion de Las Universidades a la Produccion*. ed. by Rosalba Casas and Giovanna

Weil, V.: 2001, 'Ethical Issues in Nanotechnology', in: M.C. Roco and W.S. Bainbridge (eds.), *Societal Implications of Nanoscience and Nanotechnology*, Dordrecht: Kluwer, pp. 193-198.

Weisenhorn, A.L., Hansma, P.K., Albrecht, T.R., and Quate, C.F. (1989) Forces in atomic force microscopy in air and water. Appl. Phys. Lett. 54(26), 2651-2653

Weiss, S. 1999. "Fluorescence spectroscopy of single biomolecules", *Science*, 283, 1676-1683.

Wenger, E.: 1998, *Communities of Practice: Learning, Meaning, and Identity*, Cambridge University Press, Cambridge, UK.

White, J.G. 1986. *Philos. Trans. R. Soc. Lond. B Biol. Sci.* 314: pp. 1-340.

Zhang, L., and A. Manthiram. 1997. *Appl. Phys. Lett.* 70(18):2469. 34 *Evelyn L. Hu and David T. Shaw.*

Zhang, Shuguang (2003) 'Fabrication of novel biomaterials through molecular self-assembly', *Nature Biotechnology*, 21, no. 10, 1171-78.

Zimmermann, U., N. Malinowski, A. Burkhardt, and T.P. Martin. 1995. *Carbon* 33:995. 67

Index

O

P

R

S

T

U

V

W